AF531614

Groundwater
Hydrogeochemical Investigations of Using Integrated Technique

S. Chidambaram
Professor
Department of Earth Sciences
Annamalai University
Annamalai Nagar-608002
Tamilnadu, India

New Delhi – 110 034

NEW INDIA PUBLISHING AGENCY
101, Vikas Surya Plaza, CU Block, LSC Market
Pitam Pura, New Delhi 110 034, India
Phone: + 91 (11) 27 34 17 17 Fax: + 91 (11) 27 34 16 16
Email: info@nipabooks.com
Web: www.nipabooks.com

Feedback at feedbacks@nipabooks.com

ISBN: 978-93-86546-47-0

Composed Designed by NIPA

Contents

Chapter 1

A Study on the Interrelationship of U and Rn in Groundwater of Hard Rock Aquifers

Adithya V.S., Chidambaram S., Tirumalesh K. Vasudevan U. and Nepolian M.

Department of Earth Sciences, Annamalai University, Annamalai Nagar Isotope Hydrology Section, Isotope Application Division, BARC, Mumbai

ABSTRACT

Radon being the radioactive daughter product of Uranium; the interrelationship between U and Rn has been established in the present study. It is found that Radon coexist with Uranium in the region. A total of 61 samples were collected spatially from hand pumps and bore wells of the study area. Major ions such as Ca^{2+}, Mg^{2+}, Na^{+}, K^{+}, F^{-}, Cl^{-}, HCO_3^-, NO_3^-, H_4SiO_4,PO_4- and SO_4^{2-} and U, Rn were analysed. The lithological influence of U and Radon occurrence is also highlighted. Hard rocks especially Granites are the source for both U and Rn in the study area whereas the lack of correlation exist between U and Rn in sedimentary terrain can be attributed to difference in flow conditions and chemical properties of these elements.

Key words: Radon, Uranium, Granite, Ground water, Source.

Introduction

Radon is a product of the radioactive decay of uranium. Concentrations of radon in ground water can be many orders of magnitude larger than the concentrations of its parent source. The decay of ^{238}Uranium through ^{226}radium is the primary source of ^{222}Radon. Because it is the immediate daughter product of ^{226}radium and has a short half life (3.82 days), the presence of ^{222}radon in ground water depends on a readily available source of ^{226}radium and the hydraulic properties of the aquifer materials (Hall et al, 1987). Therefore, higher radon amounts are commonly detected in areas underlain by granites and similar rocks that usually contain more uranium than do other rock types (Faure, 1986).

The association of U and Radon in granitic rocks has been reported by many authors (Knutsson and Olofsson, 2002; Armands, 1967; Akerblom and Lindgren, 1997; Vinson et al., 2009). The highest Rn value ever found in Sweden, 57,000 Bq/L was measured in a well drilled in the Blomskog granite (Akerblom and Lindgren, 1997).The highest value in Norway, 31,900 Bq/L was found in a similar type of granite, the Iddefjord granite, on the other side of the border between Sweden and Norway (Banks et al., 1998).

Uranium can be highly concentrated in intrusive igneous rocks, primarily granitic rocks. In felsic igneous rocks, uranium is concentrated in the micas and amphiboles. The large ionic radius and high valence state of uranium make it incompatible with the minerals created during the early stages of crystallization; therefore, uranium remains in the liquid phase of magma until the final stages of crystallization (Adams et al; 1959). Minerals formed during later stages of crystallization (such as biotite, muscovite, potassium feldspar, sodium feldspar, and quartz) can incorporate up to several percent of uranium in their crystal structure (Churchill, 1991). Uranium also is found in flat-lying bedded deposits in sedimentary rocks. Black shales, coals, asphalt, clays, silts, soils above limestone, and phosphate rocks also can contain large amounts of natural uranium. Concentrations of uranium in ground water also can differ as a result of Local factors, such as the availability of uranium minerals, ground-water flow characteristics, and geochemistry. Thus, the variation of uranium and ^{222}radon concentrations in ground water depends, in part, on the source of the aquifer materials. Studies found that ^{222}radon concentrations were highest in granitic and high-grade metamorphic bedrock aquifers and were related to hydrologic and hydraulic factors including well pumping (Brutsaert et al., 1981; Hall et al., 1987; Hall et al., 1985; Hess et al., 1979; Maloney, 1988).

Radon moves from its source in rocks and soils through voids and fractures. It can enter buildings as a gas through foundation cracks or dissolve in the ground water and be carried to water-supply wells. (Lindsey and Ator, 1996). Because ground water moves slowly relative to Radon decay (half-life of 3.825 days); anomalously high Radon concentrations should occur proximal to parent radionuclides. The amount of radon in air or water commonly is reported in terms of activity with units of picocuries per liter of air or water. An activity of 1 pCi/L (picocuries per liter) is about equal to the decay of two atoms of radon per minute in each liter of air or water (Otton, 1992).

Exposure to radon has been recognized as a health risk, primarily as a cause of lung cancer. A study of miners found that inhaling the decay immediate parent to radon. Only a fraction of the radon produced in rock will be found in ground water or soil gas. The presence of ground water enhances the release of radon from the solid ground. The water, which hydrates mineral surfaces, is able to absorb the kinetic energy of the radon atom formed during radioactive decay and carry it away instead of remaining adsorbed on rock surfaces (Tanner, 1980). Radon concentrations in water have been found to be thousands of times greater than dissolved uranium concentrations (Hess et al., 1985).This is due to uranium absorption on the host rock, usually on fracture walls.

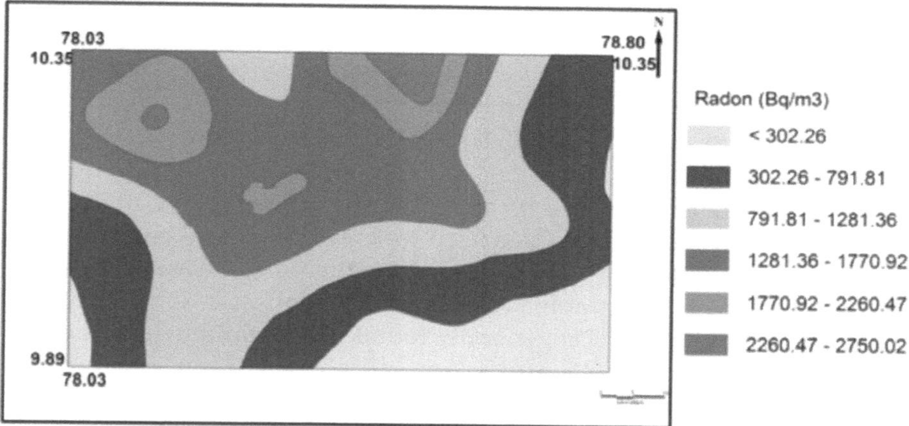

Fig. 3. Spatial distribution of Radon

It is also observed from fig 5 that the spatial distribution of Uranium also follows the same trend with higher U concentrations reported in Central Granitic terrains followed by hard rock formations in the North and North western side. Uranium might have been dissolved into ground water through rock water interaction in the U mineralised zones of the host rocks and during decay of radioactive U and its daughter products; the radon formed also dissolves into the groundwater (Arabi et al 2010). It's also reported that U is the primary source of radon is geological bedrock (Appleton, 2005, Akerblom and Lindgren, 1997). Hence it is inferred that U is the main source of ^{222}Rn in groundwater of this region (Skeppstrom and Olofsson 2007). Thus it is evident that ^{222}Rn evolves from the bedrock source of U that has been concentrated in fracture zone within the granite and gneiss (Choubey et al. 2000, Nashine et al., 1982).

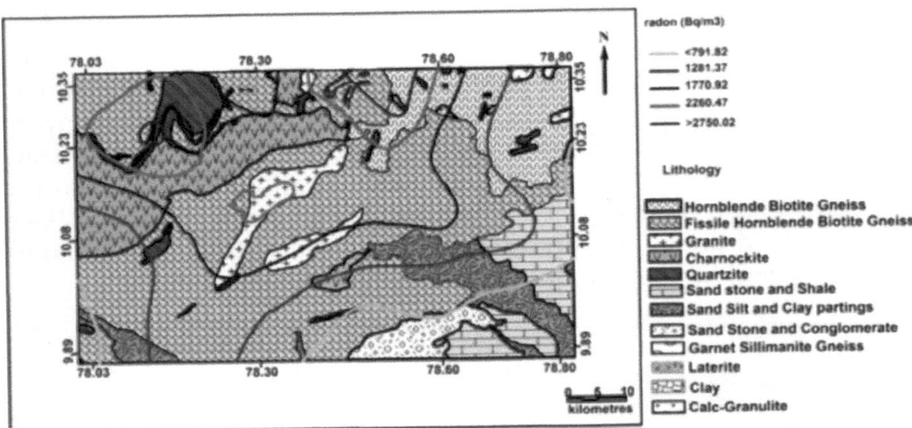

Fig. 4. Spatial distribution of Radon over lithology

It is noted that lithology controls the distribution of ^{222}Rn in groundwater. Its also inferred that hard rocks with higher U concentration should be the sources for higher ^{222}Rn content in groundwater (Langmuir 1997). In the study area granitic rocks have the highest U and ^{222}Rn (Hess et al., 1979). In granites U is generally found as Uraninite which is easily weathered to release Ra and from which ^{222}Rn can escape into groundwater with high efciency (Kraemer and Genereux, 1998).

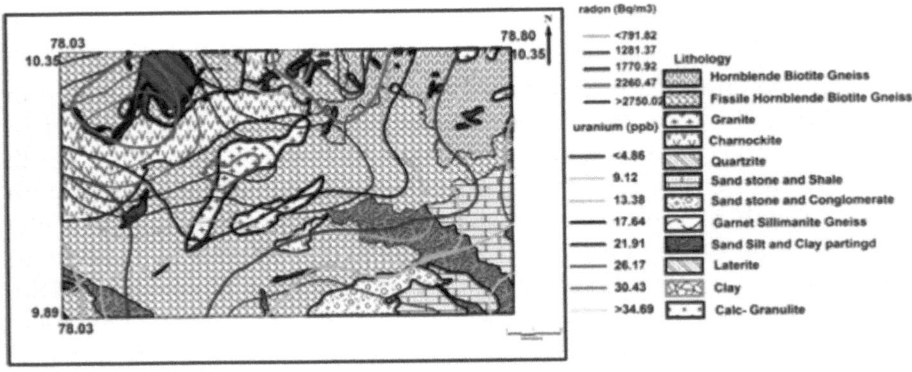

Fig. 5. Spatial distribution of U and Rn over Lithology

Fig 6 depicts the influence of land use over radon levels in ground water. In the study area Land use is primarily agricultural followed by Forest and Built- up where communities have become established. It is observed that higher Radon concentration are located in the agricultural lands and forest areas. Groundwater from wells in agricultural settings show greater concentrations of U than groundwater from wells in urban settings, probably owing to greater pH in the agricultural wells. (Ayotte et al., 2011). This could be an indirect reason behind elevated concentrations of Radon in agricultural lands.

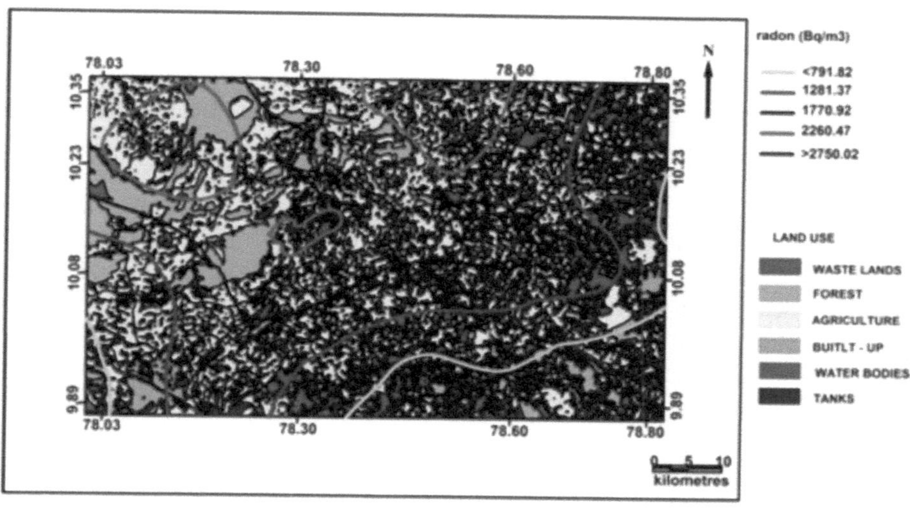

Fig. 6. Spatial distribution of Rn over Land use

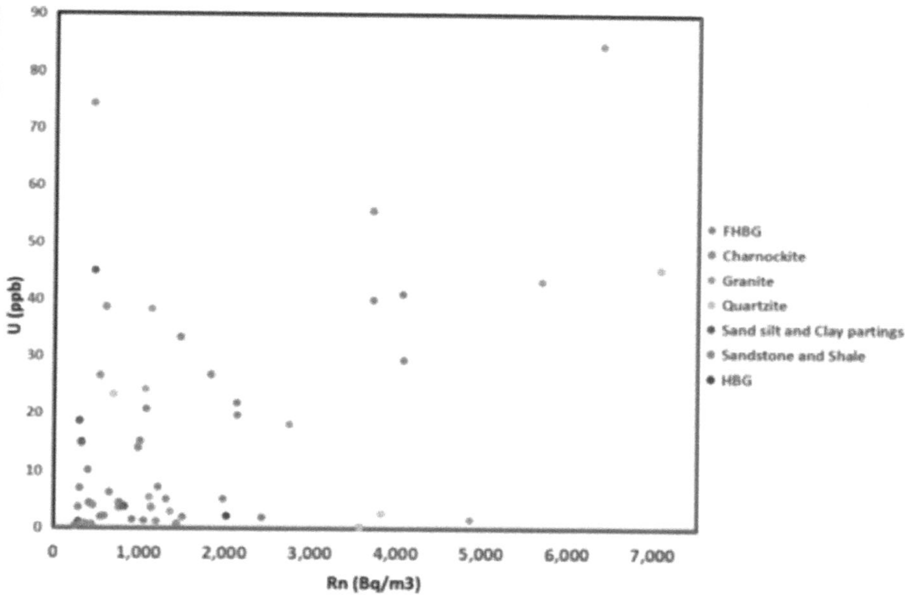

Fig. 7. Scatter plot showing inter relationship of U and Rn

The scatter plot (fig 7) between Rn and U shows that in granitic samples there is a progressive increase in Radon concentrations with higher U. similar trend is observed in FHBG and Charnockites indicating the lithological influence in U and Rn distribution in ground water. The sedimentary formations of the study area do not show any trend of U and Rn occurrence in the region. The reason for radon concentrations depends on geological structure of the area, depth of the water, source and also difference in the climate, due to the nature of place, where the specific concentration of the natural uranium and radium-226 in the soil, which is said to be the reason of high concentration. The concentration and mobility of Rn is related to the concentration of U in the sediment, specifically the sources of U near the well screen. The solubility of Rn is not dependent upon the oxidation/reduction potential of the water, whereas the mobility of U is dependent on the oxidation/reduction potential and pH. Under similar hydrologic flow conditions and favorable oxidizing alkaline conditions, U can travel substantial distances relative to Rn because the half-life of U is so much longer than that of Rn. Thus, under oxidizing conditions where U is mobile, it is possible that U can be found at higher concentrations in ground water where the aquifer matrix has little U in the aquifer matrix than in areas where the water is reducing, even if plenty of U is in the aquifer matrix. The Rn concentrations, which are primarily proportional to the U content of the aquifer matrix, may be low in the areas of rapidly flowing oxidizing water because of the potential long-term leaching of U from the aquifer matrix, especially that in soluble grain coatings. Conversely, under reducing conditions, Rn may be present in high concentrations in an area of uraniferous deposits, whereas U would not be present in the water because U is immobilized, possibly as a precipitate in the aquifer matrix (Morrow, 2001). Szabo and Zapecza (1991) noticed a lack of correlation between U and Rn concentrations in water from fractured rock aquifers indicating that long distance transport of U occurs in oxidizing waters, resulting, on occasion, in elevated concentrations of U in areas with moderate or low U content in the aquifer matrix and relatively low Rn concentration in the water.

Conclusion

The study summarizes the interrelationship between U and its daughter product. Radon distribution in ground water in the study area underlain by igneous, metamorphic and sedimentary rocks follows a general geographical pattern related to rock type. It is noted that both U and Radon are associated with granitic formations in general. This observation indicates that Rn was co-transported with U into U enriched zones whereas in other rock types concentrations of U and Rn did not significantly correlate. This result is expected because of the different chemical properties and half-lives whereby, U, but not Rn, is available for long-distance transport. It is found that Rn is not always in equilibrium with U, which could be indicative of higher flow velocities causing the increased transport of daughter products away from parent radio nuclides.

Acknowledgement

The authors express their sincere thanks to Board of Research in Nuclear Sciences (BRNS), India, for providing the necessary nancial support to this research project wide ref. No.012/35/12/BRNS/1918. dated 2012.

References

Adams J AS, Osmond JK, Rogers JJW (1959) The geochemistry of thorium and uranium, in Ahrens, L.H., Press, F., Rankama, K., and Runcorn, S.K., eds., Physics and chemistry of the Earth: Oxford, Pergamon Press, p. 298–348.

Akerblom G, Lindgren J (1997) Mapping of groundwater radon potential., European Geologist 5, 13-22.

Appleton JD (2005) Radon in air and water. In: Selinus O, Alloway B, Centeno JA, Finkelman RB, Fuge R, Lindh U, Smedley P (eds) Essentials of medical geology, impacts of the natural environment on public health. Elsevier Academic Press, Burlington, MA, USA, pp 227–262

Arabi SA, Funtua II, Alagbe SA, Zabosrki P, Dewu BBM (2010) J Am Sci 6(12): 664–672

Armands G (1967) Geochemical prospecting of a uraniferous bog deposit at Masung sbyn, northern Sweden. In Kvalheim, A. (ed.) Geochemical prospecting in Fennoscandia. Int erscien ce Publ ishers. John Wiley, Lond on, 127-154.

Ayotte JD, Gronberg JM, Apodaca LE (2011) Trace Elements and Radon in Groundwater Across the United States: U.S. Geological Survey Scientific Investigations Report 2011-5059, 115 p.

Banks D, Frugstad B, Midtgard A, Krog J R, Strand T (1998) The chem ist ry of Nor wegian groundwaters: I.The di stribution of rado n, major and minor eleme nts in 1604 crystallin e bedrock groundwaters.TheScienceofthe Total Environment 222,7 1-91.

Brooks J R (1988) Radon and your home: Maryland Department of Natural Resources, Geological Survey, 4 p.

Lindsey D B, Scott W. Ator (1996) Radon in ground water of the lower Susquehanna and Potomac river basins; U.S. Geological Survey, Water Resources Investigations Report, 4156)

Brutsaert WF, Norton SA, Hess CT, Williams JS (1981) Geologic and hydrologic factors controlling radon- 222 in ground water in Maine: Ground Water, v. 19, p. 407–417.

Choubey VM, Bartarya SK, Ramola RC (2000) Environ Geol 39(6):523–530

Churchill R (1991) Geologic controls on the distribution of radon in California: California Department of Health Services, 41 p.

CGWB (2008) Ground water resources and development prospects in Madurai region.

Vinson DS , Avner Vengosh, Daniella Hirschfeld, Gary S. Dwyer (2009) Relationships between radium and radon occurrence and hydrochemistry in fresh groundwater from fractured crystalline rocks, North Carolina (USA) Chemical Geology 260 159–171

Faure, G., 1986, Principles of isotope geology: New York, John Wiley and Sons, Inc., 589 p.

Knutsson G, Olofsson BO (2002) Radon content in groundwater from drilled wells in the Stockholm region of Sweden; N GU -BULL 4 3 9 , pp 79- 85

Hall FR, Boudette EL, Olszewski WJ, Jr, (1987) Geologic controls and radon occurrence in New England, in Graves, B., ed., Radon, radium, and other radioactivity in ground water: Somerset, N.J., Lewis Publishers, p. 15–30.

Hall FR, Donahue PM, Eldridge AL (1987) Radon gas in ground water of New Hampshire, in Proceedings of the Association of Ground Water Scientists and Engineers, Eastern Regional Ground Water Conference: Portland, Maine, National Water Well Association, p. 86–101.

Hess C T, Michel J, Horton T R, Prichard H M, Coniglio W A (1985) The occurrence of radioactivity in public water supplies in the Unites States. Health Physics 48:553–586.

Hess CT, Norton SA, Brutsaert WF, Casparius RE, Coombs EG, Hess AL (1979) Radon222 in potable water supplies in Maine—The geology, hydrology, physics, and health effects: Orono, Maine, Land and Water Resources Center, University of Maine, 119 p.

Hurlburt S (1989) Radon - A real killer or just an unsolved mystery?: Water Well Journal, v. 43, no. 6, p. 34-41.

Kraemer T, Genereux D (1998) Application of uranium- and thorium series radionuclides in catchment hydrology studies. Elsevier, Amsterdam, pp 679–722

Langmuir D (1997) Aqueous environmental geochemistry. Prentice Hall, Englewood Cliffs, NJ, p 600

Maloney TJ (1988) Maine ground-water quality, in Moody, D.W., Carr, J., Chase, E.B., and Paulson, R.W., eds., National Water Summary 1986—Hydrologic events and ground water quality: U.S. Geological Survey WaterSupply Paper 2325, p. 279-286.

Morrow W S,(2001) Uranium and Radon in Ground Water in the Lower Illinois River Basin Urbana, Illinois Water-Resources Investigations Report, U.S. GEOLOGICAL SURVEY; 01-4056

Nashine SK, Dhanraju R, Bhatnagar GR, Narayandas GR (1982) Him Geol 1:305-316

Otton J K (1992) The geology of radon: U.S. Geological Survey, General Interest Publications of the U.S. Geological Survey, 28 p.

Robillard P D, Martin K S, Sharpe W E (1991) Reducing radon in drinking water: Pennsylvania State University, Agricultural and Biological Engineering Fact Sheet, SW-135, 4 p.

Skeppstrom K, Olofsson B (2007) Eur Water 17(18):51-62

Szabo Z, Zapecza O (1991) Geologic and geochemical factors controlling uranium, radium-226, and radon-222 in ground water, Newark Basin, New Jersey in Gundersen, L.C.S., and Wanty, R.B., eds., Field studies of radon in rocks, soils, and water: U.S. Geological Survey Bulletin 1971, p. 243-265.

Tanner A B (1980) Radon mitigation in the ground: A supplementary review. In T. F. Gesell and W. M. Lowder (eds.), Natural Radiation Environment III, volume 1. Technical Information Center/U.S. Department of Energy, Houston, Texas

U.S. Environmental Protection Agency, 1992, Consumers Guide to Radon Reduction - How to reduce radon levels in your home: EPA 402-k92-003, 17 p.

Zapecza OS, Szabo, Zoltan (1986) Natural radioactivity in ground water - a review: U.S. Geological Survey Water-Supply Paper 2325, p. 50-57.

Chapter 2

Hydrogeochemical Quality of Groundwater in and Around Neyveli Lignite Mining Area, Cuddalore District-Cuddalore, Tamil Nadu

Anandhan P., Chidambaram S., Manivannan R. Paramaguru P. and Ganesh N.

Departmen of Earth Sciences, Annamalai University, Annamalai Nagar Tamil Nadu

ABSTRACT

Quality of water is affected by natural and man induced contamination to groundwater system. Area chosen for study has gained its own importance due to the location of two lignite mines and a Thermal power plant in operation. Extensive large scale pumping of water is done for lignite mining. The mine drainage and drainage from thermal power plant along with agricultural return flow from lignite charged waters increases the severity of problem. In order to achieve a panacea for the above-mentioned situation, the study has incorporated and integrated in all related factors to get a meaningful interpretation for groundwater quality and quantity management in and around Neyveli area. The frequent failure of monsoon has added to the existing problem of over exploitation. To attain a panacea for water chemistry groundwater samples were collected for four different seasons. Higher EC and TDS values are observed in Southeast and Southwest part of the study area. In Southwest monsoon and Summer seasons HCO_3 Cl dominates the anions, with few representations for SO_4. In Northeast monsoon and Post monsoon season Na + K – Ca + Mg is the dominant. Thermodynamic, studies reveal that groundwater of the region are stable with Kaolinite stability field in all the silicate systems. Bicarbonates derived from silicate mineral weathering are noted in the NW, Northern and in the mining region. These regions are located in the NW and NE part of the study area. Comparison of water quality to standards shows that the water can be used for drinking and for irrigation purposes except in few locations.

Key word: Geochemistry, Monsoon, Gibbs boomerang, USSL, Thermodynamic

Introduction

Water is an important and essential resource of on earth. In modern world all developmental activity, urban services, food production drinking water, cooling, heating, industrial growth are connected to water resources. Groundwater is a precious and the most widely distributed resource of earth. The world total water resource is estimated to be 1.36X10^8 million h.m., of these global water resources about 97.2% is salt water, mainly in oceans, and only 2.8% is available as fresh water, out of this 2.8% about 2.2% is available as surface water at 0.6% as groundwater. At present nearly 1/5th of all water used in world is obtained from groundwater. Availability of water for usage is mainly controlled by the quality of groundwater. Quality is the resultant of all biogeochemical processes and reactions, which have acted on water from the moment it has condensed in atmosphere to the time it's discharged by a well or spring. Hydrogeochemical composition of groundwater can also be indicative of its origin and history of passage through underground materials which water has been in contact with, in shallow and deep seated conditions (Anandhan et al. 2008; Chidambaram et al. 2013). Natural and anthropogenic sources along with chemical and biogeochemical constituents have been considerably altering groundwater quality in recent days. Coal mining plays a vital role in determining economic status of our country. The lignite rank of coal with a deposit of about 3, 300 million tonnes of reserves is estimated in Neyveli region of Tamilnadu. To excavate the reserves extreme quantity of groundwater is pumped from underlying aquifer so as to keep peizometric water level below lignite seam. The present study is aimed at determining hydrogeology and hydrogeochemistry of groundwater in and around Neyveli region to obtain a baseline data regarding groundwater quality. When mixed with lignite it gets altered considerably and hence leaches/infilters into the aquifer.

Location

The area chosen for study is Neyveli and adjoining region, which is located in Cuddalore district of Tamilnadu. The area is bounded between north latitudes 11°40′ and 11°25′ and east longitudes 79°20′ and 79°40′ (Fig. 1). It falls in survey of India toposheet 58 M/6, M/7, M/10 and M/11, with a total area of about 835km^2. It includes taluks of Vriddhachalam, Panruti and Chidambaram. There are two lignite mines in the proposed study area. 1st and 2nd mine, which extends from South of Neyveli Township to North of Vellar river.

Geology

The exploratory work carried out by Oil and Natural Gas Commission (ONGC) and agencies like, Neyveli Lignite Corporation (NLC) etc., was also significant. In study area (Fig.2), the basement Archean rocks consist predominantly of bluish grey granites and gneisses, this older formation are overlain by Cretaceous and Tertiary Sediment towards East. Southern part of study area is underlain by recent alluvium of 50m thick. Cuddalore formation of Mio-Pliocene age consists chiefly of sandstone and extensive clay layers with a thick band of lignite seam crops out in middle of study area, striking Northeast and Southwest and dipping towards Southeast.

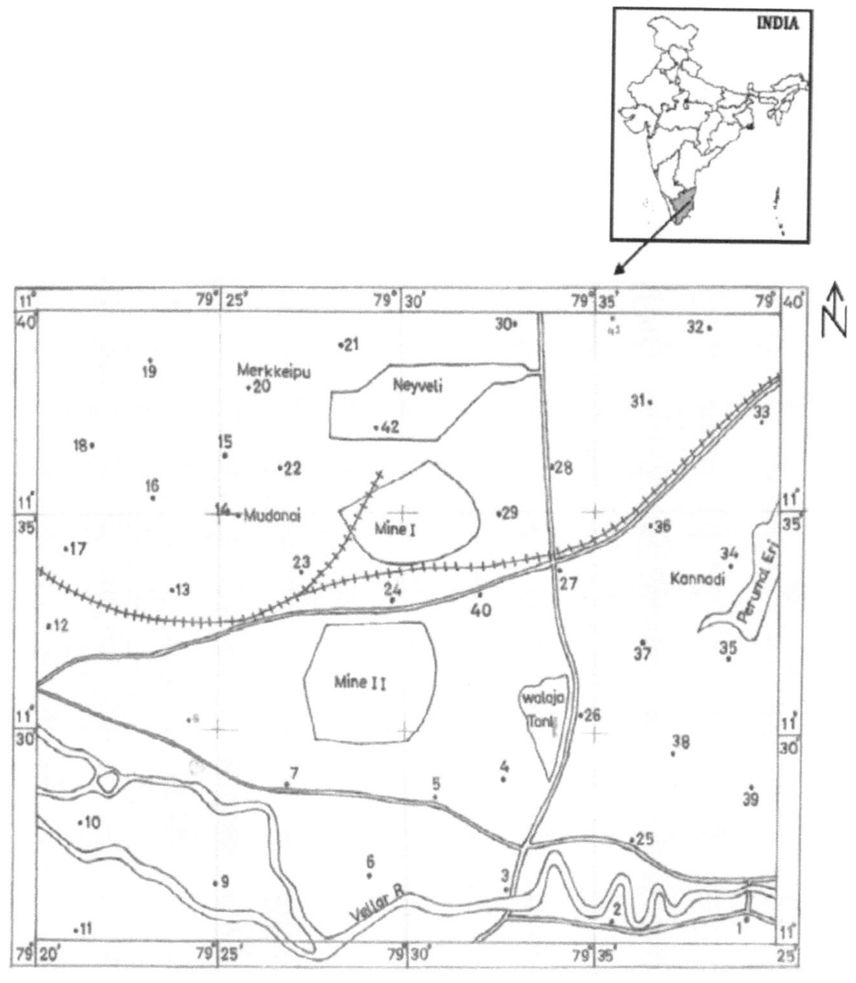

Fig. 1. Location map of the study area

Hydrogeochemical research requires proper site selection for collection of water samples and appropriate method of analysis. Sampling sites were located taking several factors into considerations like lithology, structure, geomorphology, and river influence, mining, urban, agricultural and availability of wells. Sampling of groundwater has been carried out in and around Neyveli region. The groundwater samples were collected for four different seasons. Post monsoon, SouthWest monsoon, North East monsoon and summer. A total of 168 samples were collected at the rate of 42 samples per season, Ground water sampling locations are shown in figure 4.2. One litre of water samples were collected in polyethylene bottle four times to broadly cover seasonal variation. Then it was sealed and brought to laboratory for analysis and stored properly (4°C) before analysis. Water samples were analysed for major and minor ion concentration by using standard procedures (APHA 1998; Ramanathan 1992; Ramesh and Anbu, 1996).

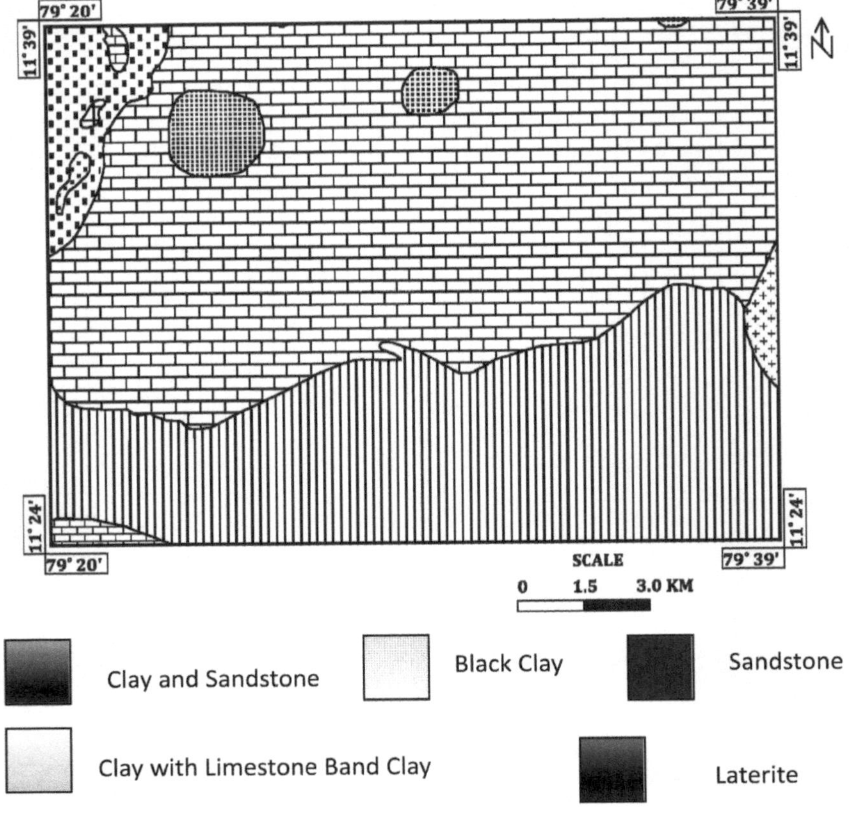

Fig. 2. Geology Map of the Study Area.

Result and Discussion

Groundwater in the study area is generally alkaline in nature with pH ranging from 5.1 to 8.65 with an average of 7.2. It was higher in post monsoon and lower in southwest monsoon period (Table.1). pH in the study area is influenced by infiltration of rain water, surface ruining water and untreated mine effluent which lowers pH. EC is an indirect measure of ionic strength and mineralisation of natural water. EC of pure water is around 0.05 μs/cm (Hem, 1991). EC ranges from 102μs/cm to 6621 μs/cm with an average of 883.5 μs/cm. Highest values of EC were observed in southwest monsoon and post monsoon and drop in EC was noted in northeast monsoon and Summer. Total dissolved solids (TDS) which is generally the sum of dissolved ionic concentration varies between 125 mgl^{-1} to 2205 mgl^{-1} with an average of 597.5 mgl^{-1}. Higher TDS was noted during southwest monsoon periods followed by summer and post monsoon season.

Table 2 shows the range of ionic concentration in groundwater of study area prescribed speciation of WHO (1996). Analytical data shows groundwater of the study area is suitable for drinking and domestic use in 60 - 70% of area. Most of the parameters are within permissible limits. TDS and EC exceed permissible limit at specific locations in southwest, summer and post monsoon indicating higher ionic concentration may be due to higher dissolution of minerals and anthropogenic impact. When SAR (alkali hazard) and specific conductance (salinity hazard) is plotted in USSL (1954) diagram, classification of water for irrigation purpose can be determined. USSL diagram (Fig.3a,b,c,d) shows most of water samples irrespective of seasons fall in C3S1,C2S3, C3S3,C2S1,C1S1 few indications of northeast, southwest and Post monsoon are also noted in C3S4 zone indicating wide range of salinity (Low-High) and low to high alkalinity. This water can be used with caution for irrigation in almost all soils with little danger to harmful effects to soil by exchanging Na into soil (Anandhan 1998).

(a)

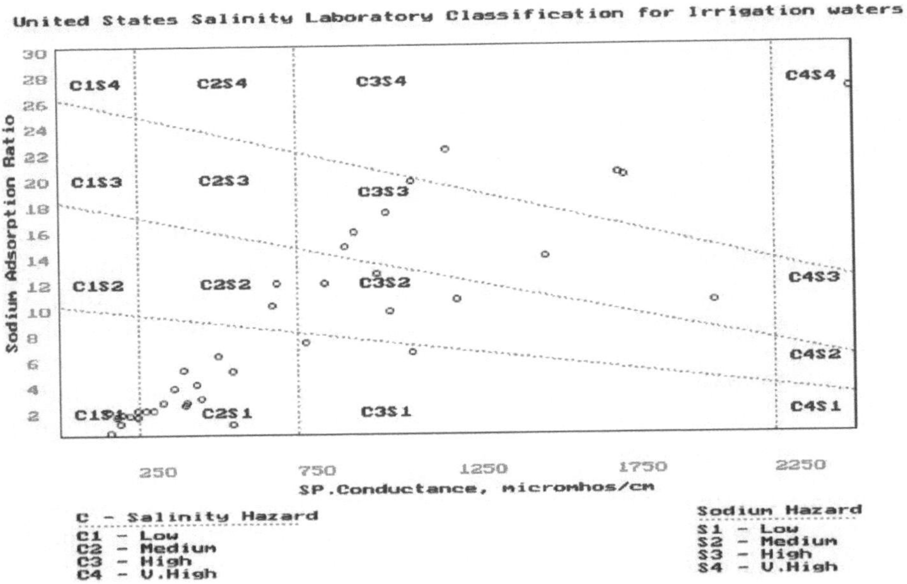

(b)

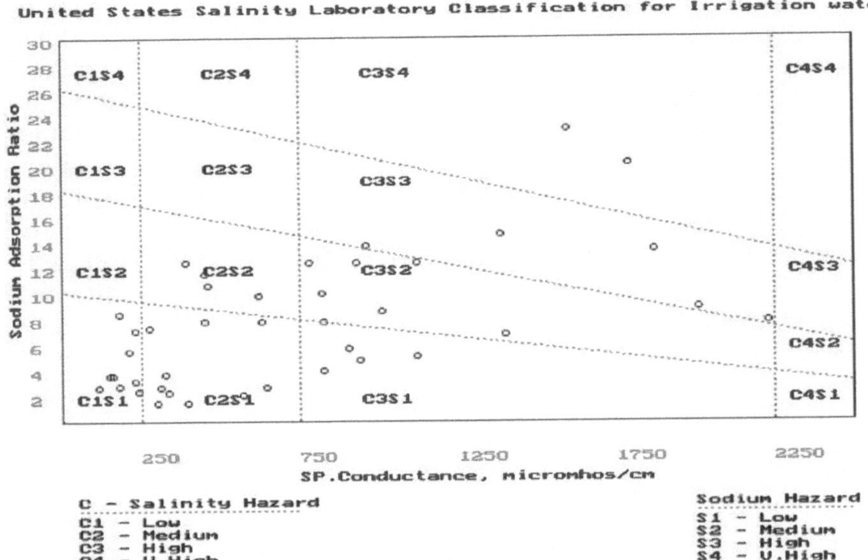

Fig. 3 (a,b). USSL diagram for the study area (NWM; SWM)

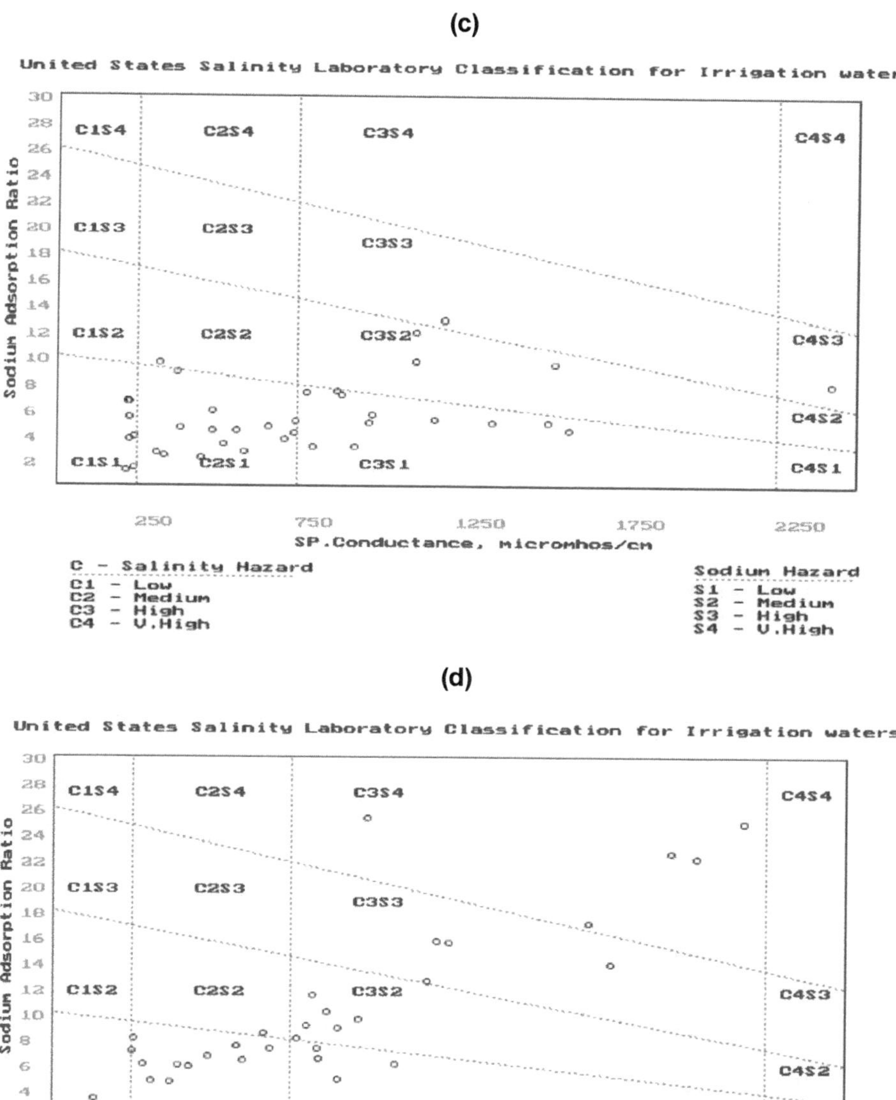

Fig. 3 (c,d). USSL diagram for the study area (POM; SUMMER)

Table 1. Maximum, Minimum, Average groundwater samples in All Seasons

		pH	EC	Cl	CO_3	HCO_3	SO_4	PO_4	NO_3	F	H_4SiO_4	Ca	Mg	Na	K	TDS
Northeast monsoon	Max	8.65	6621	487.4	60	347.7	610	1.8	5	1.5	30	30	20	936	38	2205.7
	Min	6.92	155	8.86	0	67.1	15	0.03	0.5	0.05	1.6	4	0	3	1	138.76
	Avg	7.98	970.43	130.3	15.1	137.7	176.0	0.467	2.30	0.26	22.41	16.14	6.15	211	9.5	666.35
Southwest monsoon	Max	7.9	2970	458	24	384.2	368	8.8	13	0.72	147.5	42	52	365	43	1313.9
	Min	5.78	118	8	0	12.2	0.01	0.01	0.01	0.01	8	3	1	22	1	125.55
	Avg	7.12	795	141	5.14	123.5	114.7	2.089	1.36	0.15	35.99	16.57	9.46	147	8.3	554.17
Post monsoon	Max	8.3	5680	584	18	312	197	1.75	6	0.82	183	60	42	345	74	1347.6
	Min	5.1	215	17	0	12.2	5	0.01	0.01	0.01	18	1	1	21	1	201.34
	Avg	6.79	974.62	155.6	3.85	85.24	81.5	0.295	2.12	0.23	72.67	25.64	10.5	131.1	14	547.28
Summer	Max	8.2	2180	895.2	24	250.1	341	1.4	2.5	1.1	84	48	53	781	28	1993.3
	Min	5.92	102	17.73	0	6.1	25	0.001	0.5	0.005	10	1	0	20	0	143.74
	Avg	6.93	795.02	171	5	81.46	131.71	0.436	1.89	0.1663	41.16	15.62	7.58	196	6.38	623.93

Table 2. Comparison of the chemical composition with Drinking water standard

	Groundwater	WHO (1996)	ISI (1983)
EC	102 - 6621	1400	-
pH	5.1 - 8.65	6.5 - 8.5	6.5 - 9.1
TDS	138.76 - 2205.73	1000	1500
Ca	1.- 60	500	200
Mg	1. - 53	-	100
Na	3 - 936	200	-
K	1. - 74	-	-
HCO_3	6.1 - 384.2	-	600
SO_4	0.01- 610	400	400
Cl	8 - 895.2	250	1000

Permeability index is an important factor which influences quality of irrigation water, in relation to soil for development in agriculture. Based on permeability index, Doneen (1948) classified groundwater as Class I, Class II and Class III to find out suitability of groundwater for irrigation purpose. In study area (Fig.4a, b, c, d) majority of water samples fall in Class II and Class III. Some of the samples fall in class I categories. It indicates water is moderate to good for irrigation purposes.

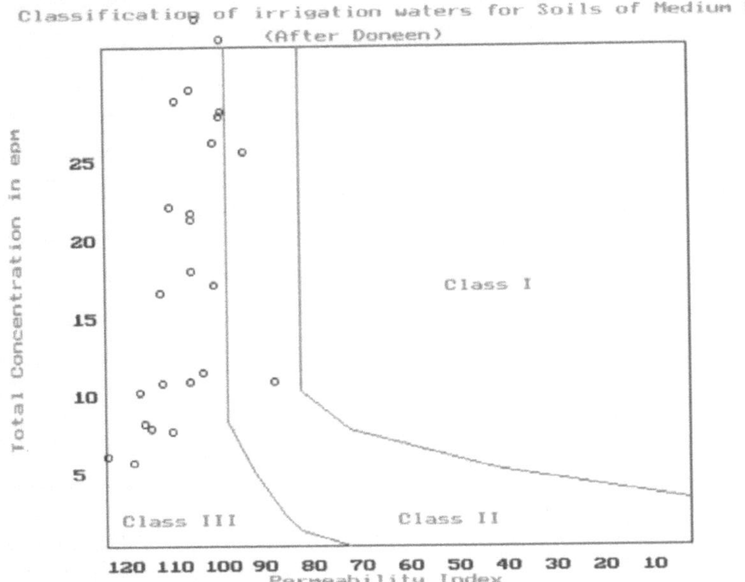

Fig. 4a. Permeability index map of the study area (NWM)

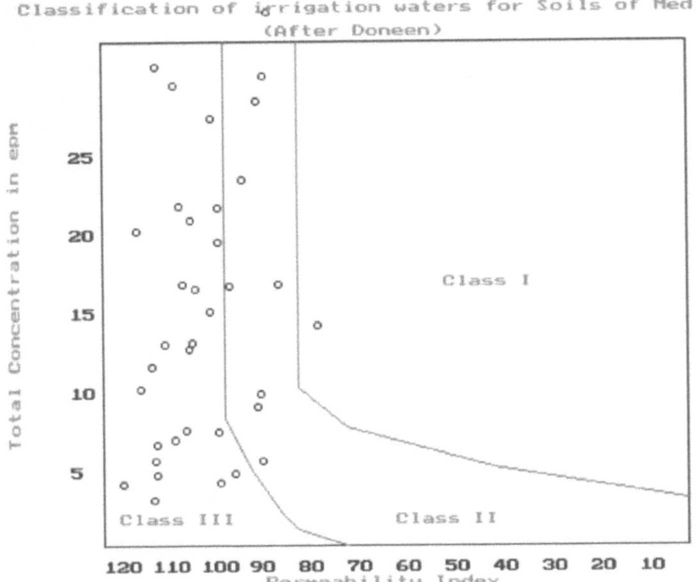

Fig. 4b. Permeability index map of the study area (SWM)

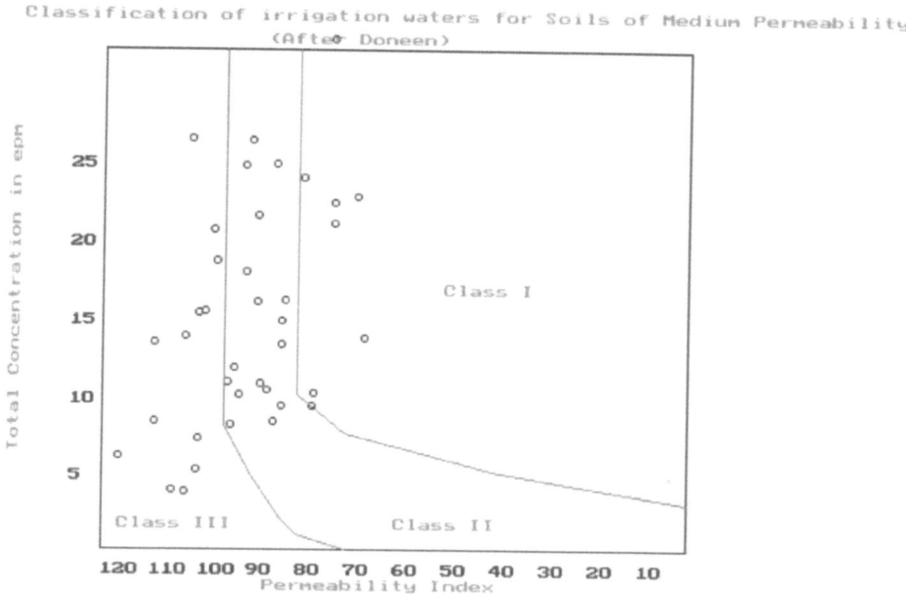

Fig. 4c. Permeability index map of the study area (POM)

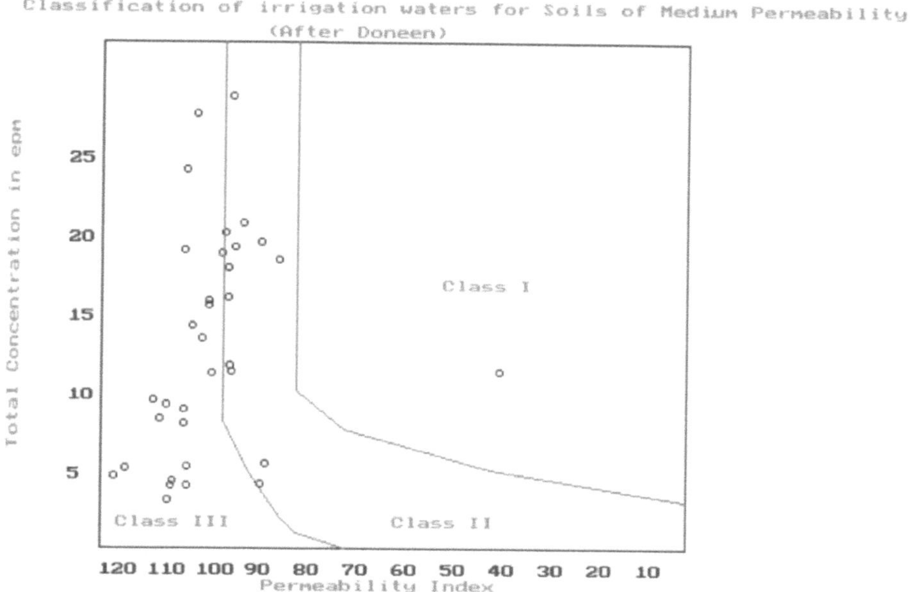

Fig. 4d. Permeability index map of the study area (SUMMER)

Piper diagram shows that in all the season (Fig.5) of groundwater samples there is a continuous mixing between Cl and HCO_3 (strong and weak acids) with few representin of Mg. The dominance of Na and Ca cation trend is noted in all season. In SouthWest and Summer season HCO_3 + Cl dominants the anions, with few representations for SO_4. In NE and POS season Na + K - Ca + Mg is the dominate. Whereas most of the POM samples show mixing trend of all cation and SO_4 + Cl + HCO_3. This indicate that the additional source of leaching from surface soil contribute to the mixed trend.

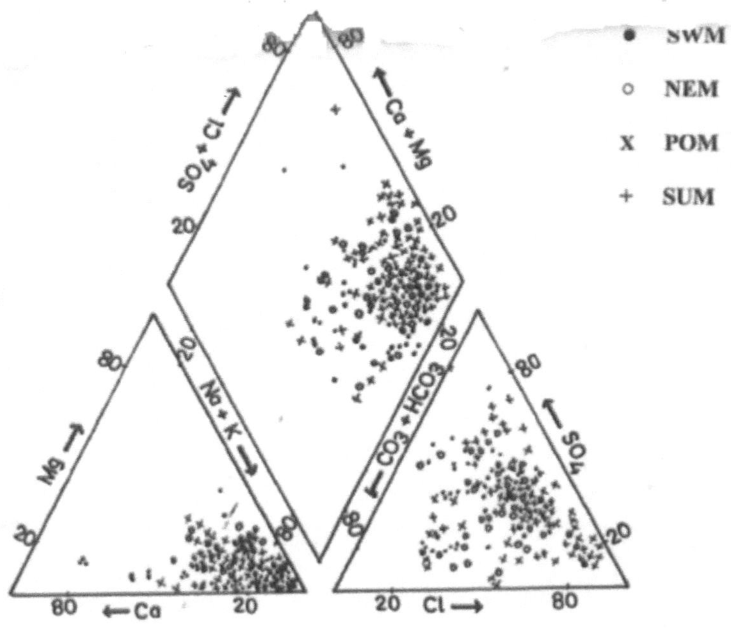

Fig. 5. Piper diagram for study in all the season

The modified Johnson's (Diamond) plot, Geochemistry of groundwater is influenced by geochemical reaction and mixing. Here cation is dominated by Na>Ca>Mg and anions by Cl>HCO_3>SO_4. Evolution of water is chiefly dependant on this relationship between rock types and water composition of anions and cations (Fig.6a,b,c,d) (Piper 1944). In figure samples fall in water contaminated by secondary precipitation and in static and dis-coordinated regimes irrespective to seasons. Single representation from SWM is also noted in high Ca+Mg+SO_4+Cl zone and HCO_3+CO_3 zones indicating weathering impact. In general alkali exceeds alkaline earth controlled by strong and weak acids.

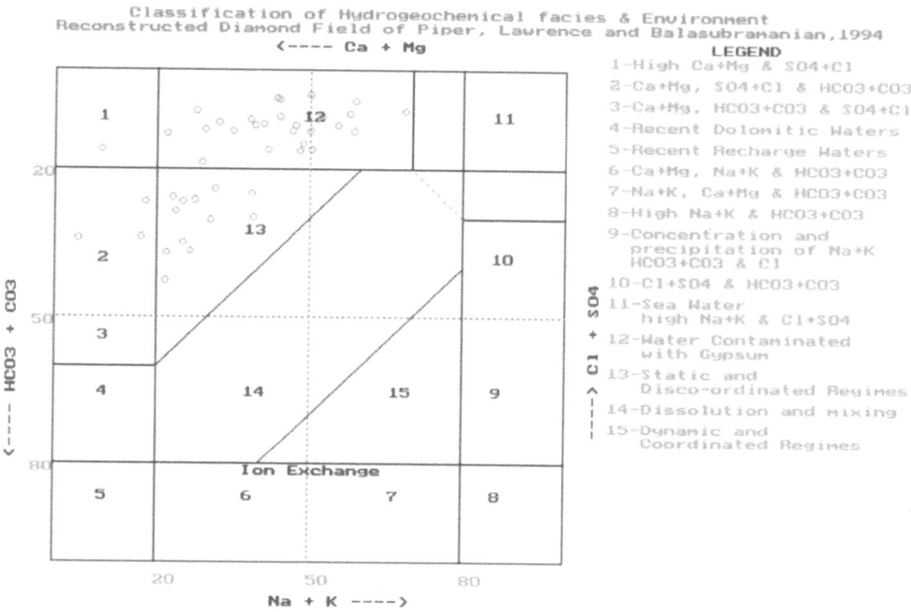

Fig. 6a. Modified diamond Johnson's plot for the study area (NWM)

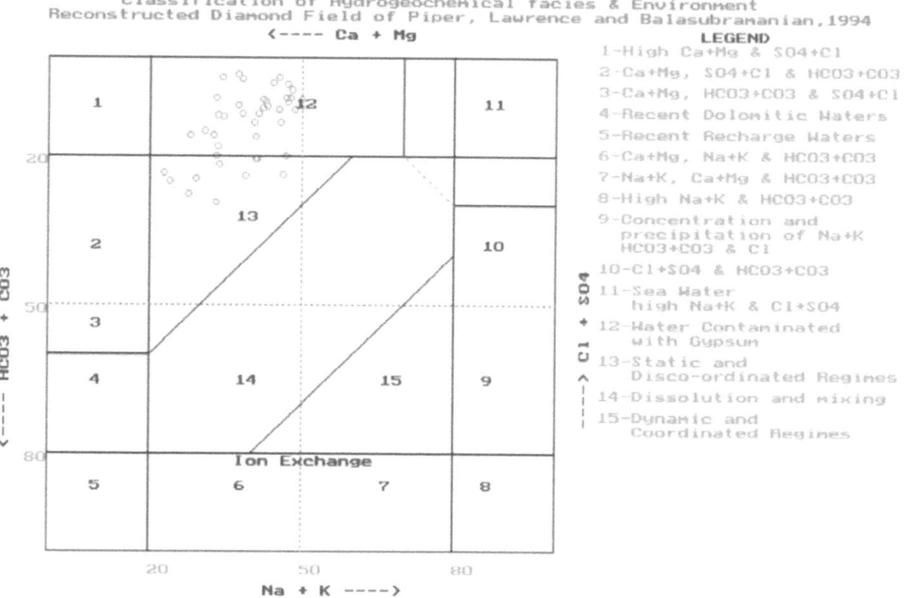

Fig. 6b. Modified diamond Johnson's plot for the study area (SUM)

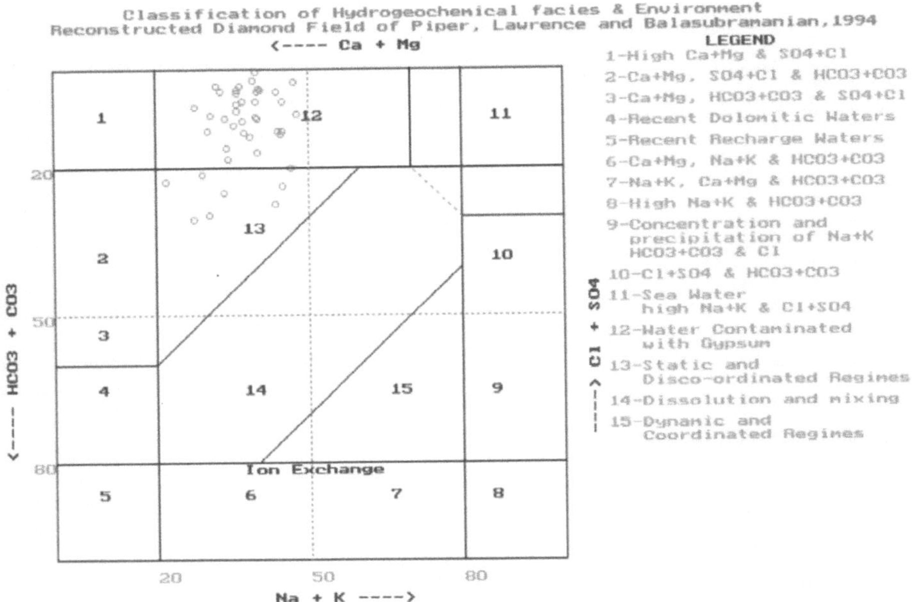

Fig. 6c. Modified diamond Johnson's plot for the study area (POM)

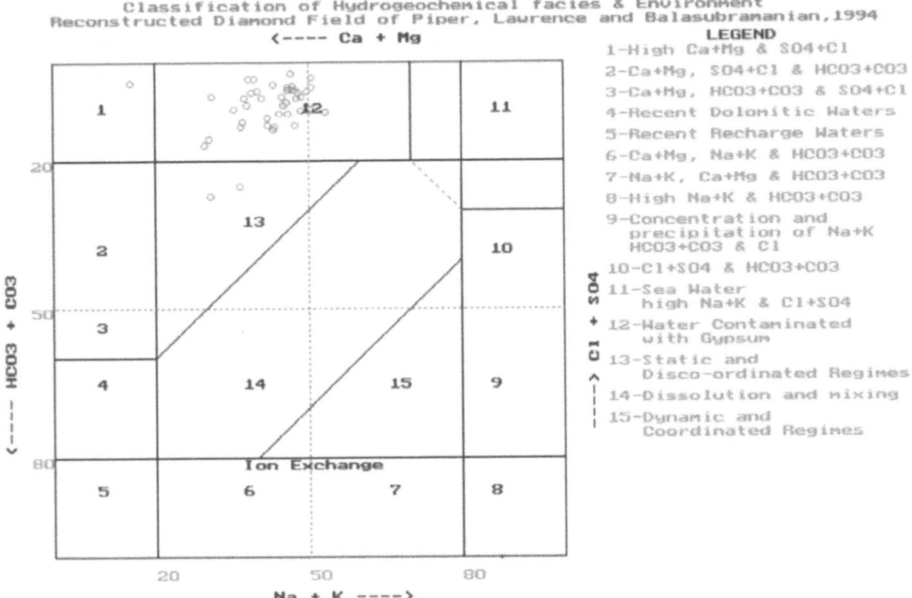

Fig. 6d. Modified diamond Johnson's plot for the study area (SUM)

Gibbs boomerang diagram shows (Fig.7a,b,c,d) mechanism controlling chemistry of groundwater in the study area (Gibbs, 1970). It is seen that majority of samples falls in weathering dominant zone in all seasons but few representations of Post monsoon fall away from boomerang zone shows impact of secondary salt precipitation and mining water controls water chemistry. The sources for these ions are secondary; hence they fall away from zone fixed by Gibbs. Further it is noted that there is no permanent evaporation zone in the study area, which control water chemistry.

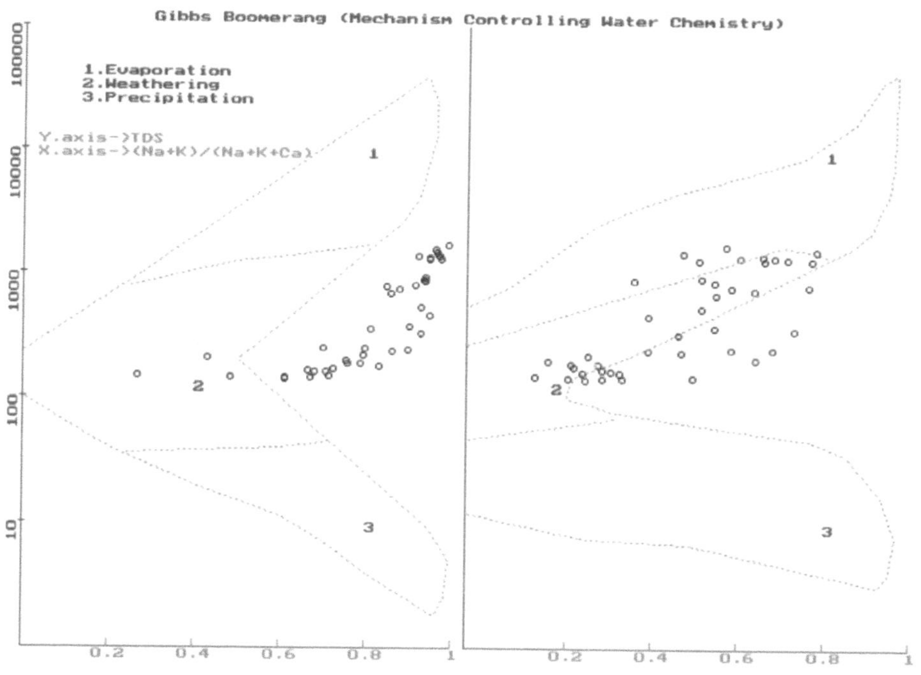

Fig. 7a. Gibbs diagram for the study area (NWM)

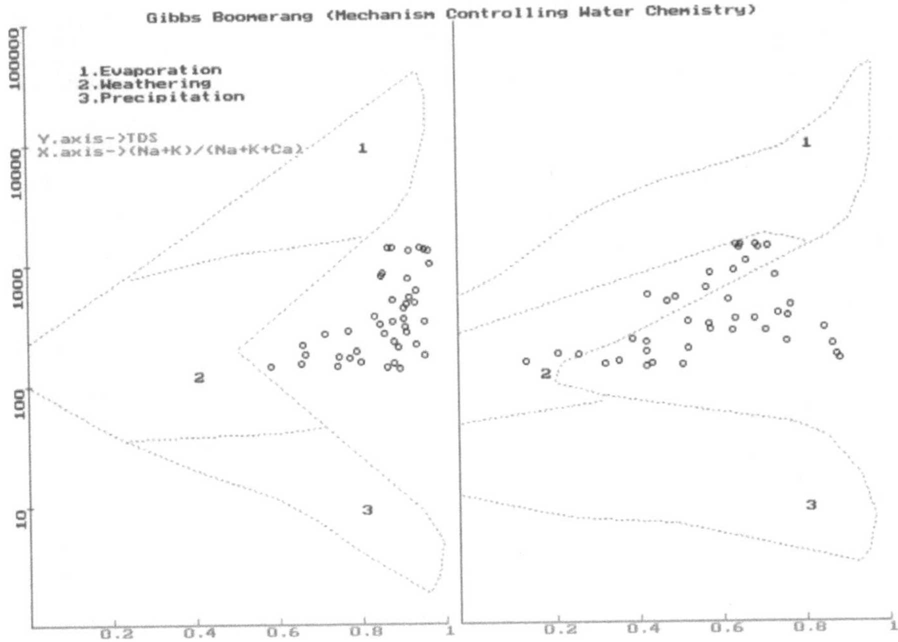

Fig. 7b. Gibbs diagram for the study area (SWM)

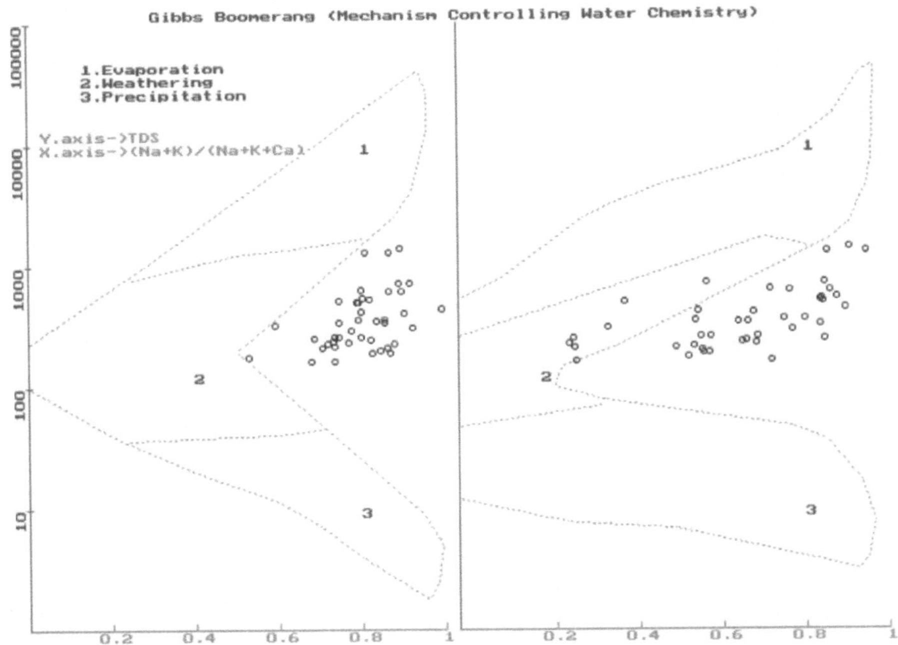

Fig. 7c. Gibbs diagram for the study area (POM)

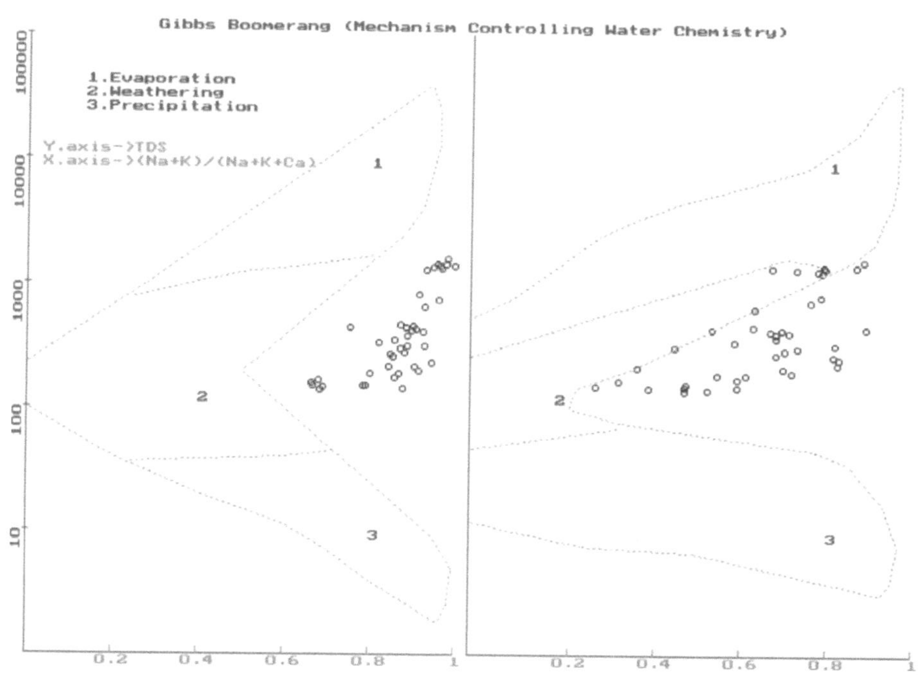

Fig. 7d. Gibbs diagram for the study area (SUMMER)

Thermodynamic stability diagram for Na System this is witnessed by samples (Fig.8a) falling in kaolinite zone and evolve ion of groundwater is noted as shift from kaolinite to Na - montmorllite boundary. In Na system samples of NWM, SWM and POM periods fall in kaolinite stability field. In summer it shifts of samples from kaolinite to montmorllionite. In decrease of silica and Na concentration is noted. This may be due to enhancement of Na dissolution summer period in groundwater. It is favoured by the reverse of reaction.

In K System the preceding relations as a plot of log [K+]/[H+] vs. log[H_4SiO_4]. Among the aspects of interest is that most ground and stream waters fall into the kaolinite field of stability. In K system majority (Fig.8b) of the samples respectively fall in kaolinite zone, but shift in stability from kaolinite to muscovite zone is also noted in few samples from SWM where reverse of reation 2 take place as K is added into the system after rainfall hence lesser K is noted in groundwater of SWM. During the other season K is related into the system and this is witness by the reaction 1 or 3 either during the conversion of K – feldspar to muscovite/kaolinite or gibbsite.

In Ca system (Fig.8c), but samples from SWM, NEM and POM periods cluster in Kaolinite zone, indicating dilution of Ca ions. In SUM it shift from Kaolinite to Montmorllionite field is noted may be due to the dissolution of Ca rich minerals in the groundwater, and Mg system (Fig.8d) majority of samples respectively of all season fall in kaolinite zone in summer it shift of samples kaolinite to chlorite zone is noted.

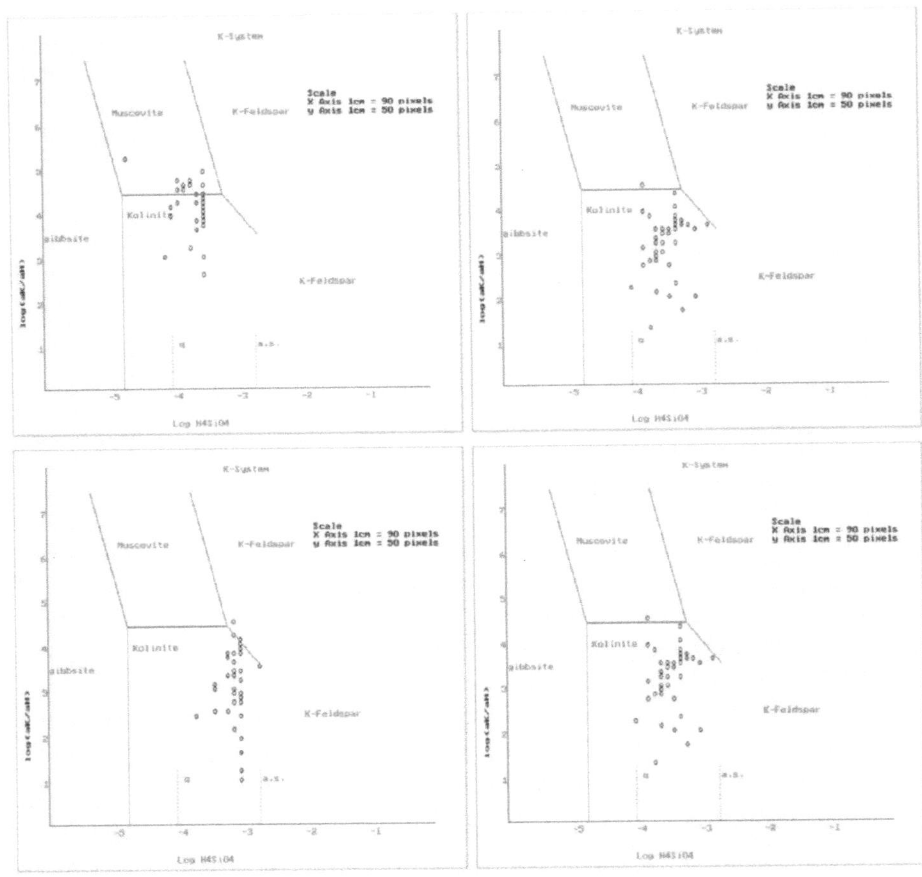

Fig. 8a. Thermodynamic stability diagram for K System for all seasons.

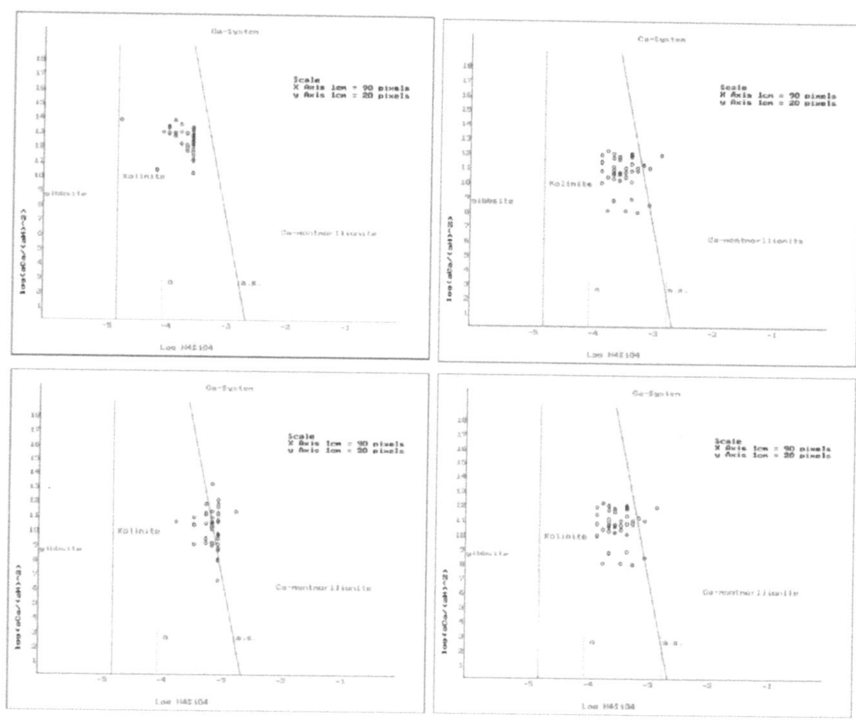

Fig. 8b. Thermodynamic stability diagram for Ca System for all seasons.

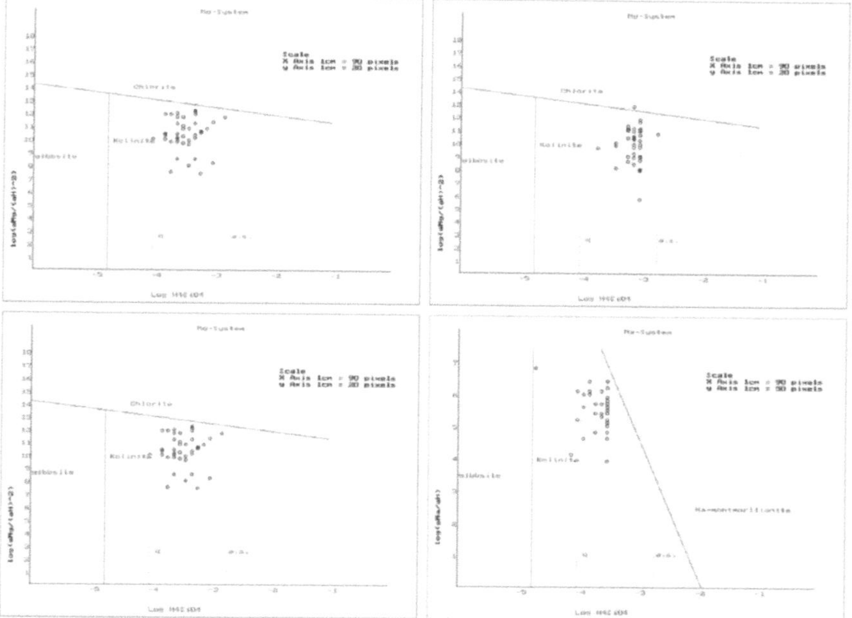

Fig. 8c. Thermodynamic stability diagram for Mg System for all seasons.

Conclusion

Groundwater in the study area is generally acidic to alkaline in nature with pH ranging from 5.1 to 8.65. It was higher in POM and lower in SWM period. Highest values of EC were observed in SWM and POM and drop in EC was noted in NEM and SUM. In SWM and SUM season HCO_3 + Cl dominates the anions, with few representations for SO_4. In NEM and POM season Na + K - Ca + Mg is the dominant. USSL classification shows most of water samples irrespective of seasons fall in C_3S_1, C_2S_3, C_3S_3, C_2S_1, C_1S_1 few indications of North East, South West and POM are also noted in C_3S_4 zone indicating wide range of salinity and alkalinity. Doneen Permeability index classification of groundwater shows majority of water samples fall in Class II and Class III. Some of the samples fall in class I càtegories. It indicates water is moderate to good for irrigation purposes. Gibbs boomerang shows that majority of samples falls in weathering dominant zone in all seasons but few representations of POM fall away from boomerang zone shows impact of secondary salt precipitation and mine water chemistry. Thermodynamic stability of Na system shows that samples of NWM, SWM and POM periods fall in Kaolinite stability field. Similar condition is noted in Ca system. In summer it shifts of samples from Kaolinite to Montmorillonite. Shift in stability from Kaolinite to muscovite zone is noted in few samples of SWM in K system. Shift in stability from Kaolinite to Chlorite zone is noted in few samples from SWM in Mg system.

References

Anandhan P (1998) Environmental impact assessment around Neyveli mining area using Hydrogeochemical, Sedimentalogical and Remote Sensing Techniques- South Eastern India. Unpublished Thesis

Anandhan P, Ramanathan AL, Chidambaram S, Manivannan R, Ganesh N, Srinivasamoorthy K (2000) A study on the seasonal variation in the geochemistry of the ground water in and around Neyveli region, Tamilnadu. In: Proceedings of International seminar on Applied Hydrogeochemistry, Annamalai University pp 86-105

APHA (1998) Standard methods for the examination of water and wastewater, 19th edition. APHA, Washington DC, USASS

Chidambaram S, Anandhan P, Prasanna MV, Srinivasamoorthy K, (2013) Arabian Journal of Geosciences 6 (9), 3451-3467 Major ion chemistry and identification of hydrogeochemical processes controlling groundwater in and around Neyveli Lignite Mines, Tamil Nadu, South India

Hem J D (1991) Study and interpretation of the chemical characteristic of natural waters. 3rd Edn. US Geol Survey, water supply paper 2254, scientific pub, Jodhpur.

Gibbs R J (1970) Mechanisms controlling world's water chemistry. Science, 170, pp.1088–1090.

Piper AM (1944) A graphic procedure in the geochemical interpretation of water analysis, Trans Geophysics Union 25: pp 914 - 923

Ramanathan AL, Chidambaram S, Anandhan P, Manivannan R, Mala V, Ganesh N (2000) Abnormalities of Certain - Dissolved major ions in Groundwaters of the coastal tract from Puduchatiram to Chidambaram, East Coast of India. In proceedings of International Seminar on applied Hydrogeochemistry, Annamalai University pp 57-75

Ramesh R, Anbu M (1996) Chemical methods for environmental analysis -Water and Sediment, 161 p

USSL (1954) Diagnosis and improvement of Saline and Alkali soils, USDA Handbook 60:147

WHO (1996) Guide lines for drinking water quality. Geneva 1: 53-73.

Chapter 3

An Attempt to Determine the Mountain Front Recharge (MFR) by Using State of Saturation Index (SI) of Carbonate Minerals Along the Foothills of Courtallam, Tamil Nadu

Banaja Rani Panda, Chidambaram S., Devaraj N. and Ganesh N.
Department of Earth sciences, Annamalai University, Annamalai Nagar Tamil Nadu, India

ABSTRACT

In semiarid climates, a significant component of recharge to basin aquifers occurs along the mountain front. In this scenario Mountain Front Recharge study has been carried along the foothills of Courtallam ,Tamil Nadu. A total of 53 groundwater samples were collected along the foothills and measured for major cations and anions using standard procedure. Saturation index (SI) of carbonate minerals like Aragonite,Magnesite, Calcite and Dolomite,LogpCO_2 and Ionic strength obtained from WATEQ4F output. pH plays a significant role in the saturation index of the minerals. Most of the Carbonate minerals shows undersaturation with the pH range 7.5. Positive relationship between SI and HCO_3 is noticed. A sharp increase of pCO_2 values towards the foothills region suggests that the groundwater has been recharged by, acquiring longer flow path during its infiltration into deeper level near foot hill region. The SI of these carbonate minerals was spatially distributed to identify the areas of higher SI and also it is compared with the water level fluctuation to understand the recharge process along the foothill region as well as riparian zone.

Key words: Saturation index, mountain front recharge, ionic strength, carbonateminerals.

Introduction

The mountain front recharge (MFR) is the contribution of mountains to recharge the aquifers in the adjacent basin. Stream channels from adjacent mountain block and mountain front (Davidson 1973; Anderson et al. 1992), can contribute a significant amount of recharge to the aquifers within semiarid catchments in comparison to the recharge through basin floor and channel at lower elevation (Huntley, 1979; Wilson et al. 1980). There are a lot of techniques adopted for the successful determination of MFR, around different regions of all over the world. James et al. 2004, Mohrbacher et al. 1984, have adopted a isotopic and geochemical approach to understand MFR, at Tucson basin, Arizona. Andrew et al. (2003) used noble gases to investigate MFR. Ping et al. (2014) have quantified groundwater recharge using the chloride mass balance method in a semi-arid mountain terrain, of South Interior British Columbia, Canada. Kambhammettu et al. (2011) evaluated three MFR estimation techniques using a numerical modeling in the Southern Jornada Del Muerto Basin, New Mexico. Although there are a lot of studies on MFR have been carried out all over the world but limited attempt found in India. This is the first attempt in India, to estimate MFR along the foot hills of Courtallam, India by using hydro geochemistry. Some previous investigations in this region focuses only on, water level fluctuation (Public Works Department (PWD), 2002),water quality and its stability to use for agriculture, drinking and domestic needs Subramani et al. 2005a, Groundwater potential assessment (Balasubramanian and Sastri 1994). These approaches does not consider, the effect of saturation index and water table fluctuation on groundwater recharge. Thus the main objective of our study is to have a detail understanding about the effect of Saturation Index (SI) of carbonate minerals and water table depth, to the groundwater recharge along the foothill.

Studyarea

The study area falls between the geographical co-ordinates (Fig. 1) 8.834°N to 9.328°N latitude and 77.155°E to77.527°E longitude. Chittar river, a tributary of Tamiraparani River form the pediment plain of the study area which arises from the hills of Courtallam and flows around 60 km through the Tenkasi and Tirunelvelli and joins with the Tamiraparani River. Several water falls in this region are the major attraction point for the tourist, which results in urbanization as well as increased water demand.The beauty of this place is due to the presence of number of hills such as AgasthiarMalai (Mountain), Agasthyamalai hills Cherumunjimottai and Naga Pothigai. Agasthiarmalai is located in the Western Ghats is a part of Anaimalai hills. Eastern part of Pothigai hills (Agasthiarmalai) falls in the Tirunelvelli district with an elevation of 1866m (6122ft).

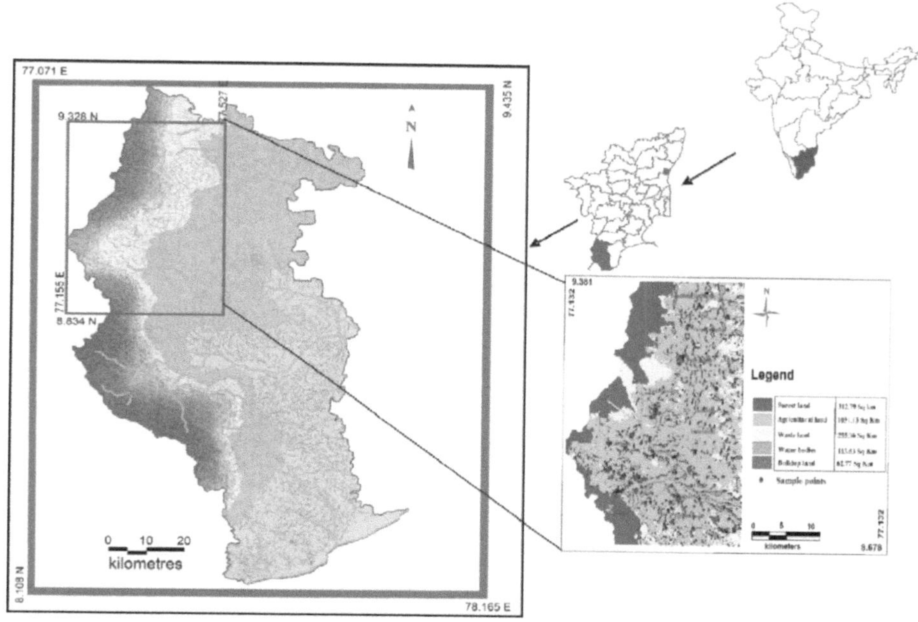

Fig. 1. Location map of the study area with land use pattern.

The studyarea receives 917.88 mm/yr rainfall annually reported by IMD, out of which NE monsoon contributes 70% and rest by SW monsoon and rainfall during transitional period. Ground water generally occurs under phreatic conditions in the weathered mantle and under semi-confined conditions in the fissured and fractured zones at deeper levels (CGWB, 2009). During November and December water table shows its maximum level and later on it starts decreasing and shows its minimum level during September or October month (Subramani, 2005). Granite, Charnockites and Biotite gneiss are the major formation of this region from which Charnockites covers the northern part and fissile hornblende gneiss the southern part of the study area (Fig. 2).

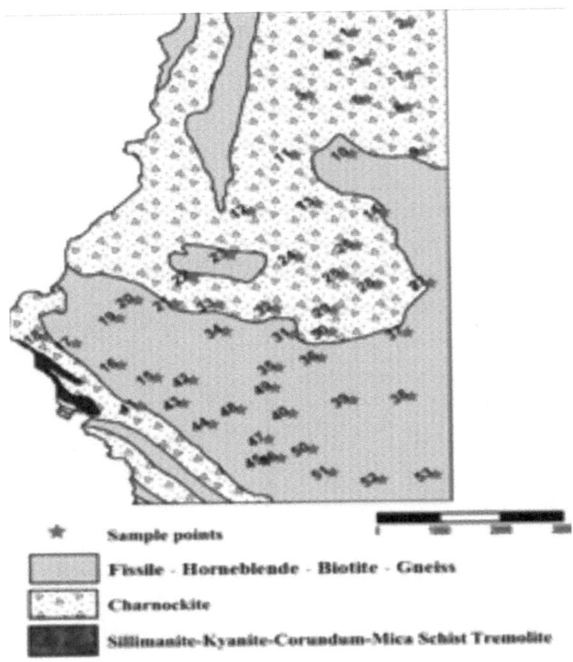

Fig. 2. Lithology map of the study area along with sampling points

Methodology

In order to understand MFR, 53 groundwater samples were collected along the foothills of Courtallam using polythene bottles during the month of April 2015, which covers the entire lithounits of the study area. Collected samples were measured for major cations and anions using standard procedure (APHA, 1998). Other physical parameter like pH, TDS, condutivity and humidity were measured using water analysis kit in field itself. HCO_3^- was determined by volumetri ctitration. WATEQ4F was used to derive the saturation index (SI) and LogpCO_2 and ionic strength were obtained from the WATEQ4F output. SI of carbonate minerals correlated with HCO_3^-, pH , Ionic strength and LogpCO_2 and plotted graphically to delineate the recharge process along the foothills of Courtallam.

Results and Discussion

The carbonate minerals Magnesite, Dolomite, Calcite and Aragonite are taken to consideration for the study. These minerals are grouped into three different categories on the basis of range of saturation index (SI). If SI of minerals <0, then groundwater is significantly under saturated. The SI value is 0 then groundwater is at equilibrium. Whereas SI >0 shows super saturated condition.

Relationship between HCO_3^- and SI of Carbonate minerals

The saturation state of carbonate minerals were plotted against the concentration of HCO_3 of all samples, it can be considered that a SI of -3.5 to 1.5 is indicative of saturation of all the minerals (Fig. 3). According to this, groundwater in all aquifer along the foothill is found to be under saturated with respect to carbonate minerals with higher concentration of bicarbonate, may be due to non-availability of cations (Ca and Mg) which might have been removed from the aqueous system due to process of cation exchange (Gomez and others, 2006). Samples away from the foothills shows that most of the minerals are under saturated in nature and shows an increasing trend of SI with increase of bicarbonate (Fig 3) whereas few mineral ranges from near saturation to saturation and also increases with increase of HCO_3. Hence there is a direct relationship in the trend of SI and HCO_3.

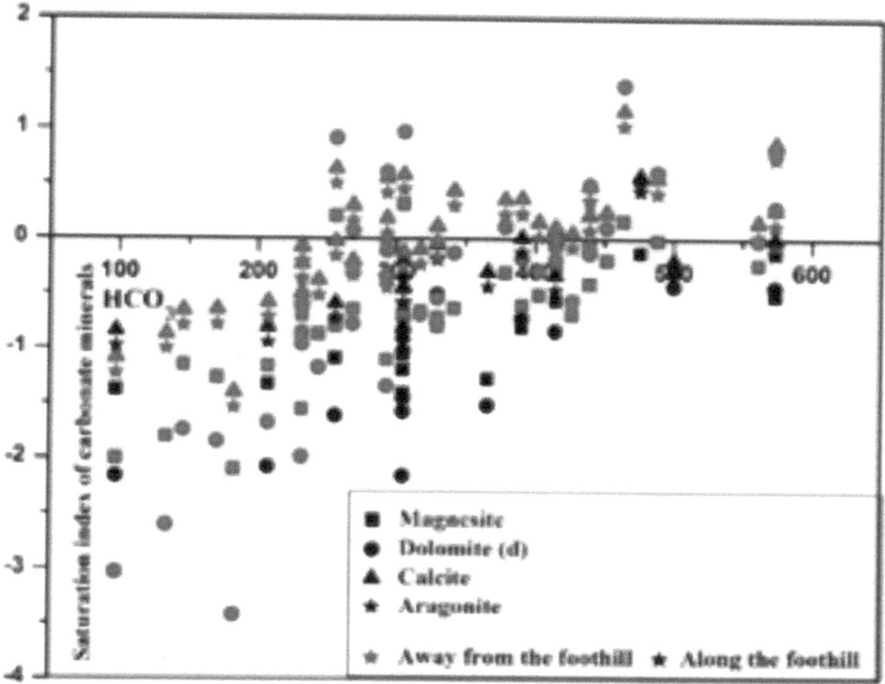

Fig. 3. The relation between the SI of carbonate minerals in all samples and their respective HCO_3.

Relationship between pH and SI of carbonate minerals

The groundwater in this region is alkaline in nature with pH ranging from 6.2 to 7.95 (Fig. 4). pH shows an increasing trend with increase of SI. According to this, groundwater in all aquifer along the foothill is found to be ranges from undersaturation to saturation. Most of the samples away from foothills, falls undersaturated category and a few samples ranges from near saturation to saturation. pH increases with increases of SI of minerals. This clearly shows that the H^+ is exchanged for the cations and these ions are made available which in turn help in the SI of carbonates (Chidambaram *et al* 2011). It is evident that the minerals which shows undersaturation mostly concentrated around the pH range of 7.5 above which it gets saturated and further it becomes over saturated.

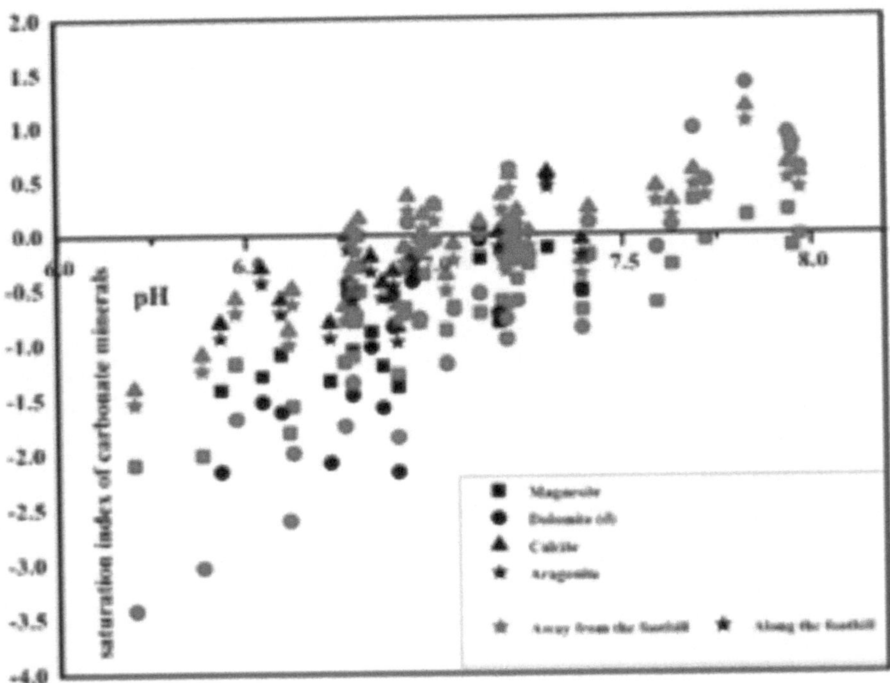

Fig. 4. The relation between the SI of carbonate minerals in all samples and their respective pH.

Relationship between logpCO_2 and SI of Carbonate minerals

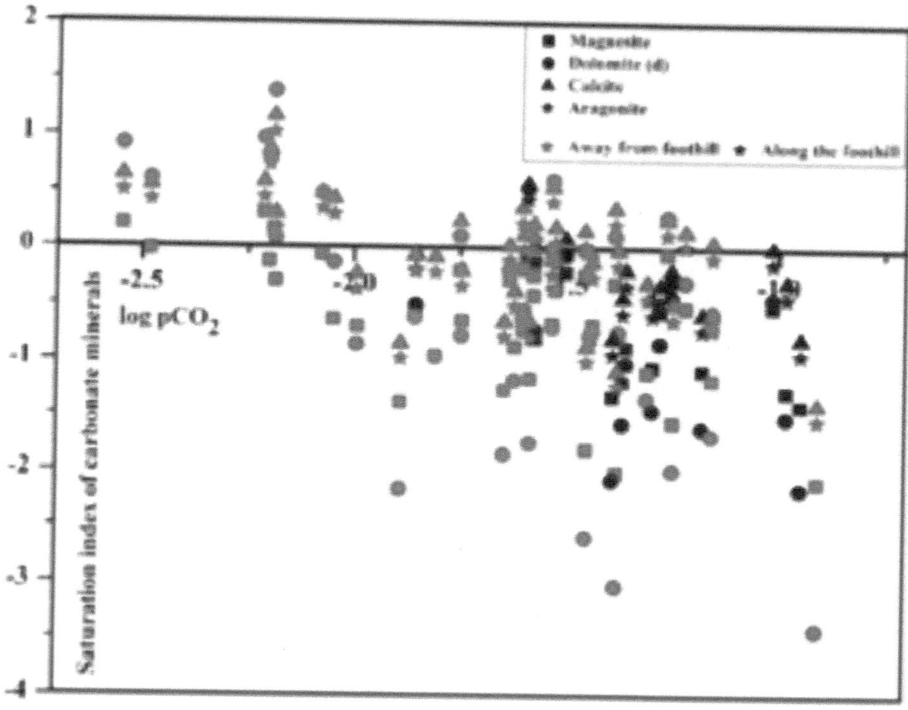

Fig. 5. The relation between the SI of carbonate minerals in all samples and their respective logpCO_2.

An increase in the carbon dioxide partial pressure during water-rock interaction influences the process of mineral dissolution. Partial pressure of CO_2 plays a significant role in the state of saturation of carbonate minerals. Considering the above fact and relationship between log pCO_2 and SI of carbonate minerals, a graph was plotted (Fig. 5).It is observed that pCO_2 ranges from -2.6 to -1. The SI of minerals generally decreases with increase of logpCO_2 values. Low log pCO_2 values show higher SI for all the minerals except a few minerals which shows near saturation. A sharp increase of pCO_2 values towards the foothills region which reveals that, the atmospheric interaction becoming lesser and recharge is predominant. Higher value of pCO_2 indicates higher residence time (Prasanna et al. 2008). It suggests that the groundwater has been recharged by, acquiring longer flow path during its infiltration into deeper level near foot hill region. A sharp decrease of pCO_2 values noted along the other regions reveals that there is more interaction of the groundwater system with the atmosphere (Chidambaram et al. 2011).Thus it can be inferred that the groundwater has been acquired a shorter flow path, and recharged from lower elevation.

Relationship between Ionic strength and SI of Carbonate minerals

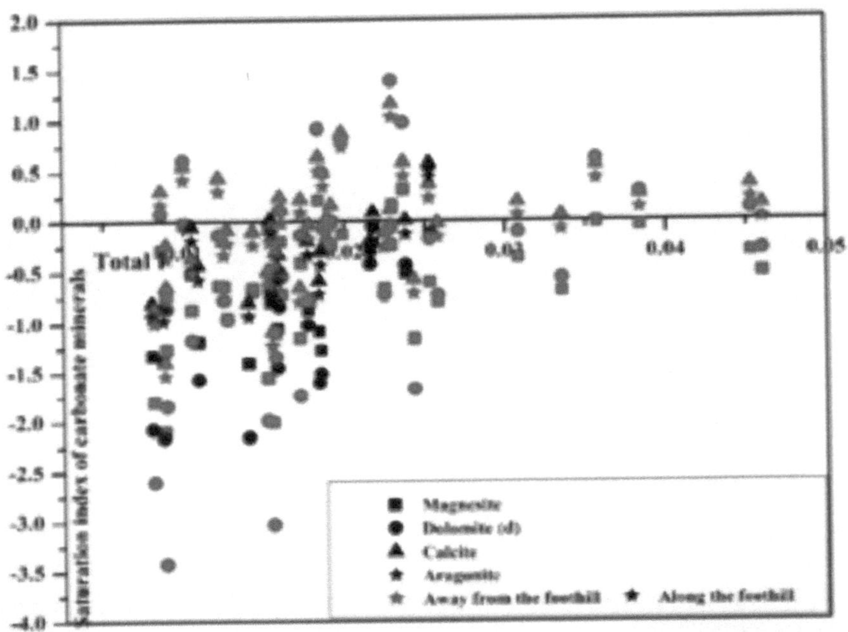

Fig. 6. The relation between the SI of carbonate minerals in all samples and their respective ionic strength.

Ionic strength (IS) is a measure of total concentrationof ions in solution which emphasizes increasedcontribution of species with charges greaterthan one to solution non-ideality (Domenico and Schwartz 1990).

$$I = 0.5\ \Sigma mz^2$$

Where m is the concentration of a given ion in moles per liter and z is the charge on that ion. The terms in the summation include one for each ionic speciespresent. The relationship of the IS of samples and SI of these minerals clearly show (Fig. 6) positive relation. The SI of all the minerals along foothill shows under saturation whereas minerals away from foothill ranges from under saturation to saturation with the increase of ionic strength. It is also evident that the under saturated minerals along the foothill mostly concentrated around the IS range of 0.025 and above which it gets saturated.

Spatial distribution of SI of carbonate minerals

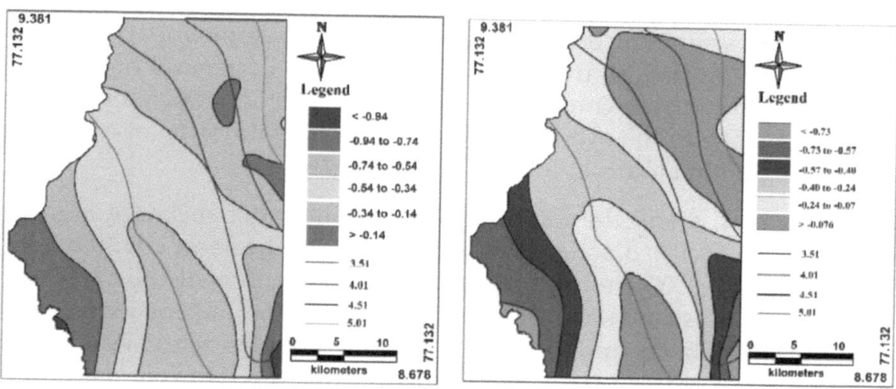

(a) Spatial distribution of SI of Aragonite. (b) Spatial distribution of SI of Calcite.

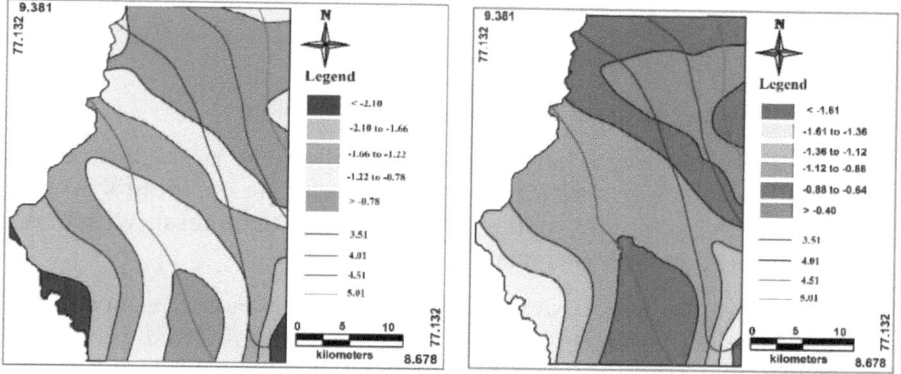

(c) Spatial distribution of SI of Dolomite. (d) Spatial distribution of SI of Magnesite.

Fig. 7. Spatial distributions SI of carbonate minerals with water level fluctuation in mbgl.

The spatial distribution of the Aragonite shows a minimum of -0.94 to a maximum of 0.049 (Fig. 7a). Higher values of Aragonite is from under saturation to near saturation is noted along the northeastern and southern part of the study area. Most of the samples show spatially under saturated nature. SI values of Magnesite (Fig. 7d) ranges from -1.85 to -0.40. Higher value noted along the northeastern and southern part of the study area. A similar trend as Aragonite is also observed in the spatial distribution of Magnesite but all the samples spatially shows under saturation though the trend remains the same. The SI ranges of Calcite ranges from -0.90 to 0.088 (Fig. 7b). All the samples spatially show under saturation to near saturation. Higher values are noted along the northeastern and southern parts of the study area. The SI values of Dolomite ranges from -2.54 to -0.54. Spatial distribution of Dolomite (Fig. 7c) shows similar trend of Calcite but all the samples spatially show under saturation though the trend remains same. In general higher SI is noted along the region comprised of Charnockites and Gneisses (Fig. 2). This may be due to the availability of the ions for saturation of carbonate minerals (Chidambaram et al., 2011).

In general, along the foothillregion groundwater recharge is much more gradual with discharge which signifies the lesser water level fluctuation. (Fig. 7). The state of saturation index for all the carbonate minerals also lesser along this region, due to the fact of increase of groundwater recharge and dilution of groundwater chemistry with respect to these minerals (Karmegam 2012). But along the riparian zone saturation index increases with higher water level fluctuation, which substantiates the fact that recharge is significantly delayed relative to discharge.

Conclusion

The study has arrived the following conclusions:

1. There is a positive relation observed in the trend of the SI to that of HCO_3 whereas groundwater in all aquifer along the foothill is found to be under saturated with respect to carbonate minerals with higher concentration of bicarbonate.
2. The minerals which shows undersaturation mostly concentrated around the pH range 7.5 above which it gets saturated and further it becomes over saturated with increse of pH .
3. pCO_2 values decreases causing increase of saturation index along the riparian zone suggesting that it is an open system and the water has been recharged from lower elevation.
4. Sharp increase of pCO_2 value along the foothill suggest that atmospheric interaction is less and recharge is predominant.
5. A positive relation is noted between the SI of carbonate minerals and IS.Under saturated minerals along the foothill mostly concentrated around the IS range of 0.025 and above which it gets saturated.

Hence the study derives the relationship of state of SI with pH, HCO_3, logpCO_2, Ionic strength and water level fluctuation which in turn helps in the identification of recharge effect along the foothill as well as riparian zone.

Acknowledgement

The author Banajarani Panda wish to express her sincere thanks to Department of Science and Technology for providing the Inspire fellowship (No: DST/INSPIRE Fellowship/ [IF150615], 27th October 2015).

Reference

Anderson, T.W., Freethey, G.W., Tucci, P. (1992). Geo-hydrology and water resources of alluvial basins in south-central Arizona and parts of adjacent states. USGS Professional Paper Reston, Virginia, USGS 1406-B.

Andrew, H., Manning, D., Kip Solomon (2003). Using noble gases to investigate mountain-front recharge.Journal of Hydrology 275: 194–207.

APHA (1998). Standard method for the examination of water and wastewater; APHA, Washington DC, USASS, 19thedn.

CGWB (2009). District groundwater brochure, Tirunelveli district, Tamilnadu.

Balasubramanian, Sastri (1994). Groundwater level fluctuation and quality monitoring. Public Works Department (PWD), Tamil Nadu, India.

Chidambaram S, Karmegam U, Prasanna M V, Sasidhar P and Vasanthavigar M (2011). A study on hydrochemical elucidation of coastal groundwater in and around Kalpakkam region, southern India; Environ. Earth Sci., doi: 10.1007/s12665-011-0966-3.

Domenico P A and Schwartz W (1990). Physical and Chemical Hydrogeology (New York: John Wiley), p.824.

Davidson, E.S. (1973). Geohydrology and water resources of theTucson Basin, Arizona. USGS Water Supply Paper (1939). E.Reston, Virginia USGS.

Gomez,P., M.J. Turrero, A. Garralon, J. Pena, B.Buil,B.de.lacruz, M. Sanchez, D.M.Sanchez, A.Quejido, C. Bajos, and L.Sanchez (2006). Hydrochemical characteristics of deep groundwaters of the Hesperian Massif (Spain) Journal of Iberian zgeology 32(1): 113-131 ISSN:1698-6180.

Huntley, D. (1979). Ground water recharge to the aquifers of northern San Luis Valley, Colorado.Geological Society of America Bulletin, Part II 90(8): 1196–1281.

James F H and Brenda E (2004). Quantifying mountain front recharge, Project report. Department of hydrology and water resources.

Kambhammettu, B.V.N.P., James, P., King, PraveenaAllena (2011). Evaluation of mountain-front recharges estimation techniques for Southern New Mexico basins. International Journal of Water Resources and Environmental Engineering 3(3): 66-72.

Karmegam, U (2012). A study on the Hydrogeochemical modelling of coastal aquifers in and around Kalpakkam. Ph.D thesis, Annamalai University.

Mohrbacher C (1983). Stable isotopes and ground water chemistry as indicators of mountain front recharge, Tucson basin, Pima County, Arizona. Hydrogeology and water resources in Arizona and the southwest. 13: 77-84.

Ping J, Nichol C, Wei X (2014). Quantification of groundwater recharge using the chloride mass balance method in a semi-arid mountain terrain, South Interior British Columbia, Canada. Journal of Chemical and Pharmaceutical Research. 6(1): 383-388.

Prasanna MV (2008). Hydrogeochemical studies in the Gadilam river basin, Tamilnadu. Unpublished Ph.D. Thesis, Annamalai University.

Subramani T (2005). Hydrogeology and identification of geochemical processes in Chittar River Basin, Tamil Nadu. Anna University, Ph.D.Thesis.

Subramani, T., Elango, L., Damodarasamy, S.R. (2005a). Groundwater quality and its suitability for drinking and agricultural use in Chithar River basin, Tamil Nadu, India. Journal of Environmental Geology 47: 1099–1110. doi:10.1007/s00254-005-1243-0.

Wilson, L.G., DeCook, K.J., Neuman, S.P. (1980). Regional Recharge Research for Southwest Alluvial BasinsTucson, Arizona, University of Arizona, Water Resources Research Center.

Chapter 4

Determination of Coastal Vulnerability Using Isotope and Geochemistry

Chidambaram S., Paramaguru P., Karmegam U. and Johnsonbabu G.[1]

Department of Earth Sciences, Annamalai University, Department of Geology
[1]Presidency College, Chennai-5, India

ABSTRACT

The geochemical signature and vulnerability of groundwater of the East Coastal aquifers of India from Pondicherry to Nagapattinam were determined. The coastal aquifers are fragile and this situation becomes more intense in aquifer systems like that of the Pondicherry to Nagapattinam region. A total of 136 water samples were collected during pre and post monsoon of 2008 and 2009 respectively. Out of which 45 samples for each pre and post monsoon from shallow groundwater and 23 samples for each pre and post monsoon from deeper groundwater.. The samples collected were analysed for major, minor cations and anions. Selected samples were also analysed for stable isotopes (d^{18}O and dD). The analysed samples were studied for; their utility for different purpose, the nature of the geochemical classification, their variations in ion ratios, extent of sea water intrusion and the percentage of mixing, The study reveals that chief mechanism controlling shallow groundwater is the salts precipitated in the pore spaces during the Pre monsoon were leached and dissolved during the subsequent monsoon apart from sea water intrusion. Southern part of the study area is more vulnerable than the Northern part of the study area.

Key words: Geochemical, groundwater, coastal aquifer

Introduction

The demand for good quality water is rapidly increasing, but the process of salinization (e.g. Petalas and Lambrakis, 2006; El Yaouti et al.,

2009; Ghiglieri et al., 2012; Sdao et al., 2012) threatens the exploitation of additional water resources such as groundwater. Salinization of aquifers in coastal areas is the result of concomitant processes related to both seawater intrusion and water–rock interaction, which in some cases are almost indistinguishable. The natural balance between freshwater and saltwater in coastal aquifers is disturbed by groundwater withdrawals and other human activities which lower groundwater levels and reduce fresh groundwater inflow to sea. This ultimately causes saltwater intrusion into coastal aquifers (Barlow, 2003). The hydrogeochemical process reveals the zone and quality of water that are suitable for drinking, agriculture and industrial purposes. Hydrogeochemical studies are to understand the changes in water quality due to rock - water interaction as well as anthropogenic influences. Many of the ground waters in the coastal aquifer are suffering from seawater intrusion both by the natural processes and anthropogenic activities like over - extraction, urbanization and agricultural activities (Ozler, 2003). Hydrogeochemical study is useful tool to identify the processes that are responsible for groundwater chemistry (Elango et al., 2003). In recent years, due to the advent of industrial growth, large-scale application of synthetic fertilizers for agriculture production and use of pesticides and insecticides for production has caused serious concern regarding susceptibility of groundwater contamination. In addition to this, various human activities, particularly those involved in land disposal of industrial efuents and sewage efuents, sludge and solid waste, septic tank efuents, urban runoff, and various unknown human activities, contaminate groundwater sources. Isotope hydrology provides complementary information on the type, origin and age of groundwater. The isotope content within the aquifer, will reflect the origin of the water, particularly the location, period and processes of recharge. If the isotope content changes along groundwater paths, this will reflect the history of the water, particularly the process of mixing, salinization and discharge A detailed study on seawater intrusion along the Cuddalore (Ramanathan et al., 1999), Chennai (Gnanasundar and Elango 1999; Satheesh Herbert Singh and Lawrence 2007), and in Pichavaram (Chidambaram et al., 2005) geochemistry of Pondicherry region (Thilagavathi et al., 2012)were carried out. Keeping all the above facts in the mind, an study has been made to understand the complexity of hydrogeochemical characters in the study area. The Pondicherry to Nagapattianam has been selected to have a holistic picture. The main objectives of this study are to understand the hydrogeochemical variation of groundwater with respect to space and time and study the vulnerable zone of this region and probable factors responsible for water chemistry of the region.

Study Area

The area chosen for study is along the Pondicherry to Nagapattinam coastal region, which is situated on the eastern margin of Tamilnadu. It includes the taluks of Cuddalore, Chidambaram, Sirkazhi, Mayiladuthurai, Tharangambadi, Nagapattinam, Kilvelur, Thirukkuvalai and Vedaranyam. The area has also been extended into the Karaikkal Union Territory so as to get a complete overview along the coast.

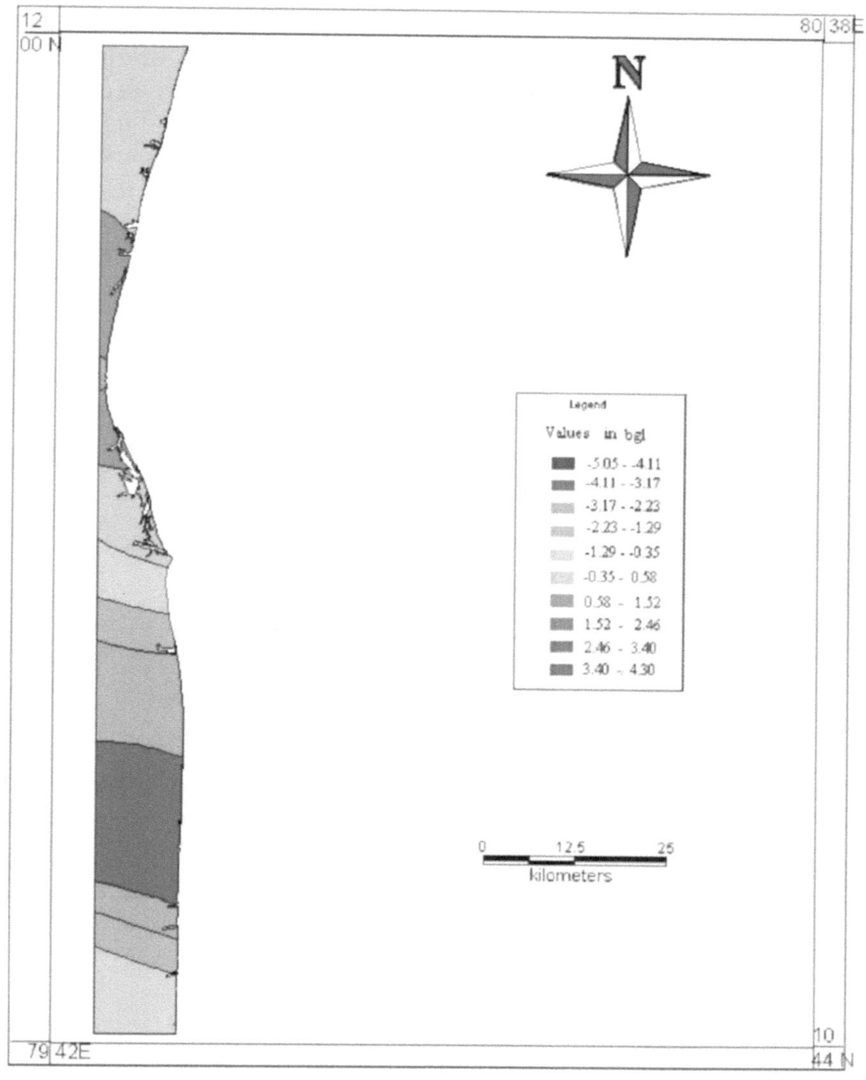

Fig. 1. Spatial distribution of reduced water level in Study area

The temperature varies from 40.6° to 19.3° C with sharp fall in night temperatures during monsoon period. The normal annual rainfall of the Study area is 1230 mm. The major part of the Study area is covered by black clay soils (matured) as per the classification of soil survey and Land Use shows that an isolated patch of brown clay loam soil in the area bordering the north-western boundary of the Karaikal Region. The Study area forms part of Cauvery river basin and is drained by a network of rivers like the Cauvery, the Kollidam, the Arasalar, the Vettar and the Ponnaiyar. In Study area ground water occurs under water table, semi-confined and confined conditions. The general descriptions of the various geological formations occurring in the study area are briefed in the following paragraphs (Table 1). A study of this map revealed that the depth to water table in the Study area ranged between 4 - 9m. In major part of the Study area (Fig. 1) the depth to water table lies within 4.0 mbgl. The reduced water level shows that the major portion of the study area falls in the negative gradients, except along the north eastern boundary of the study area, indicating the probability of sea water intrusion.

Table 1. Main Branch & Rivers of Cauvery in the Study Area

Irrigation Rivers	Irrigation cum	Drainage Rivers
I. CAUVERY SUB-BASIN		
1. Cauvery	1 Nattar	1 Arasalar
2. Kudamurutti	2 Krittamani river	2 Nular
	3 Virasolanar	3 Vanjijar
	4 Vikramanar	4 Nandalar
	5 Cholasundaramaniar	5 Manjalar
	6 Patharasanar	6 Mahimalayar
	7 Mudikondan	7 Manniyar
	8 Puttar	8 Pudumanniar
	9 Palayar	
	10 Thirumalairajanar	

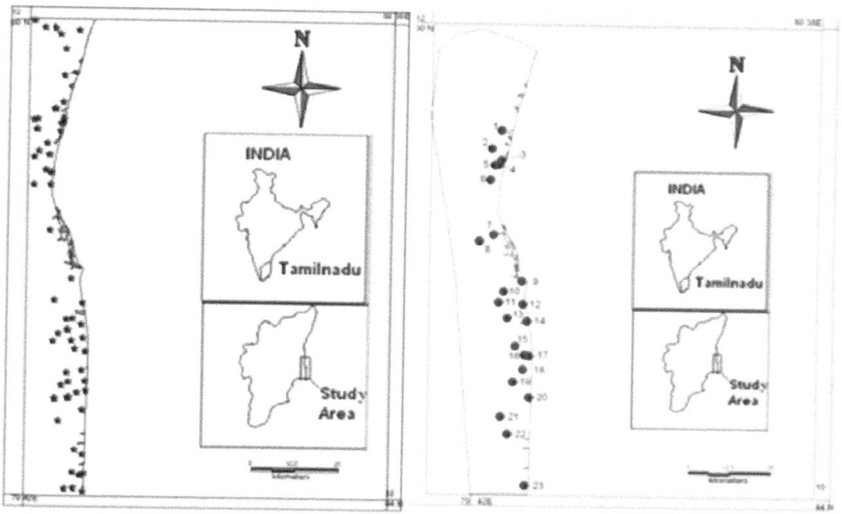

Fig. 2. Sampling location for the surface and groundwater samples

Methodology

A total of 136 water samples were collected during pre and post monsoon of 2008 and 2009 respectively. Out of which 45 samples for each pre and post monsoon from shallow groundwater and 23 samples for each pre and post monsoon from deeper groundwater. The Shallow and Deep groundwater sampling locations are shown in (Fig 2). The quality of ground water is affected by the over-extraction and natural discharge. One liter of water sample was collected in a polyethylene bottle. The samples were then brought to the laboratory for analysis and stored properly (4°C) before analysis. The samples were analysed for major cations, anions, and also analyzed the stable isotopes of oxygen-18 ($d^{18}O$), deuterium dD and carbon-13 (^{13}C) analyses, using standard procedures (APHA, 1998). Groundwater samples were collected in a 60mL capacity polyethylene bottles after pumping out the stagnant water, in order to get representative samples. The sample bottles were filled completely and preserved airtight in order to avoid evaporation. Deuterium and oxygen-18 were measured by mass spectrometer (Geo 20 Europa) using gas equilibration method (Coplen et al., 1991) with a precision of ± 0.5 ‰ and ± 0.1 (2s criterion) respectively. Stable isotope results were expressed with respect to VSMOW, where

$$\delta(‰) = (R_{sample} - R_{SMOW} / R_{SMOW}) X 10^3 \quad \text{(Eq- 1)}$$

$$R = D/H \text{ or } ^{18}O/^{16}O \quad \text{(Eq- 2)}$$

Carbon stable isotope (^{13}C) analysis was performed on CO_2 samples generated from dissolved inorganic carbon through acid hydrolysis with phosphoric acid. Carbon isotope (^{13}C) values are reported in permil (‰) notation against VPDB. The overlay analysis was carried out with selected parameters, which includes EC, Electrical conductivity, Water types, stable isotope and tritium. The layers were prepared for surface and groundwater with respect to seasons. In mapinfo, the vertical mapper tool was used to plot the variation of parameters. Maps were prepared for with three categories viz, Vulnerable, Moderate Vulnerable and Less Vulnerable. Inverse Distance Weighing method was applied to generate the raster graphical output. The raster graphical output is suitable for the overlay analysis which gives high spatial variability of multi-attributes. The Inverse Distance Weighting (IDW) algorithm is a moving average interpolator that is usually applied to highly variable data. For certain data types it is possible to return to the collection site and record a new value that is statistically different from the original reading but within the general trend for the area.

Results and Discussion

The average, minimum and maximum values for different seasons are given in (Table 2). The groundwater in study area is generally odorless and colorless in most of the places. The average temperature at the time of sampling varies from 25° C to 31° C.

Table 2. Maximum, Minimum and Average Value of hydrogeochemical parameters for Shallow and Deep groundwater (all values in mg/l) except EC µs/cm and pH

	pH	EC	TDS	Ca	Mg	Na	K	Cl	HCO_3	SO_4
Shallow Pre Monsoon										
Maximum	8.11	9848	5908	180	185	1898	1012	2065	597	51
Minimum	6.19	253.6	127	12	7.2	21	6.4	64.2	10.2	0.01
Average	7.41	2638.9	1152.6	45.9	39.7	295.9	174.0	404.2	200.6	11.1
Shallow Post Monsoon										
Maximum	8.68	8380	8860	148	108	1139	572.6	2517	720	69
Minimum	6.02	215	99	4	0	53.5	32.1	35.45	24.4	0
Average	7.46	2293.8	1944	45	24	338	162	476	257	16
Deep Pre Monsoon										
Maximum	7.60	6522	2543.6	516	164	672	250	1567	732	954
Minimum	5.50	230	112.9	11	8.7	16.2	1.3	12.9	87.5	8.6
Average	6.89	1969	892	89	39	254	41	397	327	128
Deep Post Monsoon										
Maximum	7.8	6910	2763	209	111	1330	154.5	1823	568	604
Minimum	60	248	181	11.3	7.95	12.7	1.8	17	88	8
Average	7.0	1916	1917	57	34	292	32.5	390	297	129

Spatial Distribution of Electrical Conductivity

Electrical conductivity (EC) measures the dissolved ionic concentration and it depends on the type of ions and their concentration and temperature at which measurement was taken. Spatial distribution of EC of samples was plotted for both shallow and deep depths for each season (Fig. 3 and 4). The distribution of the Electrical conductivity of samples collected during the PreMonsoon at Shallow depths (SPRM) shows that higher values near the southern part of the study area covered by Karaikal, Thirunallar, Thirumurugal and Nedungadu. Lesser EC values are noted in the northern part of the Coleroon River. The spatial representation of the samples collected during the Post Monsoon at Shallow depth (SPOM) indicates that the same regions represent higher EC values as that of the SPRM. The deeper aquifers of the study area during the Pre Monsoon (DPRM) shows higher values in the south eastern part of the study area falling near the Karaikal region. The PostMonsoon samples (DPOM) also show representation of higher values in the same region as that of premonsoon. In all seasons irrespective of depth the higher values are concentrated around the southern part of the study area.

Based on EC, the groundwater has been classified into (1) fresh (<1,500 µS/cm), (2) brackish (1,500-3,000 µS/cm), and (3) saline >3,000 µS/cm (Saxena et al. 2003), (Table 3).

Table 3. Classification of Electric conductivity of groundwater (Saxena et al. 2003)

Values in µs/cm		SPRM (n=62)	SPOM (n=62)	DPRM (n=22)	DPOM (n=22)
Fresh	<1500	39	39	12	10
Brackish	1500 - 3000	19	27	7	10
Saline	>3000	27	19	4	3

Based on this pattern of classification, the percentage of groundwater samples in each group shows that 30% of the samples of SPOM and 43% of SPRM are saline. The salinity after monsoon may indicate the dissolution and leaching of secondary salts by infiltrating rain water (Chidambaram et al. 2005). Higher representations were noted in the fresh water category in both the seasons. The higher representation of samples in brackish category during SPOM and an increase in saline category is noted during SPRM, which causes deterioration of groundwater quality. The Deep aquifers show lesser representation of

Saline water with 17% during DPRM and 13% during DPOM, but good representations of samples were noted in the brackish to saline category irrespective of season (Fig 4).

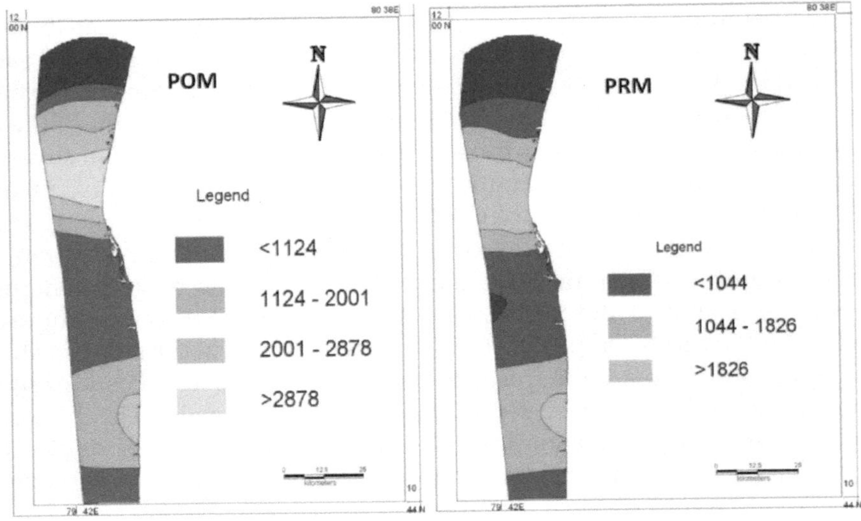

Fig. 3. EC for deep groundwater samples

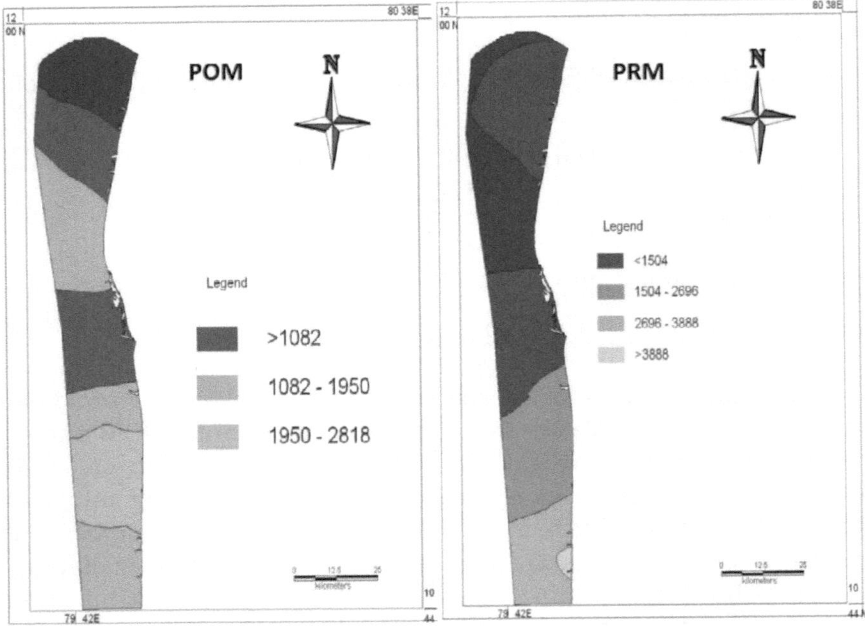

Fig. 4. EC for shallow groundwater samples

Geochemical Classification

The hydrochemical facies reflect the chemical processes operative in a specific lithological and geochemical environment. The general classification and trend of variation in the groundwater of the study area is exhibited in the Piper - diagram (Thilagavathi et al., 2012). Groundwater of few samples are dominated by alkali (Na^+ and K^+) over alkaline–earth cations (Ca^{2+} and Mg^{2+}) reflecting predominance of salinity. Weathering also results in the release of Na^+ and K^+ from the K-feldspar and plagioclase alteration. On the other hand, the position of the groundwater in the anion triangle indicates dominance of the Cl^- and HCO_3^-, but a less abundance of sulfate in these waters. Hydrogeochemical facies interpretation is an useful tool for determining the flow pattern, origin of chemical histories of groundwater masses (Chidambaram et al. 2012; Karmegam et al. 2010). Most of the samples irrespective of formation fall in the mixed zone facies of Ca-Na-HCO_3 and Ca-Mg-Cl facies indicating the transition zone (Fig 5).

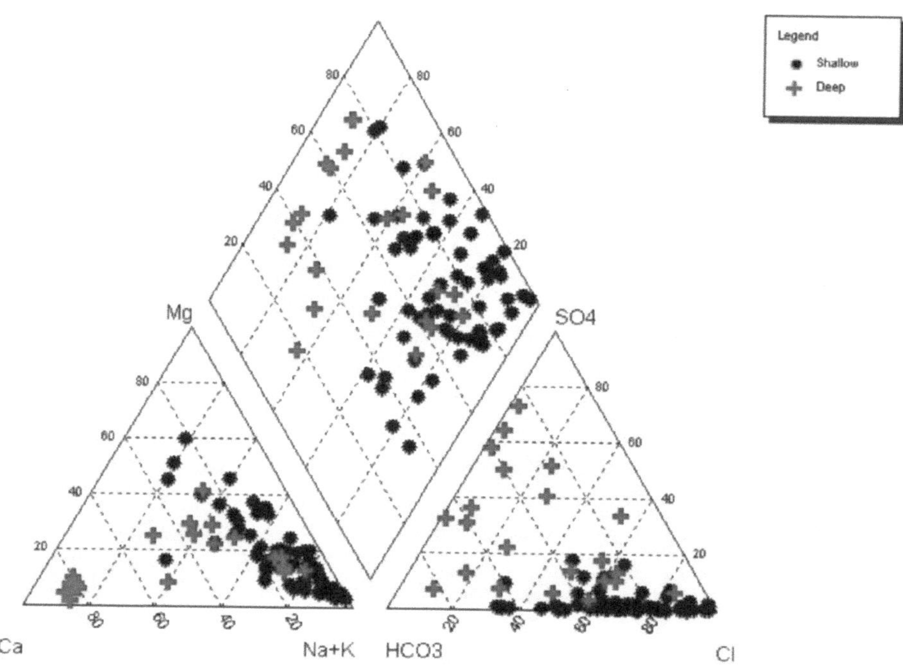

Fig. 5. Shallow ground water samples shows the migration of the facies from during Pre monsoon and Post monsoon season 2008

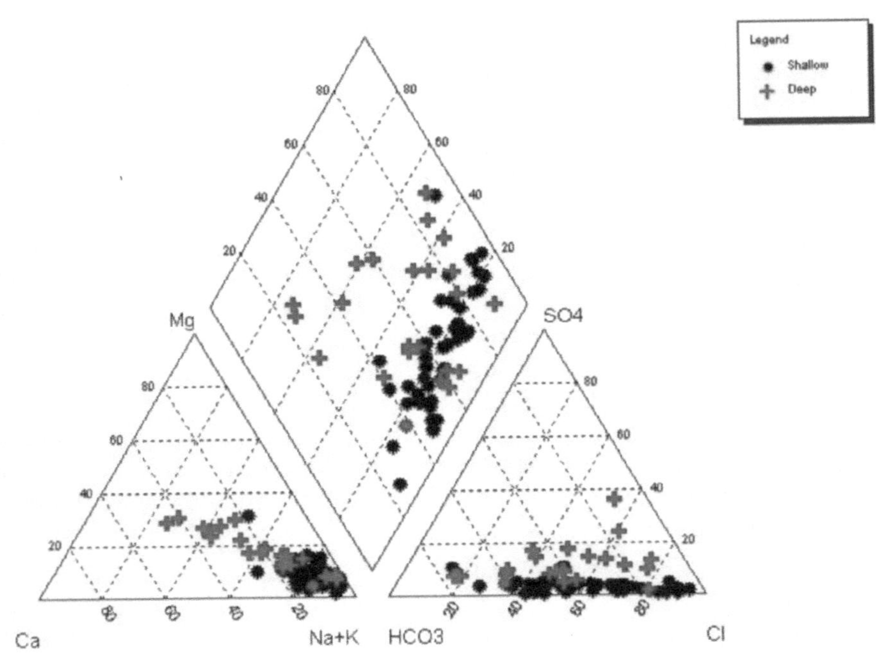

Fig. 6. Deep groundwater samples shows the migration of the facies from during Pre monsoon and Post monsoon season 2009

In general most samples are Na-Cl or Na-mixed-anion type, although the low salinity waters are Na-HCO_3 or mixed-cation-HCO_3 type. The high salinity waters are Na-Cl type (maximum Cl concentration is 2,150 mg/l). The Piper diagram shows (Fig. 5 and 6) the migration of the facies from Ca-Na-HCO_3 type during Pre monsoon to Na-Cl type during Post monsoon season irrespective of the depth (Fig 5 and 6). The monsoon adds Cl to the groundwaters by dissolution and leaching of ions precipitated during the premonsoon from unsaturated zone. It is also to be noted that the Cl^- concentration increases by the agricultural return flow. The processes of ion exchange also results in the Na^+ HCO_3^- Cl^- water type. There are substances which absorb and exchange their cations, with cations present in the groundwater. Variation of anions from recharge to discharge varies from $HCO_3^- \rightarrow Cl^-$ type due to the dissolution of ions in the pathway or due to precipitation and removal of HCO_3^- from the aqueous system (Tirumalesh et al., 2010) Low TDS, high Ca^{2+}/Mg^{2+} ratio, low SO_4^{2-} and high HCO_3^-occurs in the recharge area and the reverse in the discharge area.

The projection of the points in the central diamond-shaped field (Fig 5 and 6) revealed the heterogeneity of the groundwater chemical composition in this area, but a general trend from Na-HCO_3^--Cl^- waters to Na-Cl waters from inland towards the coast.

These patterns indicate that the groundwater chemistries are changed by cation exchange reaction, as well as simple mixing (Appelo and Postma, 1999). The region of the Ca-Cl type water in few samples may be a leading edge of the seawater plume (Vengosh et al., 2002; Appelo and Postma, 1999;). Large proportions of the groundwaters showed Na-Cl^- type and Na-HCO_3^--Cl^-, which generally indicates a strong seawater influence. The Na-HCO_3^--Cl^- type of water represents partly flushed remains of an ancient, confined saline water body.

A hydrochemical diagram proposed by (Chadha, 1999) has been applied in this study to interpret the hydrochemical processes occurring in the study area. The same procedure was successfully applied by (Karmegam et al., 2010; Vandenbohede et al., 2010) in a coastal aquifer to determine the evolution of two different hydrogeochemical processes. Data were converted to percentage reaction values (milliequivalent percentages) and expressed as the difference between alkaline earths (Ca+ Mg) and alkali metals (Na+ K) for cations, and the difference between weak acidic anions (HCO_3+ CO_3) and strong acidic anions (Cl + SO_4). The hydrochemical processes suggested by Chadha (1999) are indicated in each of the four quadrants of the graph. These are broadly summarized as:

Field 1: Ca-HCO_3 type of recharging waters

Field 2: Ca-Mg-Cl type of reverse ion-exchange waters

Field 3: Na-Cl type of end-member waters (seawater)

Field 4: Na-HCO3 type of base ion-exchange waters

The plot shows that majority of samples of study area fall in Field -7 (Sea water intrusion type) with few representation in Field -6 and 8, indicating Reverse Ion Exchange and Base Exchange respectively (Fig 7). Very few samples fall in the recharge region. It is interesting to note that greater impact of sea water intrusion is noted in the Shallow aquifers than the deeper ones irrespective of season.

In general the nature of the sample scatter in the plot shows that Na^+ + K^+ is predominant in all seasons irrespective of depth, but during post monsoon season at Shallow depth representation of the samples are noted with dominance of Ca^{2+} + Mg^{2+} cations, indicating reverse ion exchange . In anions only few of the samples shows dominance of HCO_3 in SPRM and more representation of SPOM is noted. Field-7 shows good representation of SPRM, SPOM, DPRM and DPOM indicating the impact of Sea water intrusion, SPOM and SPRM shows Reverse Ion Exchange and Base Ion Exchange processes.

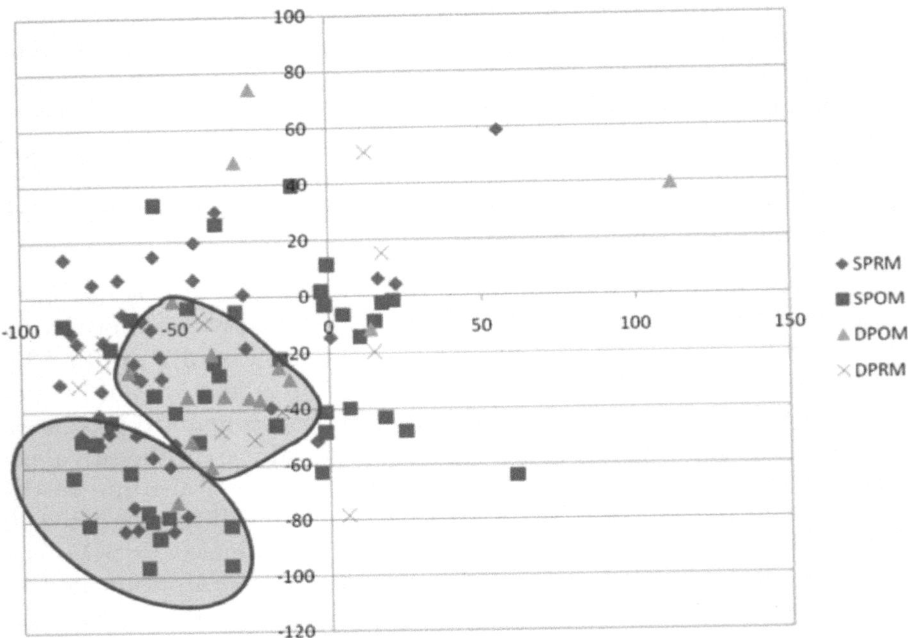

Fig. 7. Chaddas plot exhibiting the geochemical nature for shallow and deep groundwater for both seasons

Seawater has distinct ionic ratios such as Na^+/Cl^- = 0.86, SO_4^{2-}/Cl^- = 0.05, $Ca^{2+}/(HCO_3^- + SO_4^{2-})$, Mg^{2+}/Ca^{2+} =5.2 (Vengosh et al., 2002; Vengosh and Rosenthal 1994). The average values of these ratios in the samples of the study area are given in (Table 4). The Na/Cl ratio of the study area is greater than unity in all seasons. The Na^+/Cl^- molar ratio could reach unity due to the mixing of seawater and freshwater, which has a Na^+/Cl^- ratio greater than unity (Vengosh and Rosenthal 1994).Relative to seawater, the saline groundwater has relatively low Na^+/Cl^- ratios (0.5–0.8), low Mg^{2+}/Ca^{2+} ratios and high $Ca^{2+}/(HCO_3^- + SO_4^{2-})$ ratios (greater than unity) (Vengosh et al., 2002).

Table 4. The comparative ionic ratios of potential salinization sources and study area groundwater parameter for seawater intrusion.

Parameter	Seawater intrusion	Deep saline upconing	Agriculture return flows	Wastewater infiltration	SPRM	SPOM	DPRM	DPOM
Na^+/Cl^-	0.86–1[a]	<0.8[b]	–	1.1[c]	2.6-0.18	13.2-0.03	2.09-0.3	1.72 – 0.44
SO_4^{2-}/Cl^-	0.05[a,c]	*0.05[c,d]	>>0.05[e]	0.09[d]	0.5-0.006	7.16-0.08	1.9-0.05	1.99 -0.44
K^+/Cl^-	0.019	<0.019	–	>>0.02	1.26-0.03	11.42-0.05	1.16-0.05	0.412-0.015
Mg^{2+}/Ca^{2+}	<5[b]	<1[a]	–	–	3.12-0.2	1.67- 0.09	1.32-0.09	1.4-0.009
$Ca^{2+}/(HCO_3^-+SO_4^{2-})$	0.35–<1	<1[a]	–	–	1.9-0.04	< 0.94	1.09-0.02	0.9-0.02

The Mg^{2+}/Ca^{2+} ionic ratio greater than 5 is a direct indicator of seawater contamination (Metcafe and Eddy, 2000). The value of these ratios in the study area is greater than unity and it may be due to deep saline upconing or ion exchange process. The chemical signature of water polluted with urban wastewater or from the reuse of treated wastewater is very distinguishable. Sewage effluent has a relatively high Na^+/Cl^- ratio (greater than unity) and comparatively with high SO_4^{2-}/Cl^- ratio (0.09) (Ghabayen et al., 2006). This water is characterized by high SO_4^{2-}/ Cl- ratios (much greater than 0.05) (Vengosh et al., 2002). A high SO_4^{2-}/ Cl^- ratio is attributed to the application of gypsum fertilizers. Lesser values of K/Cl and SO_4/Cl^- are noted in the deeper aquifers indicating relatively lesser anthropogenic impact.

The comparison of the Na^+/Cl- ratio shows that fresh water mixing with sea water is responsible for high salinity in the area (Table 4). However in the post-monsoon ratio of Shallow aquifers are found to have higher values in comparison to pre-monsoon samples which give a hint of migration of anthropogenic impact or dissolution and leaching of precipitated salts during monsoon. This is also evident by the high SO_4^{2-}/Cl^- ratio in the Shallow aquifers than the deeper aquifers which substantiate that there is some impact from agriculture return flows. It is also true in the background that salt patches can be easily observed in the area at the time of high evapo-transpiration. Nevertheless this temporary phase has been released in post-monsoon and once again inland salinity seems to be the governing factor. K^+ seems to be contributed from anthropogenic sources, this finding get strengthen by the comparison of K^+/Cl^- ratio that found quite higher and equivalent to waste water infiltration. $Ca^{2+}/(HCO_3^- + SO_4^{2-})$ ratio depicts Sea water intrusion process is mostly responsible in both seasons except Mg^{2+}/Ca^{2+} ratio in the post-monsoon, that may be result of several complex reactions in the aquifer.

Stable Isotopes

The isotope $\delta^{18}O$ and äD compositions are sensitive tracers and widely applied to studying the natural water circulation and groundwater movement.

THE $\delta^2H - \delta^{18}O$ RELATIONSHIP IN PRECIPITATION

The LMWL is helpful for interpreting local water movements, sources and processes to which the water has been subjected (Dansgaard, 1964). The LMWL was constructed by analysing the stable isotopes of rainwater samples in the present study area. The samples from 14 stations were collected and plotted along with stable isotope composition of the Shallow groundwater samples (Fig 8). The trend of LMWL is different from GMWL line, with $\delta D = 0.15\ x + \delta\ ^{18}O\ 0.87$ this deviation. Such deviation result from differences in climatic factors such as air temperatures secondary evaporation, seasonality of precipitation and moisture source (Clark and Fritz 1997), which occurs in precipitations globally. The groundwater's fall within a cluster either close to or to the right of the LMWL, indicating that these samples have undergone some evaporation prior to infiltration.

Isotopic Ratio in Groundwater Samples

About 37 Shallow groundwater samples and 15 deep groundwater samples were collected in total of about 52 samples were analyzed for stable isotopes of oxygen-18 and Deuterium in PRM. It is observed that there is a clear demarcation of isotopic signatures in ground water of different depths in the study area. The isotope ratio shows enrichment of heavier isotope in the deeper aquifers than the Shallow aquifers. More over there is a definite trend noted in the Shallow aquifers than in the deep aquifers. The δD and $\delta\ ^{18}O$ data provides information on the secondary processes acting on the water as it travels into the subsurface.

a) Shallow groundwater

The collected groundwater samples of Shallow formation have a range from -7.6% to -0.5% for $\delta\ ^{18}O$ and from -55.6% to -1.8% for δD. The mean isotope values of groundwater are -4 % and -31 % for $\delta\ ^{18}O$ and $\delta\ D$ respectively. The δD and $\delta\ ^{18}O$ values of the groundwater samples fall nearer to LMWL indicating recharge from meteoric source (Chidambaram et al., 2007b; Prasanna et al., 2009). The ground water has slightly less $\delta^{18}O$ and δD than the weighted mean for precipitation. Minor deviation from the weighted average precipitation can also be attributed to the land use practices Groundwater samples of this shallow depth fall parallel to the LMWL indicating recharge has taken place by the local precipitation (Chidambaram et al., 2007b). There are two groups of sample clusters with heavier isotopes and lighter ones. The heavier ones represent the sea water intrusion process and the lighter ones with fresh water influx.

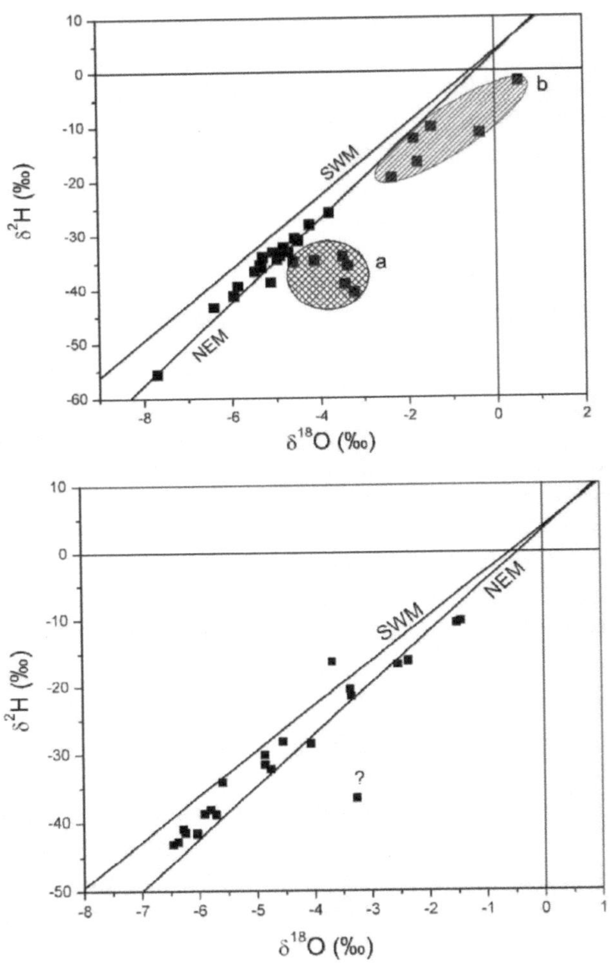

Fig. 8. Deuterium Vs O18 shallow and deep groundwater samples

Deeper groundwater

The groundwater samples of the deeper depth range from -6.47% to -1.4% for $\delta^{18}O$ and from -47.5‰ to -2.7% for δD. The mean isotope values of groundwater are -3.6% and -22.6% for $\delta^{18}O$ and äD respectively. The slope of $\delta^{18}O$- $\delta^{2}H$ relation was only ~2, much lower than the range for the evaporation from the open water surfaces. Deep waters of pre-monsoon season show a clear contribution from SWM source (Fig. 8) indicated by their alignment along the LMWL of SWM. The SWM isotope signatures are dominantly seen in groundwater isotopic values. During post monsoon period, the isotopic signatures are relatively enriched and fall between the LWMLs of both the monsoons (Fig. 8). Groundwater sample from Caper Quarry which is collected from a depth of 215m bgl doesn't show any fluctuation in stable isotopic composition indicating single source of recharge and probable derived from long flowpath, which is further supported by presence of negligible tritium.

Stable isotopes (δ ^{18}O, δ D) Vs. Chloride

A comparison of the deuterium and chloride data provides greater understanding of groundwater-surface water interaction processes in the study area. There is no significant variation of isotopic characters with respect to Cl^- in the formations with depth. In general, deep aquifer has lower Cl^- values and higher value in the Shallow aquifers. The chloride - $\delta^{18}O$ plot (Fig. 9) indicates that there are two major mechanism operative in the study area i)direct sea water incursion into the aquifer and ii) recharge of the mixed sea water and the saline backwaters. The chloride-deuterium plot suggests that two types of groundwater occur in the study area namely: (i) groundwater samples from the Shallow aquifers are characterized by low chloride and depleted δD, (except in certain locations near the coast and the those near the coleroon river mouth in the north part of the study area) represent areas that are recharged frequently by rainfall. (ii) Groundwater samples along the coast with high chloride and enriched äD indicating rarely receive recharge from meteoric water and recharge of the evaporated waters from a different source nearby or due to the evaporation taking place along the coast. Higher concentration of chloride was attributed to the leaching of secondary salts from the formation (Chidambaram et al., 2007) or due to sea water intrusion.

It has been reported that few regions are characterized by low chloride and relatively enriched δD signatures whereas diffuse recharge would tend to be enriched in chloride and depleted in δD (Fig. 9). Majority of Shallow groundwaters with low Cl^- and depleted deuterium, representing are that are recharged more frequently by local rainfall and less frequently by river water. (2) Some groundwater samples characterized by low Cl and relatively enriched deuterium that is frequently recharged by river waters. (3) Certain groundwater samples, with very high Cl^- and lower deuterium is due to the diffusive recharge along the flow path.

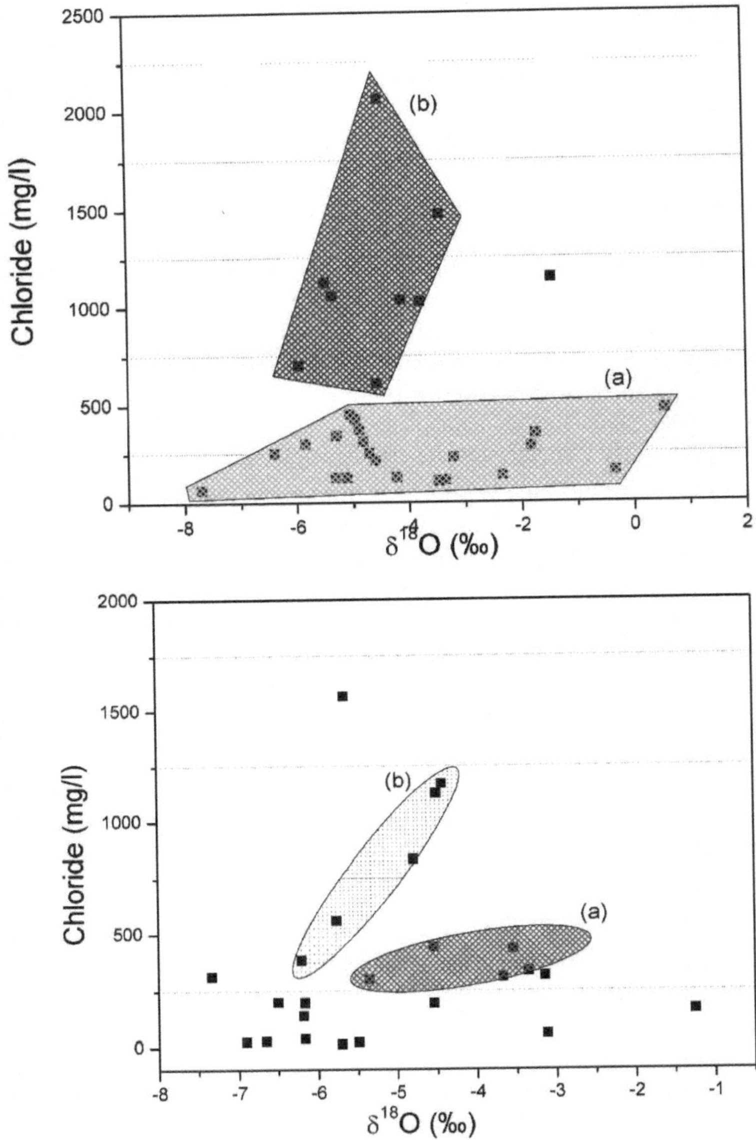

Fig. 9. $\delta^{18}O$-chloride correlations shallow and deep groundwater samples

Environmental Radioisotopes [Tritium (^3H) and Carbon-14 (^{14}C)]

Water samples collected during premonsoon season from the investigated area were measured for tritium and carbon-14 contents and the data has been given in the Table.

The samples collected from deep zones have tritium values in the range of 2 to 5 T.U., except three samples viz., Caper Quarry, Periankuppam and Thorurai. The tritium values in these samples are < 1 T.U. indicating these are old groundwaters. These samples are mostly collected from very deep zones (150 to 200m bgl) and from Lower Tertiary formation. There is no variation observed in the chloride values, stable isotope values in different seasons. These findings infer that these groundwaters are probable recharged from distant areas.

In the ^{3}H versus Cl plot (Fig. 10) most of the groundwaters fall in the rain dominated area. Group (a) represents groundwaters recharged from distant areas and fresh in quality. Group (b) samples are modern and are brackish to saline in nature. The source of salinity is mostly surface water sources like backwater. These trends are further verified by using ^{3}H versus $\delta^{18}O$ plot. (Fig. 11) shows almost same trends shown in (Fig. 11). Old groundwaters show depleted isotopic values [group (a)] indicating distant recharge mostly from western parts of the study area. Group (b) samples show enriched isotopic composition as well as high tritium indicating the source of salinity in these samples is surface waters that have undergone evaporation and backwaters are most plausible sources in this region. Rests of the samples are modern and fresh in quality.

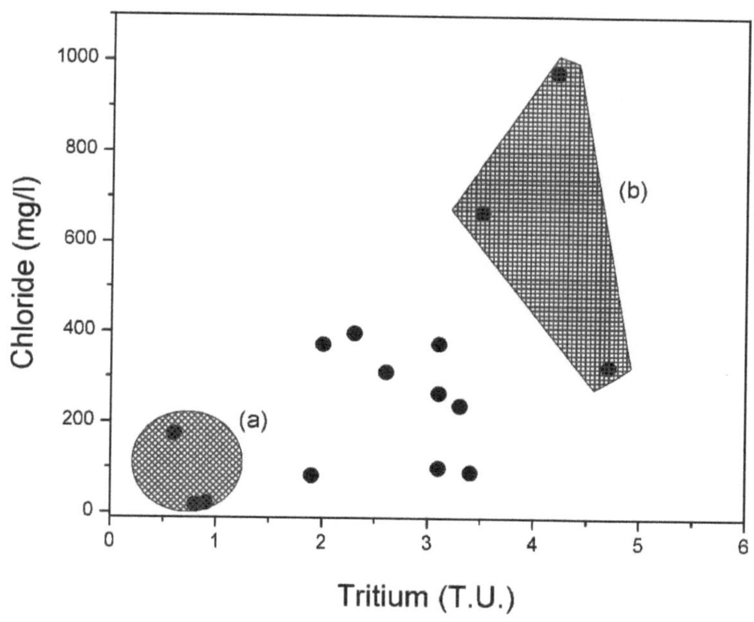

Fig. 10. Tritium Vs Chloride

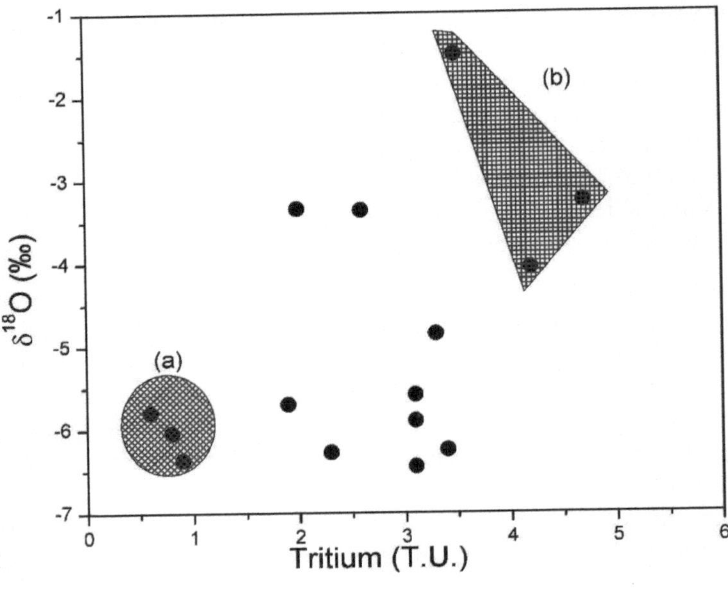

Fig. 11. Tritium Vs O18

Vulnerability Assessment of groundwater

Electrical conductivity, Water types, stable isotope and tritium of groundwater were taken into an account to identify the physio chemical parameters. Groundwater vulnerability maps have become a standard tool for protecting groundwater resources from pollution. After preparing the spatial water quality index maps for various parameters as explained earlier, overlay was carried out during PRM in shallow and deep groundwater Similar procedures were adopted, for other season of POM and overlay analysis was attempted. The maximum value in both case were considered as good/permissible and the region representing low value indicate that they are sensitive to pollution.

Spatial Index =
1. Zone I (Vulnerable)
2. Zone II (Moderately Vulnerable)
3. Zone III (Less Vulnerable)

The vulnerability index map of the study area indicates that the Three zones of vulnerability was obtained of which Zone I is more vulnerable and covers about 30% of the southern part of the study area consisting of Kuzhaiyar and Karaikal regions. In the northern part of the study area this zone is represented near the velar river mouth and in the SIPCOT industrial estate region. The northern most part of the study area comprising of Caper Quarry, Periyaganganankuppam etc., fall in Zone III with lesser vulnerability. Less vulnerability zone is represented in the Northern part of the study area. Pollution of groundwater here is mainly due to industrial effluents. Municipal wastes in water bodies is another major concern for the this zone The remaining part of the study area is covered by zone (Zone II) of medium vulnerability based on the above mentioned factors (Fig. 12). The flow of groundwaters is initiated from the topographic high to the low which is witnessed by the water table conditions and the integrated output reveals that major portion of study area are represented with moderate/Vulnerable groundwater quality spatially in the Northern and South Eastern part of the study area (Fig. 12) surrounded by the river. The over extraction of groundwater from the coastal aquifers, result in reduction of groundwater resource and lowering of water level. Further anthropogenic processes like agriculture activities, industries and domestic sources result in less vulnerability zone.

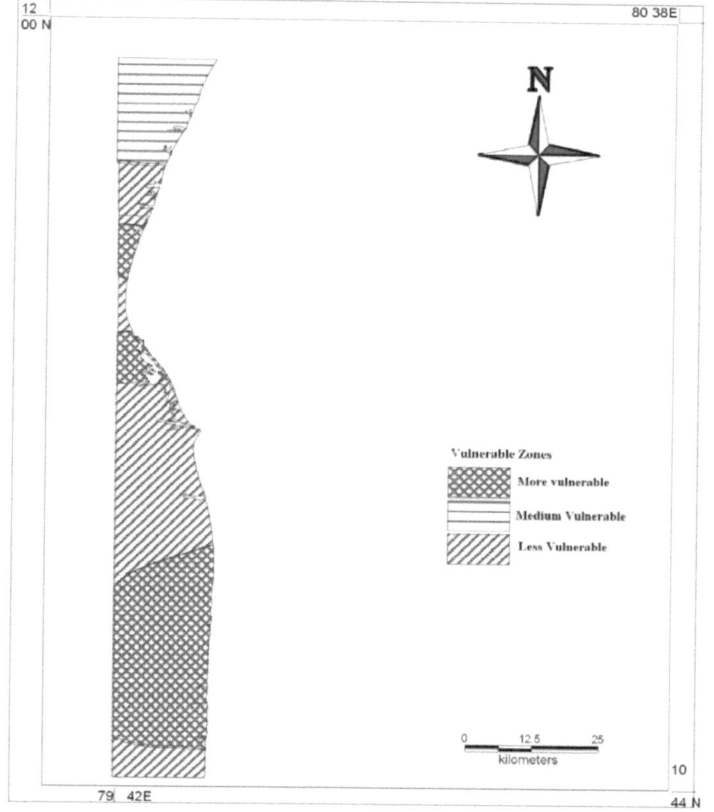

Fig. 12. Groundwater vulnerability index map of the study area

Conclusion

The water level shows negative gradient in the southern part of the study area promoting the incursion of sea water. Higher values of electrical conductivity were noted near the southern part of the study area covered by Karaikal, Thirunallar, Thirumurugal and Nedungadu. Lesser EC values are noted in the northern part of the Coleroon river. The ratio of Na^+/HCO_3^- is close to unity from low to high salinity indicating that ion exchange is not a controlling hydrogeochemical process. The Cl^- concentration in the $Na\text{-}HCO_3$ groundwaters is also low, with plots of Na^+ vs. HCO_3^- and Na^+ vs. Cl^- showing distinctive Na–Cl-saline and $Na\text{-}HCO_3$ sodium rich trends. These ^{18}O-enriched groundwaters are mainly Na–Mg–Cl end member waters. This slight enrichment may show isotopic exchange with the rock mass through slow water–rock interaction reactions. The Na–Cl end member waters are more $\delta\ ^{18}O$ depleted than the other groups. These groundwaters lie along a line tangential to the GMWL. The chloride concentration in these waters shows that the Na-Cl type groundwaters generally have increased TDS and isotopic enrichment along the trend. The tritium in groundwaters of the study area shows that recent waters are noted along the river course of Coleroon and near the Perumal Eri reservoir located in the north western part of the study area. Later by integrating all the layers (Water level, Electrical conductivity, Water types, stable isotope and tritium) the area vulnerable for contamination was derived. Three zones of vulnerability was obtained of which Zone I is more vulnerable and covers about 30% of the southern part of the study area. In the northern part of the study area fall in Zone III with lesser vulnerability. The remaining part of the study area is covered by zone (Zone II) of medium vulnerability based on the above mentioned factors. Groundwater vulnerability and risk mapping is a relatively new scientific approach for facilitating planning and decision making processes in order to protect this valuable resource.

Acknowledgement

The authors express their sincere thanks to Board of Research in Nuclear Sciences (BRNS), India, for providing the necessary nancial support to this research project.

References

APHA (1998) Stanard methods for the examination of water and wastewater 19th edn. APHA, Washington, DC.

Appelo CAJ, Postma D (1993) Geochemistry, groundwater and pollution. Balkema Press.

Barlow PM (2003) Ground water in fresh water-salt water environments of the Atlantic Coast. U.S. Geological Survey Circular 1262.

Chadha DK (1999) A proposed new diagram for geochemical classication of natural waters and interpretation of chemical data. Hydrogeol J 7(5): 431–439.

Chidambaram S, Anandhan P, Prasanna MV, Srinivasamoorthy K, Vasanthavigar M (2012) Major ion chemistry and identification of hydrogeochemical processes controlling groundwater in and around Neyveli Lignite Mines, Tamil Nadu, South India Arab J Geosci DOI 10.1007/ s12517-012-0589-3.

Chidambaram S, Prasanna MV, Vasu K, Shahul Hameed A, Unnikrishna Warrier C, Srinivasamoorthy K, John Peter A, Anandhan P (2007B) Study on stable isotopic signatures in ground water of gadilam river basin, Tamilnadu, India.Indian journal of Geochemistry.

Chidambaram S, Ramanathan, A L, Anandhan P, Srinivasamoorthy K, Prasanna M V (2005) A comparative study on the coastal surface and ground water in and around Puduchattiram to Coleroon, Tamil Nadu. International Journal of Ecology and Environment Sciences, 31(3), 209–306 Special Issue.

Clark ID, Fritz P (1997) Environmental isotopes in hydrogeology. Lewis publishers, boca raton, new york, pp: 1–328.

Coplen TB, Wildman J, Chen J, Anal. Chem. 63 (1991) 910-912.

Dansgaard W (1964) Stable isotopes in precipitation. Tellus 16, 436–468.

El Yaouti F, El Mandour A, Khattach D, Benavente J, Kaufmann O. Salinization processes in the unconfined aquifer of Bou-Areg (NE Morocco): a geostatistical, geochemical and tomographic study, Appl. Geochem., 24, 16–31, 2009.

Elango L, Kannan R, Senthil Kumar M (2003). Major ion chemistry and identification of hydrogeochemical processes of ground water in a part of Kanchipuram District, Tamil Nadu, India. Environmental Geosciences, 10(4): 157- 166.

Ghabayen MS, McKee Mac, Kemblowski Mariush (2006) Ionic and isotopic ratios for identication of salinity sources and missing data in the Gaza aquifer. J Hydrol 318:360–373

Ghiglieri G, Carletti A, Pittalis D. Analysis of salinization in the coastal carbonate aquifer of Porto Torres (NW Sardinia, Italy), J. Hydrol., 432-433, 43–51, doi:10.1016/j.jhydrol.2012.02.016, 2012.

Gnanasundar D, Elango L (1999). Groundwater quality assessment of a coastal aquifer using geoelectrical techniques. Journal of Environmental Hydrology, 7(2), 1–8.

Karmegam U, Chidambaram S, Sasidar P, Manivannan R, Maikandan S, Anandhan P (2010) Geochemical Characterization of groundwater's of Shallow Coastal Aquifer in and Around Kalpakkam,South India. Research Journal of Environmental and Earth Sciences 2(4) ; 170-177,2010 ISSn: 2041-0492.

Metcafe, Eddy (2000) Integrated aquifer management plan: nal report. Gaza Coastal Aquifer Management Program, USAID Contract No. 294-C-00-99-00038-00.

Ozler HM (2003). Hydrochemistry and saltwater intrusion in the Van aquifer, east Turkey. Environ. Geol., 43: 759-775.

Petalas C, Lambrakis N. Simulation of intense salinization phenomena in coastal aquifers – the case of the coastal aquifers of Thrace, J. Hydrol., 324, 51–64, 2006.

Prasanna M V, Chidambaram S, Shahul Hameed A, Srinivasamoorthy K. (2009). Study of evaluation of groundwater in gadilam basin using hydrogeochemical and isotope data. Environmental monitoring and assessment, doi 10.1007/s 10661-009-1092-5.

Ramanathan A L, Subramanian V, Ramesh R, Chidambaram S, James A (1999) Environmental geochemistry of Pichavaram ecosystem (tropical), southeast coast of India. Environmental Geology, 37(3), 223–233.

Satheesh Herbert Singh D, Lawrence J F (2007) Groundwater quality assessment of shallow aquifer using geographical information system in part of Chennai city, Tamilnadu. Journal of Geological Society of India, 69, 1067–1076.

Saxena V K, Mondal N C, Singh V S (2003) Identification of seawater ingress using Sr and B in Krishna delta. Current Science, 86(4), 586–590.

Sdao F, Parisi S, Kalisperi D, Pascale S, Soupios P, Lydakis Simantiris N, Kouli M. Geochemistry and quality of the groundwater from the karstic and coastal aquifer of Geropotamos River Basin at north-central Crete, Greece, Environ. Earth Sci., 67, 1145–1163, 2012.

Tirumalesh K, Ramakumar K L, Shivanna K, Chidambaram S (2010) An over view on NETPATH geochemical modeling- a case study from coastal aquifer of Pondicherry, Groundwater Monitoring and Management through Hydrogeochemical Modeling Approach, IK International Publishing House, New Delhi, pp.33-47.

Thilagavathi R, Chidambaram S, Prasanna MV, Thivya C, and Singaraja C (2012) A study on groundwater geochemistry and water quality in layered aquifers system of Pondicherry region, southeast India. Applied Water Science. doi:10.1007/s13201-012-0045-2.

Vandenbohede A, Courtens C, William de Breuck L (2010) Fresh-salt water distribution in the central Belgian coastal plain: an update. Geol Belg 11(3):163–172.

Vengosh A, Gill J, Davisson ML, Hudson GB (2002) A multiisotope (B, Sr, O, H, and C) and age dating study of groundwater from Salinas Valley, California: hydrochemistry, dynamics, and contamination process. Water Resour Res 38(1):1–17.

Vengosh A, Rosenthal E (1994) Saline groundwater in Israel: its bearing on the water crisis in the country. J Hydrol 156:389–430.

Chapter 5

Hydrogeochemical and Inverse Modeling of Coastal Aquifers of Kalpakkam Region, Tamil Nadu

Chidambaram S., Karmegam U., Prasanna, M.V. and Sasidhar P.[1]

Department of Earth Sciences, Annamalai University, Annamalai Nagar
[1]SRL, IGCAR, Kalpakkam, India

ABSTRACT

A total of 30 groundwater samples were collected in four different seasons (Pre Monsoon, South West Monsoon, North East Monsoon and Post Monsoon) from the coastal aquifers of Kalpakkam region of Tamilnadu. The samples were collected covering the seasonal and lithological variations to infer the influence of natural geochemical processes and anthropogenic sources on the groundwater chemistry. The collected samples were analysed for the parameters like EC, pH and Major ions. The saturation index of silicates and carbonates shows that state of saturation to over saturation, but the sulphate minerals is dominantly under saturated. In the inverse modeling three models are obtained using PHREEQC and the error percentage is noted as 7.45%.

Key words: Groundwater, Monsoon, Saturation Index, Geochemical

Introduction

Hydrogeochemical composition of groundwater can also be indicative of its origin and history of passage through underground materials which water has been in contact with, in shallow and deep-seated conditions. Natural and anthropogenic sources along with chemical and biogeochemical constituents have been considerably altering groundwater quality in recent years. The computerized geochemical models are powerful tools for understanding the chemical state of natural waters and for predicting the behavior of such waters under variety of hypothetical conditions. Global problems, approaches and priorities of chemical modeling were discussed in detail by Jenne (1978). Implication of computer programme in calculation of equilibrium distribution of inorganic species of major and minor elements in natural waters using chemical analysis and insitu measurements of Eh, pH were discussed in detail by Truesdell and Jones (1973). Mixing of water from different aquifer with water from deep bedrock aquifer in groundwater discharge area having an intermediate composition was explained by dilution of calcite precipitation using Mix2 chemical equilibrium model, in drainage basin of Holocene age, Canada. Simulation of calcite dissolution and porosity changes in salt water mixing zone in coastal aquifers was done by Sanford and Konikow (1989). They found that there is an increasing porosity and permeability by enhanced dissolution, using geochemical model PHREEQC. The same model was used by Elangovan et al (1999) to find out di-equilibrium indices of calcite along with other minerals such as Calcite, Aragonite, Dolomite, Hematite, Goethite, $Fe(OH)_3$ and Strontionite. Either Davis or extended Debye Huckel equation can be used for the calculation of activity coefficients (Garrels and Christ, 1965; Drever, 1988). Inverse modelling using PHREEQC was also attempted by Anandhan (2005) for delineating the geochemical process from the recharge to discharge. Two specific flow paths were selected to study the hydrogeochemical variations in the Neyveli aquifer (Chidambaram et al., 2012). Several earlier workers have also used geochemical models, for solubility equation studies in groundwater (Sanford and Konikow 1989; Plummer et al, 1990; Elangovan, 1997 and Murphy and Shock, 1999). Saturation index of minerals were also adopted in studies of groundwater in Salem district (Srinivasamoorthy, 2004). Hence the present study has been carried out in Hydrogeochemical processes of Kalpakkam region of Tamilnadu using inverse modeling.

Study Area

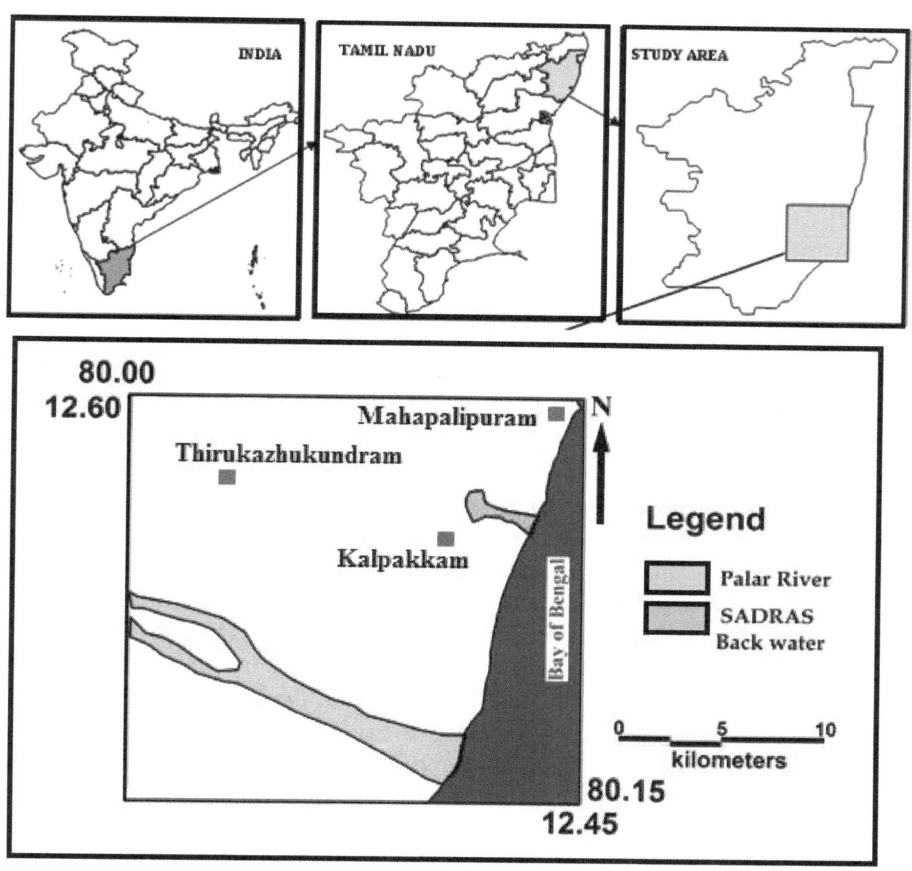

Fig. 1. Location map of the study area

The study area Kalpakkam falls in the Kanchepuram district it comprises of hard rock and sedimentary formation located in northern part of Tamilnadu between 12p 37′ and 12p 25′ North latitudes, 80p 00′ and 80p 12′ East longitudes (Fig.1). The study area is bounded by Bay of Bengal in east. The SADRAS back water of the Bay of Bengal in the east and the landfill of Mahabalipuram town is noted in NE part of study area. The study area reveals that there are two major lithologies, i.e., Charnockite (hard rock) and alluvium formation (fluvial and marine). The Palar River flows from the northwest to southeastern part of the study area fringed by this Quaternary alluvium (flood plain). The famous Indira Gandhi Atomic Research (IGCAR) station is located in this Kalpakkam region. The region depends on the South-West and North-

East Monsoons for rainfall. Normal and actual rainfall recorded during the year 2005-06 is 1252.1 mm and 1825.5 mm respectively (CGWB, 2007). River Palar, a major river course, which drains this region originates from Western Ghats in Karnataka state, and discharges in Bay of Bengal near Pudupattinam. The Buckingham canal runs parallel to the coast and presence of SADRAS back water is noted in NE part of the study area.

Methodology

A total of 30 groundwater samples were collected for post monsoon, Northeast monsoon, Summer, southwest monsoon, in the study area (Fig. 1). Sampling and analysis were carried out using standard procedures (APHA, 1995). Five hundred milliliters of water samples were collected in polyethylene bottle. Then, it was sealed and brought to the laboratory for analysis, stored properly (4°C), and filtered with 0.45-ìm filter paper before analysis. pH and electrical conductivity (EC) were determined in the field using Field Analysis Kit (Eutech Handheld Instruments). The collected groundwater samples were analyzed for major cations and anions. Calcium, magnesium, bicarbonate, and chloride were determined by titrimetric method. Sodium and potassium were analyzed through flame photometry (ELICO CL 378). The range of flame photometry is 1–100 ppm with an accuracy of ±1 digit. Silica, phosphate, and sulfate were determined by spectrophotometry (ELICO SL 171 minispec). The range of spectrophotometry is 340– 1,000 nm with an accuracy of ±2.5 nm. The reliability of the results was determined by the ionic balance of groundwater samples and a 5–10 % of percentage error was noted. Saturation index and Inverse modeling were performed by PHREEQC software. Saturation indices (SI) are indicators if the mineral will precipitate or dissolve in the aqueous condition. SI value <-0.5 indicates that groundwater is undersaturated with respect to the mineral and if the value is zero, groundwater is at equilibrium, and when the values are between -0.5 and 0.5, it indicates near saturation, while values > 0.5 indicate a supersaturated condition.

Results and Discussion

The dominance of anion shows that HCO_3 > Cl > SO_4 > NO_3 > F > PO_4 and that of cation is Na > Mg > Ca > K. In all the seasons the NE part of the study area has higher EC due to the presence of back water region near SADRAS. Higher concentration of EC was observed in all major litho units and along their contacts and along the river alluvium. The landfill sites for the Mahabalipuram Township it also located along the NE part of the study area (Karmegam et al., 2010).

Saturation Index of Carbonate Minerals

Saturation index of groundwaters for minerals such as Magnesite ($MgCO_3$), Aragonite ($CaCO_3$), Calcite ($CaCO_3$), Dolomite ($CaMg(CO_3)_2$) were calculated. Bicarbonate derived by the pressure of weathering or from other source remains in the aqueous medium trying to equilibrate with different ions like Ca, Mg or both to attain saturation with different minerals like Magnesite (SI_M), Dolomite(SI_D), Calcite (SI_C) and Aragonite(SI_A). The SI of Carbonate minerals shows under saturation to saturation state in all four seasons. The saturation state of Carbonate minerals are in the following orders $SI_C > SI_M > SI_A > SI_D$. Saturation index of these carbonate minerals were plotted against the concentration of HCO_3 (in mg/l). During the POM (Fig. 2a) and SUM (Fig.2c), an increasing trend of SI is noted with the increasing HCO_3. More number of groundwater samples were found fall between saturation to oversaturation during POM and SUM may be due to significant evaporation concentration in these samples. The Magnesite and Aragonite minerals are near the state of equilibrium showing an increasing trend with HCO_3. During NEM (Fig. 2b) and SWM (Fig. 2d) similar trend is noted for all carbonate minerals. They are nearly saturated to under saturated state and this indicates that state of saturation decreases during NEM and SWM it is due to the fact that increase of groundwater recharge and dilution of the groundwater chemistry with respect to these minerals (Calcite and Dolomite) after the subsequent monsoon period or due to the non-availability of Ca. In general there is an increase or fall of saturation index of the Carbonate minerals with increasing HCO_3. During NEM and SWM the excess of HCO_3 results in stoichiometric dissolution of Calcite and ion exchange (Gomez et al., 2006). During POM and SUM the saturation index of the carbonate minerals in almost 80% of the sample are saturated with respect to all Carbonate minerals. When HCO_3 concentration increases, there is an increase of SI of all the Carbonate minerals and they form near saturation to saturation state. Few samples show under saturation though with high HCO_3, may be due to non availability of cations (Ca and Mg) which might have been removed from the aqueous system due to process of cation exchange (Gomez et al., 2006). The carbonate minerals generally attain saturation when the concentration of HCO_3 reaches nearly 300 mg/l.

Saturation Index of Silicate Minerals

The different states of Silicate minerals like crystalline, cryptocrystalline and amorphous forms are represented by Crystobalite, Chalcedony and Silica Gel respectively. The saturation index of this three different minerals are represented as, SI_{cr}, SI_{ch} SI_{sg}. It is observed that the saturation index of these minerals in the samples are in the following trend SI_{cr} >SI_{ch} >SI_{sg}. The SI_{cr} and SI_{ch} are in over saturation and SI_{sg} is in the state of undersaturation to saturation. It is also observed that there is a linear increase in saturation index of these minerals with the increase in concentration of dissolved Silica, in all the seasons. SI_{cr} and SI_{ch} shows near equilibrium to over saturation state in all the four seasons, but it also shows few samples in the state of under Saturation during NEM (Fig.3b) and SWM (Fig.3d) seasons This indicate that availability of dissolved Silica is more during POM (Fig.3a) and SUM (Fig.3c) and in all the seasons there is an exponential increase of saturation index of these minerals with increasing dissolved silica concentration. It is observed that SI_{Ch} and SI_{Cr} attains equilibrium with H_4SiO_4 around 25ppm and that of silica gel at 90ppm during POM and in NEM , SUM and SWM the saturation of SI_{Ch} and SI_{Cr} is around 20ppm and for that of Silica gel is around 65ppm. At lower concentration of dissolved silica, SI_{cr} and SI_{ch} attain equilibrium, but still higher concentration of dissolved Silica is essential for SI_{sg} to attain saturation Hence, It is clearly evident that the gel form of silica needs higher concentrations for reaching the state of saturation.

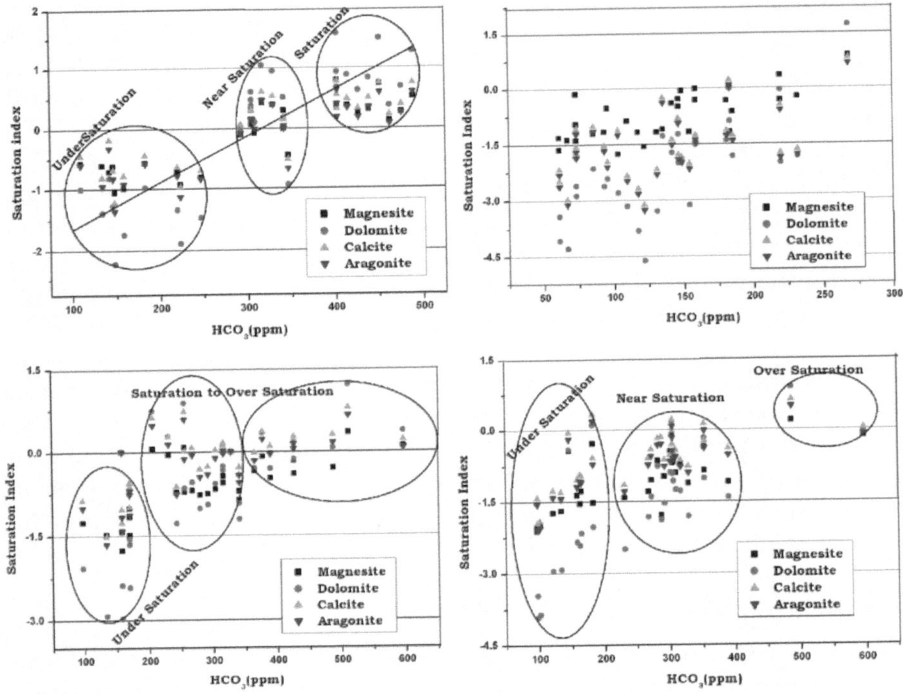

Fig. 2. Variation of Saturation index of different Carbonate minerals with dissolved HCO_3 for samples collected during (a) POM; (b) NEM; (c) SUM; (d) SWM

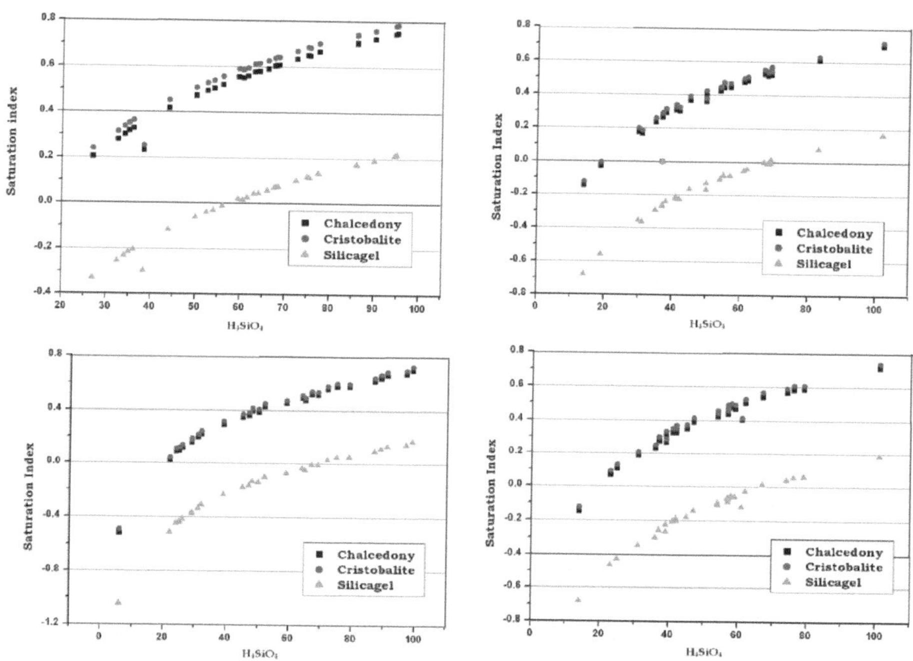

Fig. 3. Variation of Saturation index of different Silicate minerals with dissolved H_4SiO_4 for samples collected during (a) POM; (b) NEM; (c) SUM; (d) SWM

Saturation Index of Sulphate Minerals

Gypsum is a common Sulphate mineral present in sedimentary formations and other sulphate mineral is Anhydrite ($CaSO_4$). The saturation index of these two different minerals is represented as SI_{Gy} and SI_{An}. It is observed that during all the seasons, the saturation index of these minerals are in the following trend $SI_{Gy} > SI_{An}$. The SI of both the minerals are in state of Under saturation. It is also observed that there is a linear increase in saturation index of these minerals with the increase in concentration of Sulphate during POM, SUM and SWM with Anhydrite (Fig. 4a, c, d). It may be due to the common ion effect or the interference of Na in the formation of these minerals, which varies in chemical composition of the groundwater with season. All the four seasons shows that there is a steep increase of SI with initial low SO_4 concentration depicting steeper slopes and the further increases of the SO_4 concentration shows a gentler slope even with a decrease increase of SI. This trend was not clearly observed in NEM samples as it had higher SO_4 concentration ranges (Fig. 4b) than all other seasons, due to flow of the pollutant load in the Palar from the upstream. The NEM season is the peak agricultural season hence the impact of fertiliser may also be prominent, which may also result in higher concentration of this ion in groundwater's of this season.

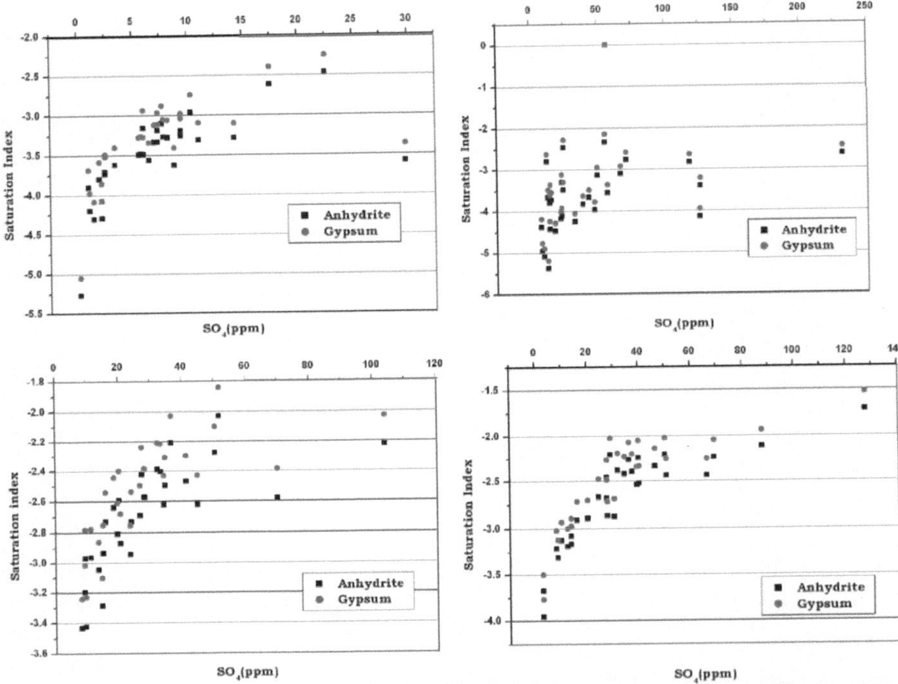

Fig. 4. Variation of Saturation index of different Sulphate minerals with dissolved SO_4 for samples representing (a) POM; (b) NEM; (c) SUM; (d) SWM

Inverse models for a specific sample

A model is a subject of the selected phases that satisfies all the selected constraints. The model is of the form.

Initial water + "Reactant Phases" $\rightarrow$ Final Water + "Product Phases"

The modeling assumes that the selected initial water mix is of some proportion, with or without further mineral water reactions. The fraction of initial water mixed to produce the final water is displayed in a mixing case. The variation of ions in a single location (sample 11) during two different seasons POM and SUM were studied in order to know the probable phase mole transfers taken place during this period. For data input to PHREEQC, identifiers in the INVERSE MODELING data block were used for the selection of aqueous solutions POM 11 and SUM 11, default uncertainty limits was selected, reactant phases include,

Magnesite, dolomite, Calcite, Aragonite, Anhydrite, Gypsum and Quartz. The phases were selected after deriving the SI of the ground waters with respect to these minerals. Range calculations were determined to 1000(-range) and minimal models were opted. Aqueous solution compositions are defined with the SOLUTION, SOLUTION SPREAD, data block and reactants were defined with PHASES or EXCHANGE SPECIES data blocks. The output of the run has evolved three different types of models giving their respective solution fractions and Phase mole transfer.

Model Calibration

Observations were made in two seasons in POM 11 and SUM 11 for inverse modeling. Samples of POM was taken as the initial solution and SUM as the final solution. The results of these models were then compared with the saturation index tables of the initial and final solution (Table.1). The saturation index of minerals like Anhydrite, Gypsum, Magnesite, Dolomite, Chalcedony, Calcite, Aragonite, Cristobalite and silica gel were observed in both the solutions. The order of the dominance of saturation is observed in the following order;

Dolomite> Magensite> Calcite> Aragonite> Cristobalite>Chalcedony

The saturation index table shows dissolution of Calcite, Dolomite and Gypsum. The precipitation was noted in Dolomite, Calcite and Chacedony. Depending upon the significant dissolutions and precipitations noted form the Saturation index of different phases the constraints and uncertainties were determined.

Table 1. Saturation index of the initial and final solution

Minerals	POMSolution 1	SUMSolution 2	Composition
Magnesite	0.7866	0.0648	$MgCO_3$
Dolomite	1.5904	0.3092	$CaMg(CO_3)_2$
Calcite	0.7727	0.1759	$CaCO_3$
Anhydrite	-3.3346	-2.6222	$CaSO_4$:1/2H_2O
Gypsum	-3.115	-2.4336	$CaSO_4$:2H_2O
Aragonite	0.629	0.0368	$CaCO_3$
Chalcedony	0.2786	-0.5134	SiO_2
Cristobalite	0.3144	-0.4898	SiO_2
Silica gel	-0.2546	-1.0422	SiO_2

Table 2. Phase mole transfer of three models

Model-1		Model-2		Model-3	
Phase	Mole transfer	Phase	Mole transfer	Phase	Mole transfer
Calcite	-5.286×10^{-4}	Dolomite	-4.192×10^{-4}	Gypsum	-2.022×10^{-4}
Gypsum	1.13×10^{-4}	Calcite	-2.658×10^{-4}	Calcite	2.684×10^{-4}
Anhydrite	-1.24×10^{-4}			Chalcedony	1.119×10^{-4}

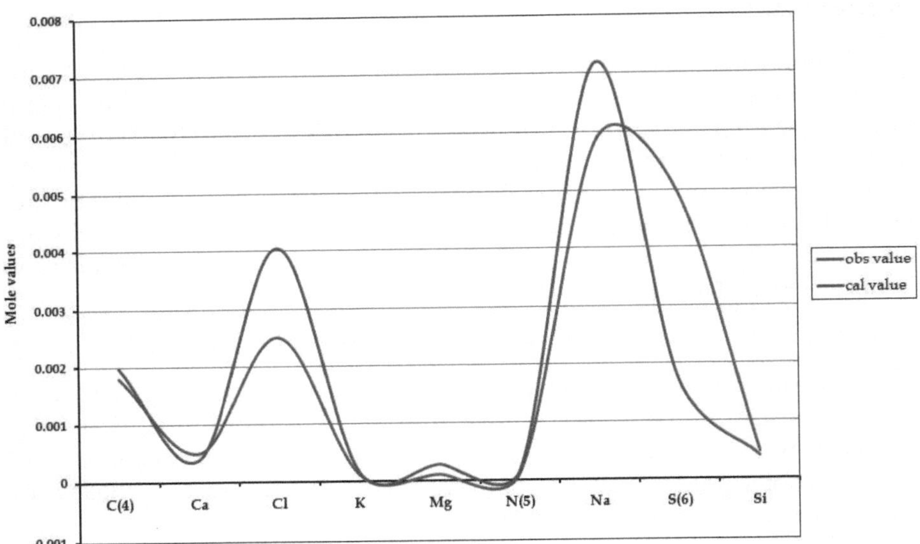

Fig. 5. Line Graph against Mole Values (Validation against the calculated and observed values)

The three models were obtained (Table.2) are calibrated to achieve the apt field model by establishing the relationships. The addition and removal are considered as dissolution and precipitation of the phases from the system. This is derived by the positive and negative symbols respectively, in the phase mole transfer. Hence, dissolution is noted by positive values in phase mole transfer.

Thus out of these models model 3 has positive representation of calcite indicating the increase in the concentration by dissolution and in model 1 has the negative representation of Anhydrite indicating precipitation and removal. Both the cases show contradictions when compared to the original saturation index of the considered solution 1 and solution 2. Hence the appropriation of these models is ruled out. The second model has higher precipitation and removal of Dolomite and Calcite which match with the original phases saturated. Hence the model this is believed to be the model which replicates the natural system to the maximum.

Any model run requires a validity and in order to check the validity of the model, the composition of the solution 2 is plotted as a line graph (Fig. 5) against the modeled value. The variations in the major and minor ions with the original solution 2 and that of the modeled value are minimal. The total error percentages was calculated to be 7.45% (the difference in the TDS/Sum of the both the solutions).

Conclusion

It is also interesting to note that the order of dominance during under saturation is $Si_C > SI_{Ar} > SI_{Mg} > SI_{Do,}$ but this case almost reverse during the state of saturation except for Calcite and Aragonite. It may be due to the preferential addition of Mg to $CaCO_3$ and formation of new $CaMg(CO_3)_2$, which results in the increase of SI of Dolomite in the saturation state than the other carbonate minerals. It is observed that in all the seasons, the saturation index of these minerals are in the following trend $SI_{cr} > SI_{ch} > SI_{sg}$ and higher concentration of Silica is noted in the non-monsoon periods. The saturation index of sulfate minerals are in the following trend $SI_{Gy} > SI_{An}$ and they are under saturated in all the seasons. For the inverse modeling samples were considered from two seasons in POM 11 and SUM 11. Samples of POM was taken as the initial solution and SUM as the final solution. Three models were obtained, in that second model is considered as apt which replicates the natural system to the maximum.

Acknowledgement

The authors express their sincere thanks to Atomic energy Regulatory Board (AERB),India, for providing necessary financial support to this research project.

References

Anandhan, P. (2005) Hydrogeochemical studies in and around Neyveli mining region, Tamilnadu, India. Ph.D Thesis, Department of Earth Sciences, Annamalai University, 189p.

APHA (1995) Standard methods for the examination of water and waste water, 19th edition, APHA, Washington DC, USASS.

CGWB (2007) Ground water resources and development prospects in Kancheepuram region.

Chidambaram S, Bala Krishna Prasad M, Manivannan R, Karmegam U, Singaraja C, Anandhan P (2012) Environmental hydrogeochemistry and genesis of fluoride in groundwaters of Dindigul district, Tamilnadu (India). Environmental Earth Science, 68(2), pp. 333–342.

Drever, IJ. (1988) The Geochemistry of Natural Waters, 2nd Edition; Englewood Cliffs NJ (Eds) Prentice Hall, pp.388.

Elangovan, K., Balasubramanian, A., Thirugnana Sambandan, R., Rengaragan, R., Satish, PN. and Janardhan, AS. (1999) Geochemical Modelling Of Crystalline Aquifers Of Salem District, Tamilnadu. Indian J Geochem, v.14: pp.1 - 17.

Garrels, RM. and Christ, CL. (1965) Solutions minerals and equilibria: New York, Harper and Row, 450 p.

Gomez P, Garralón A , Buil B, Turrero M J, Sánchez L, De La Cruz B, (2006) Modeling of geochemical processes related to uranium mobilization in the groundwater of a uranium mine. Science of the Total Environment, In Press, Corrected Proof.

Jenne, EA.(1978) Chemical Modeling In Aqueous Systems ,Speciation Sorption, Solubility And Kinetics. US Geol Sury Water Resources Division, Meto Park

Karmegam U, Chidambaram S, Prasanna M V, Sasidhar P, Manikandan S, Johnsonbabu G, Dheivanayaki V, Paramaguru P, Manivannan R, Srinivasamoorthy K, Anandhan P (2010) An study on the mixing proportion in groundwater samples by using Piper diagram and Phreeqc model, Chin. J. Geochem, Vol. 30, pp.490–495.

Murphy W M, Shock E L (1999) Environmental aqueous geochemistry of actinides. In: Uranium: Mineralogy, Geochemistry and the Environment (Eds. BURNS, P.C. and FINCH, R.). Reviews in Mineralogy, 38, pp. 221-254.

Plummer L N, Busby J F, Lee R W, Hanshaw BB (1990) Geochemical Modeling of the Madison Aquifer In Parts Of Montana, Wyoming, And South Dakota. Water Resources Research, v.26, pp.1981 - 2014.

Sanford WE. and Konikow. (1989) Simulation of Caicite dissolution and porosity change in salt water mixing zones in coastal aquifers. Water Resources Res, v.25, pp.655-667.

Srinivasamoorthy (2004) Hydrogeochemistry of groundwater in salem district, Tamilnadu, India. Unpublished Ph. D thesis, Department of Earth Sciences, Annamalai University, India.

Truesdell, A.H. and Jones, B.F. (1974) WATEQ, A computer program for calculating chemical equilibria of natural waters, U.S. Geological Survey, 2 233-274.

Chapter **6**

A Study on the Removal of Fluoride in Groundwater by Using Natural Materials

Manikandan S.[1], Chidambaram S.[1], Ramanathan A.L.[2] Johnsonbabu G.[1] and Karmegam U.[1]

[1] Department of Earth Sciences, Annamalai University. Annamalai Nagar, India
[2] School of Environmental Sciences, Jawaharlal Nehru University, New Delhi India.

[3] School of Engineering and Science, Department of Applied Geology Curtin University, Sarawak Malaysia, CDT 250, 98009 Miri, Sarawak, Malaysia

ABSTRACT

Deduction of overindulgence fluoride (F) from the water has been endeavor by numerous investigators by using different materials both natural and artificial. The objective of this paper was to attempt the fluoride removal by using the locally available natural materials such as red soil, untreated charcoal, local powdered brick, fly-ash from Neyveli (thermal power plant) and mineral serpentine were used, adopting column method. Maximum defluoridation occurs immediately after the experiment has started. The morality of fluoride removed per gram is more in red soil when compared to the other materials. Variation of fluoride removal with time was studied after the erection of the column /unit, the defluoridation setup was then connected to the public water supply from the overhead tank of 50,000L capacity. Higher F removal was noted when flow rate was lesser. An attempt on the regeneration of the fluoride adsorbed soil was also made.

Key words: Adsorption, Defluoridation, Lateritic soil, Serpentine, Red soil

Introduction

Removal of excess fluoride from the natural water supply is still a challenging task. Fluoride removal was done by passing through columns, as earlier studies on stirring methods had confirmed that an increase in stirring rate increase the fluoride for a given temperature (Killedar and Bhargava, 1993). In general aluminium compounds are found to be good fluoride removers because of reaction between Al and F molecules, several fluoride removal methods were carried out using the aluminium based compounds (Bulusu and Nawlakhe, 1992). Few studies have been attempted on the possibility of using clay vessels for water defluoridation (Agarwal et al. 2003). Such vessels, usually made up of clay-dominated soil, are quite common in rural India and are used for storing water, especially during summers. On account of their porous nature, the clay vessels facilitate water evaporation and thereby act as effective water coolants. Such vessels, having a high F sorption ability, are also expected to effectively defluoridate the stored water. Silty clay (C) fraction exhibiting maximum F^- sorption was further amended chemically to enhance its potential for water defluoridation—a method when extrapolated to fabrication of vessels would have enhanced potential of utilizing vast naturally occurring adsorbent (i.e. C) for defluoridation of water. Yadav et al. (2006) proved that adsorption of fluoride on Brick Powder (BP) from aqueous solution was found to be first order reaction and mechanism of fluoride removal on adsorbent was found to be complex. The surface adsorption as well as intra-particle diffusion contributes to the rate-determining step. Deflouridation capacity of BP can be explained on the basis of the chemical interaction of fluoride with the metal oxides under suitable pH conditions. The optimum pH was found to be in range of 6.0–8.0 for maximum adsorption of fluoride, which makes it very suitable for use in drinking water treatment, especially in rural areas.

During the last decade the removal of low-molecular weight compounds and the removal of salts with different valences from ground and surface waters for drinking water production were carried out by nano filtration (NF) increasingly (Rautenbach and Groschl 1990). Generally, nano- hydroxyapatite (n- HAp) as synthetic apatite was found to be a suitable sorbent for fluoride because of their low costing, availability and higher adsorption efficiency as previously reported by the authors (Simons 1993). This review of the literature shows that the study of fluoride removal by natural materials/soil is less. One such attempt was made (Chidambaram 2000; Chidambaram et al. 2003) for removal of fluoride in groundwater by using natural materials and showed that, untreated charcoal seems to have lesser effect on the concentrations of fluoride with time and it adsorbs only 0.5–1 mgl of fluoride.

The present study aimed to find a suitable low-cost environmental friendly method for the removal of fluoride in the groundwater that is used by common man. Therefore, few natural materials such as red soil, untreated charcoal, local powdered brick, fly-ash from Neyveli (thermal power plant) and mineral serpentine were used. The ability of soil to absorb fluorine from solution has been studied by earlier researchers (Bower and Hatcher, 1967; Gupta et al. 1982). The capacity of fluoride removal by the individual materials was studied and accordingly two columns were set up and studied for their defluoridation capacities. The results obtained may indicate a possible solution for the removal of fluoride from a rural water supply scheme.

Methodology

Defluoridation was done by using Red soil, serpentine mineral, charcoal, brick powder and fly ash. Based on these observations Red soil was attempted for defluoridation Chidambaram et al., (2003). It is noted that the red soil is fine grained substance and with clay size particles, with good ion exchange capacity but lesser permeability, Hence a column was set up for the study by using mixture of red soil and sand with a ratio by weight of 4:1 to increase the permeability. The removal rate of fluoride was tested using 250gm of mixture soil (200g of red soil and 50g of sand). The standard solution of 10mg/l of fluoride was prepared by using DDW and passed through the column. The samples were collected at definite intervals. 11 different Batch experiments were conducted by passing Fluoride standard of 10 ppm. 10 ml of sample was collected after the treatment and analysed. A total of about 62 samples were collected in different batches during the entire experimental period. Out of these samples collected different parameter were observed along with Fluoride concentration. Rate of flow of water through the column, pH, percentage of fluoride adsorption, adsorption capacity, bicarbonate concentration and EC were also observed. pH and EC were analysed by ion electrodes HCO_3 was estimated by titrimetry method using standard procedures (APHA, 1998).

The adsorption capacity was calculated by using the formula,

Qe (mmol/g) = (Co - Ce) x V/W x 1000

Ce = equilibrium adsorbate (fluoride) concentration (mmol/l)

Co = initial adsorbate (fluoride) concentration (mmol/l)

Qe = amount of fluoride adsorbed per unit gram of the adsorbant at equilibrium (mmol/g)

V = volume of solution (l)

W = weight of adsorbant (g)

Defluoridation Using Natural Materials

The study reveals that of the five materials were used for Defluoridation, red soil has a good defluoridation capacity followed by brick, serpentine, fly-ash and charcoal. Near equilibrium is attained after about 30 min of the experiment (Table.1). Maximum defluoridation occurs immediately after the experiment has started. The morality of fluoride removed per gram is more in red soil when compared to the other materials. Red lateritic soil has the highest fluoride removal capacity because it has oxide of aluminium and iron as its major components. Both are good removers of fluoride but the red soil is found to be very effective.

Table 1. Fluoride in time intervals (mg/L)

Time (mins)	Red soil	Charcoal	Fly-ash	Brick	Serpentine
15	0.64	9.05	3.4	4.7	3.7
30	0.09	9.05	3.7	4	4.1
60	0.06	9.1	4.7	3.7	4.6
90	0.035	9.05	6	4	4.8
120	0.029	9	7.1	3.9	5.8

The column study reveals that use of the amount of materials according to fluoride removal capacity seems to be very effective. This study lacked detailed monitoring of data over a periodic time interval of 1 min, where sequential changes can occur rapidly; further, the soils are not studied in detail for textural character and are not penetrated. The observation reveals that after a period 30 min removal of fluoride is reduced in the laboratory experiment and for the specific volume of the chromatography column. Perhaps due to saturation of the anion-exchange sites the concentration of fluoride shoots up. The removal of fluoride in this lateritic soil is not only due to Gibbsite but also due to Kaolinite. The kaolinite structure includes highly accessible hydroxyl groups positioned on the gibbsite layers which allow for anion exchange (Kau et al., 1997). Reaction was assumed to be a direct indication of the reaction progressing with the concentration change of Al^{3+} in both gibbsite and kaolinite, which was difficult to monitor (due to lack of quantitative characterization of this heterogeneous material). The observed kinetics can thus be treated as pseudo-first order kinetics.

This is also evident in the increase of the HCO_3 content in water, which increases along with pH. It is noted that the red soil has greater EC than the other natural materials used. Different combination of these materials was studied in the laboratory in sequential batch experiments. The regeneration of the materials used in all these studies were also attempted. The monitoring was carried out in batches and during this Rate of flow pH and % of adsorption, adsorption capacity, HCO_3 and EC was observed. The statistical representation of these values is given in Table 2.

Table 2. The maximum, minimum and average values of rate of flow, pH, adsorption, HCO_3 and EC

	Rate of flow	pH	% Adsorption	Adsorption capacity	HCO_3 (mg/L)	EC (μs/cm)
Maximum	0.83	9.50	96.00	0.0038	707.60	4354.00
Minimum	0.02	6.63	0.00	0.0000	24.40	291.00
Average	0.22	7.69	86.45	0.0025	182.80	723.52

Table 3. Fluoride removal efficiency and rate of flow

	Rate of flow	pH	Removal efficiency	Adsorption capacity	HCO_3	EC
Rate of flow	1.00					
pH	0.58	1.00				
Removal efficiency	0.03	-0.17	1.00			
Adsorption capacity	-0.22	-0.53	0.59	1.00		
HCO_3	-0.50	-0.57	-0.04	0.30	1.00	
EC	0.22	0.03	0.04	0.42	0.04	1.00

The study shows that Rate of flow ranges maximum from 0.02 to 0.83 with an average of 0.22. The pH shows maximum from 9.50 to minimum 6.63 with an average of 7.69, the percentage of adsorption shows maximum from 96 to minimum 0 with an average of 86.45, (Table.3) Adsorption capacity maximum from 0.0038 minimum to 0.00 with an average of 0.0025, the HCO_3 maximum from 707.60 mg l^{-1} minimum to 24.40 mg l^{-1} with an average of 182.80 mg l^{-1} and the EC maximum from 4354 μs/cm minimum to 291 μs/cm with an average of 723.52 μs/cm. The correlation analyses of the parameters were carried out and these parameters show that the rate of flow of water has positive correlation to pH and negative to HCO_3, pH has negative correlation to all other parameters, with strong negative correlation to fluoride adsorption capacity and HCO_3 (Table 3). This indicates that the efficiency of fluoride removal increases when HCO_3 concentration decreases. The efficiency of fluoride removal is higher when the adsorption capacity is more. The adsorption capacity shows a poor positive correlation to electrical conductivity of the solution.

Transfer of Technology to the field

The above said results of the preliminary study was considered and red soil of silt size fraction mixed with sand which are locally available was used for the present study as the chief fluoride removal agent in the Fluoride Removal Tank (FRT). Earlier experiments were tried with following options : 1) red soil + sand mixture, 2) pebbles and red soil + sand mixture, 3) Pebbles + Red soil + Sand mixture and Charcoal, 4)Pebbles + Red soil + Sand mixture + Charcoal and Brick powder. These were experimented to avoid clogging, prevention of color in the treated water, prevention of movement of red soil materials from the FRT along the supply line and to optimize the flow in the FRT. Among all these mixture the fourth sequence was found to be fruitful.

The defluoridation tank erected was 6′ height and 1.5′ diameter rubber coated and the sequence of the material filled in is given in the figure (1).The water flows against gravity first flows through pebbles, followed by charcoal, then through red soil then by brick and then goes to the supply. The size fraction arrangements are in such a way that the charcoal and brick powder are sand sized and red soil was silty sand. The red soil was comparatively finer than the other two fractions which increase the retention time to the water passing through the medium. It is also observed that after the introduction of the FRT, of water supply along the normal distribution pipe line reduces the pressure, hence the consumers at the far end of the water supply line receives a lesser quantity of water and a few places dependent people at the end of the pipeline are even devoid of water supply.

Variation of fluoride removal with time was studied after the erection of the column /unit, the defluoridation setup was then connected to the public water supply from the overhead tank of 50,000L capacity and it was allowed to flow through the defluoridation column (Fig. 1 and 2).

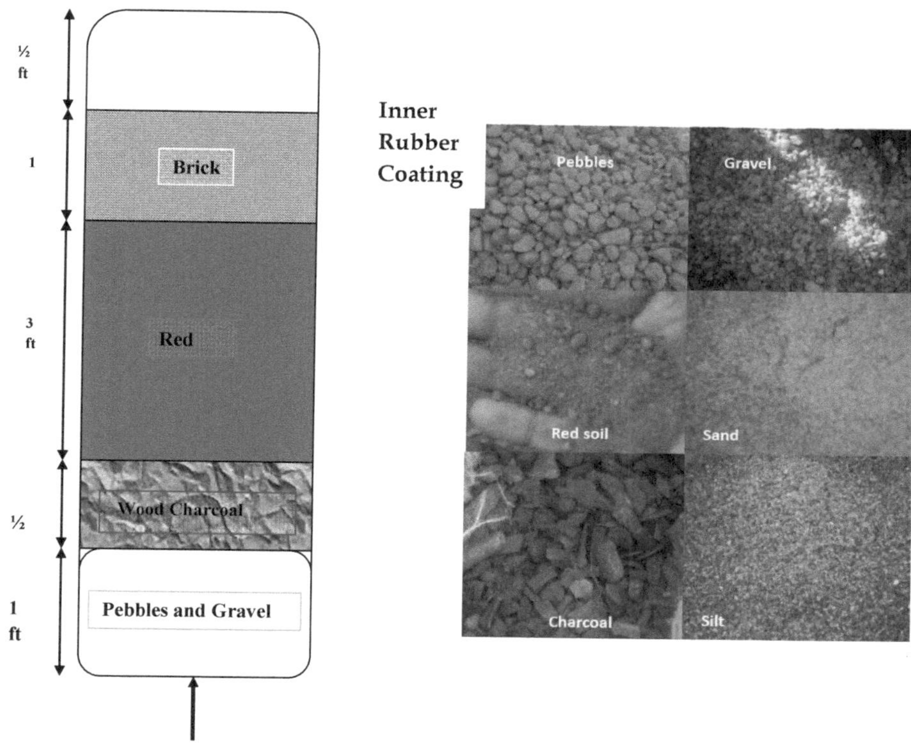

Fig. 1. Model of FRT erected

Fig. 2. Natural materials used in the tank

The flow through the column was observed for 24hrs after the erection in 2hours time interval. The water was continuously pumped to the OHT (by normal process) and it was passed though the defluoridation column erected through gravity. The continuous observation shows that the initial concentration of 4.5 mg/l was reduced to 2.5 mg/l in the first sample collected later there was a decrease in the fluoride concentration around 0.5mg/l after 14hrs of erection (Fig 3). Later it was almost stable around 0.68 mg/l.

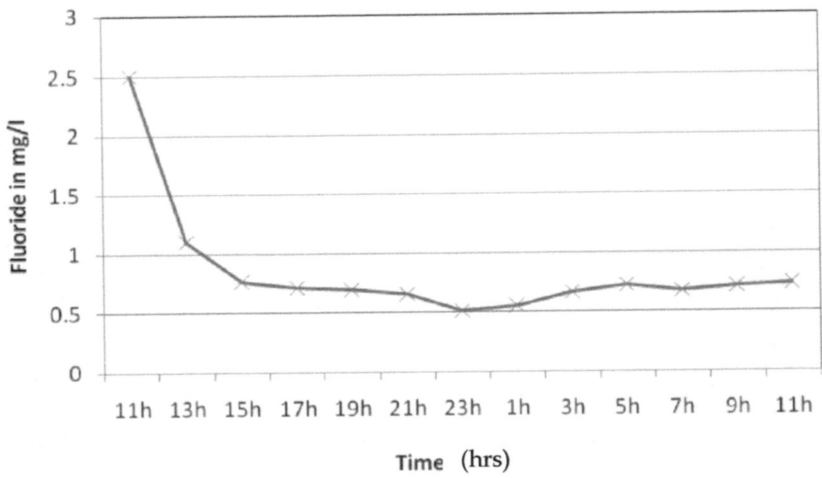

Fig. 3. Rate of fluoride removal in 24 hours monitoring immediately after the erection of FRT

Water supply to the dependent local village community was provided for four hours approximately two hours in the morning and two in the evening. The material capacity was observed from 3rd day 7th day 10th and 15th day (Fig 4). Later new material was refilled and further similar observations were made, four similar refilling and similar time interval observations were made. An average of the time period observations show that during morning the water supply was from 6.00 am to 8.00 am.

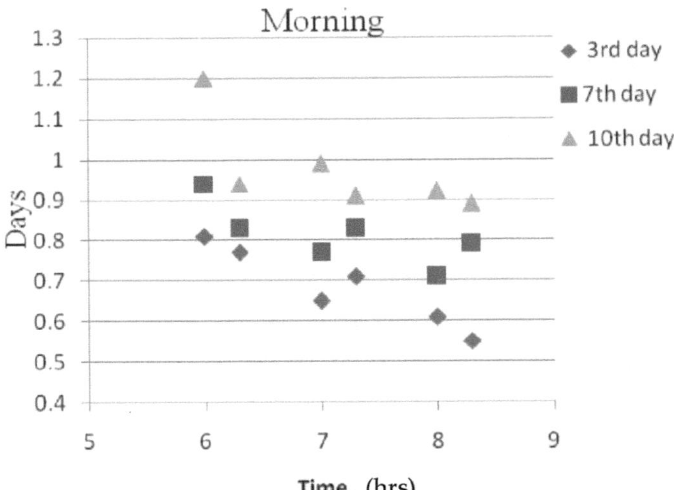

Fig. 4. Four times average of fluoride removal from FRT (Morning)

Half an hour interval observation shows that there is a decrease in fluoride concentration with time. It is also due to the variation in water level in the OHT; at initial time period the water level in the full and hence the rate of flow is higher which results in lesser contact time and the removal is lesser. But as the time passes the water level in the tank reduces and contact time increases which results in further removal of the fluoride ions in water.

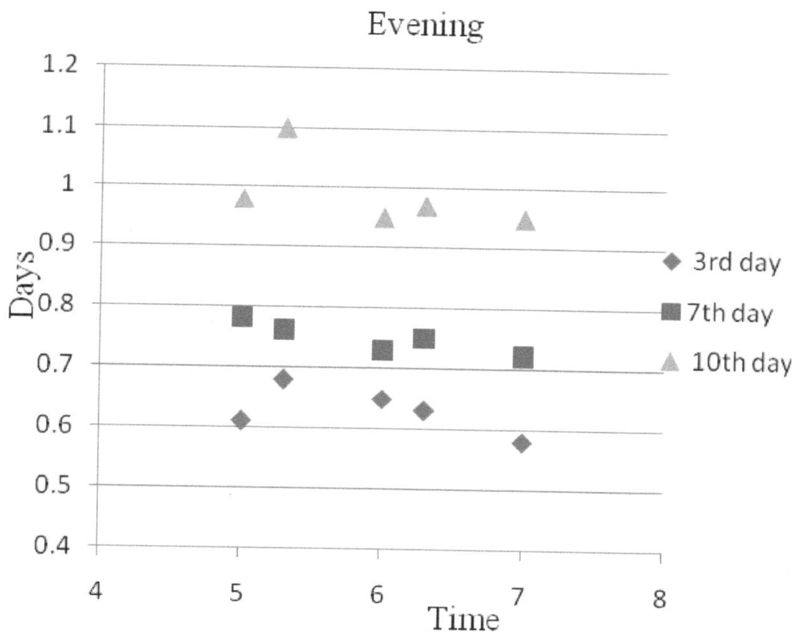

Fig. 5. Four times average of fluoride removal from FRT (Evening)

It is also observed that the efficiency of the fluoride removal decrease with days and at the 10th day it's almost near the permissible limit, (Fig 5) Hence, the material were refilled again. It is interesting to note that the trend of the fluoride removal remains the same. Similar trend was also noted in the evening 2 hrs of delivery. It is also to be noted that the F removal capacity may also vary with the variation in the groundwater pH. Though the source of groundwater had remaining the same it may vary with time.

Characterization of Red Soil Used in the FRT

The red soil was characterized by powder X-ray diffraction (PXRD) using Shimadzu 600 X-Ray Diffractometer. The PXRD patterns were collected in continuous scan mode with monochromatic CuKa (l=1.5418 Å) radiation that was selected using a curved germanium (111) monochromator. X-rays which were produced at the X-ray source (copper radiation) were reflected at the germanium monochromator (111 planes) giving pure Cu, Ka radiation which was diffracted by the sample. The data were collected in the range from 10.0 to 80.0° (2θ) by a linear PSD which was set at a step size of 0.5° (2θ) and counting time of 60s per step. Compound identification was made using a search-match computer program supported by the Joint Committee on Powder Diffraction Standards database JCPDS, (1997) together with diffraction profiles obtained from standards of red soils and clay minerals as reported in literature (Mineral database, 2002).

The mineralogical characteristics of the lateritic soil used in this study are shown by the PXRD patterns in (Fig 6). The raw phase profiles qualitatively indicate presence of gibbsite ($Al(OH)_3$) as shown by intense reflection at 2θ angle of 17.93° (d= 4.94 Å), and double at 2θ angle of 20.51° probably combining the literature reflections at d= 4.34 Å and 4.30 Å of a crystalline gibbsite. Intense kaolinite ($Al_2Si_2O_5(OH)_4$) reflections are also shown at 2θ angles of 11.58° and 24.80°.

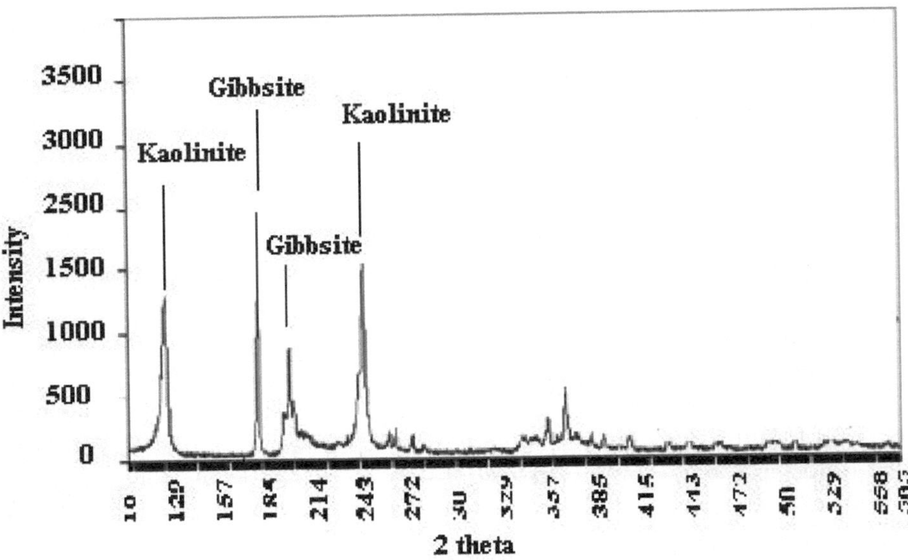

Fig.6. PXRD profile of the red soil used for Defluoridation in the study. Gibbsite and Kaolinite reflection are indicated within the profile.

Table 4. Fluoride concentration and defluoridation capacities of red soil at 2.5g/ 200ml solution of 8 Mg/L F^-

	Initial fluoride Conc.(Mg/L)	Equil.Fluoride Conc.(Mg/L) (Mg/L)	Fluoride removed	Defluoridation capacities (%)
Red soil	8.00	0.52	7.6	95

Table 5. Correlation coefficients between increase in concentration and decrease in fluoride concentration in defluoridation with bauxite.

Ions	PO_4^{3-}	SO_4^{2-}	CO_3^{2-}	NO_3^-	Cl^-	Ca^{2+}	K^+	Na^+
Correlation	-0.29	-0.61	+0.91	-0.05	+0.75	-0.91	-0.15	-0.11

The removal of fluoride in this lateritic soil is not only due to gibbsite but also due to kaolinite. The kaolinite structure includes highly accessible hydroxyl groups positioned on the gibbsite layers which allow for anion exchange (Kau et al. 1997). Reaction was assumed to be a direct indication of the reaction progressing with the concentration change of Al^{3+} in both gibbsite and kaolinite, which was difficult to monitor (due to lack of quantitative characterization of this heterogeneous material). It was also assumed that the bauxite concentration was significantly larger than that of fluoride. The observed kinetics can thus be treated as pseudo-first order kinetics.

In order to determine the maximum dose of the lateritic soil in a 200 ml solution of 8 mg/L F^-, different masses of this soil were mixed with the solution and the equilibrium residual fluoride concentration was determined. The equilibrium concentrations decreased quickly with increase in mass of the soil used in defluoridation upto about 10g dose. After this point the equilibrium concentrations were within 0.05 to 0.2 mg/L. (Table 4 and 5) This implied that increasing the mass of material per volume of water would only increase defluoridation efficiency up to a dose of 10g/200ml (1g/20ml).

Environmental Benefits

Fluoride from water can be divided into two categories: precipitation and adsorption. Precipitation of fluoride with calcium and aluminium salts (Saha, 1993, Hichour et.al., 2000) has been used to remove fluoride from industrial wastewater. Typically, lime is used as a calcium source, which first reduces the fluoride concentration down to 10–20 ppm. The Ca^{2+} ions released from calcium salts interact with fluoride and form CaF_2 precipitate.

Gas concrete (Oguz, et al., 2005) which is a building material, is included in the light concrete category. It is produced from different mixtures from silica, sand, cement, lime, water, and aluminium cake, which produce gas. One of the main advantages of $F^{''}$ removal by using concrete over the other chemical treatment methods is that it does not produce any chemical sludge.

The results of Oguz et al., (2005) study indicate that gas concrete is an effective adsorbent for removal of fluoride from aqueous solutions. Fluoride removal up to 96% was obtained. It was found that the amount of fluoride adsorbed on gas concrete efficiently depends on pH, and weakly agitation rate and temperature of the solution. It also shows that the adsorption of fluoride is approximately maximum at pH 6.9 and 8.9. At a pH above 6.9, fluoride removal was not changed because of stronger competition with hydroxide ions on adsorbent surface. Adsorption is also not favoured in the acidic range (Tokalýoglu et.al., 2001); this may be as a result of the formation of weakly ionised hydrofluoric acid.

Hence keeping these observations in mind the materials used in the FRT were made into concrete bricks in order to further remove the adsorbed fluoride by the calcium rich compounds in the concrete. It is thought that the removal of fluoride by gas concrete at the values of pH 6.9 and 8.9 has taken place both adsorption and precipitation of Al^{3+} and Ca^{2+} salts ($F^{''}$) (Oguz 2005). The predominant product on the surface of as concrete materials is AlF_3 because the solubility value (7.08×10^3) for AlF_3 is lower than that of CaF_2 (2.57×10^{10}). So the concrete bricks made by the materials used for the Defluoridation in the FRT were made into concrete bricks and later studied for leaching of Fluoride ions in to the environment so that the commercial viability of these bricks can be understood (Fig. 7). The stability of the bricks were also confirmed by the TLCP test (toxicity characteristics leaching procedure).

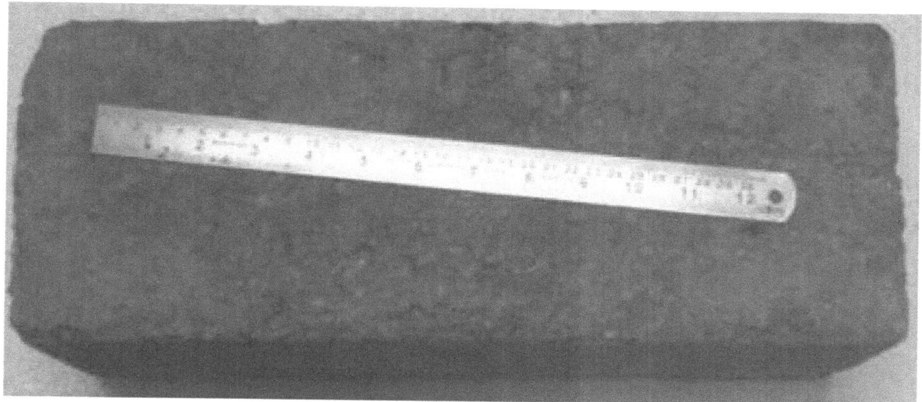

Fig. 7. A Cemented brick made out of the materials used in FRT

Conclusion

The Variation of fluoride removal with time was studied after the erection of the column/unit, the defluoridation setup was then connected to the public water supply from the overhead tank of 50,000L capacity and it was allowed to flow through the defluoridation column. The study reveals that of the five materials used for defluoridation red soil has a good defluoridation capacity followed by brick, serpentine, fly-ash and charcoal. Near equilibrium is attained after about 30 min of the experiment. The column study reveals that use of the amount of materials according to fluoride removal capacity seems to be very effective. This study lacked detailed monitoring of data over a periodic time interval of 1 min, where sequential changes can occur rapidly. The column technique is going to be very significant. In this respect our study will be of great relief to those drinking-water supply systems where F is reported to be very high. The results obtained indicate this as a possible method for the removal of Fluoride for an effective rural water supply scheme. Further due to its control over the rate of water ow, column experiments is proved to be an easy method for transferring the technology to the eld. This method requires no power supply since water moves down by the gravitational force hence it becomes more cost effective and easy to handle by local community as a suitable green and clean technology for rural drinking water supply in Fluoride affected areas.

Acknowledgement

The authors wish to express their thanks to Ministry of Environment and Forest (MoEn&F), India for providingthe necessary nancial support to carry out this study vide letter No.16/2007-CT Dated 07.10.2009.

References

Agarwal M, Rai K, Shrivastav R, Dass S. (2003) Defluoridation of water using amended clay. J. Cleaner Produc. 11, 439–444.

APHA (1998) Standard methods for the examination of water and wastewater, 19th edn. APHA, Washington, DC.

Bower CA, Hatcher J T (1967) Adsorption of fluoride by soils and minerals. Soil Sci 103: 209-218.

Bulusu KR, Nawlakhe WG (1992) Defluoridation, National environmental engineering Research Institute, Nagpur. Oct.

Chidambaram S, Ramanathan AL, Srinivasamoorthy K (2003) Lithological influence on the groundwater chemistry-Periyar district a case study In: Conference on coastal and freshwater issues 173p.

Chidambaram S (2000) Hydrogeochemical studies of groundwater in Periyar district, Tamilnadu, India, unpublished Ph.D thesis, Department of Geology, AnnamalaiUniversity.

Gupta RK, Chhabra R, Abrol I P (1982) Fluoride adsorption behavior in alkaline soils: relative roles of pH and sodicity. Soil Sci. 133, 364-68.

Hichour M, Persin F, Sandeaux J, Gavach C (2000) Fluoride removal from waters by Donnan dialysis. Sep. Purif. Technol. 18, 1–11.

Kau PMH, Smith DW, Binning P (1997) Fluoride retention by kaolin clay. J. Contam. Hydrol. 28, 267–288.

Killedar D J, Bhargava DS (1993) Effect of Stirring Rate and Temperature on Fluoride Removal by Fishbone Charcoal, Indian Journal of Environmental Health, 35(2), 81-87.

Oguz (2005) Adsorption of fluoride on gas concrete materials, J. Hazard. Mater. B 117, 227–233.

Rautenbach R, Gröschl (1990) A Separation potential of nanofiltration membranes, Desalination. 73-77.

Saha S (1993) Treatment of aqueous effluent for fluoride removal. Water Res. 27, 1347–1350.

Simons R (1993) Trace element removal from ash dam waters by nanofiltration and diffusion dialysis, Desalination 89 (3) 325–341.

Tokalioglu S, Sahin U, Kartal S (2001) Determination of fluoride and some metal ion levels in the drinking water in Kayseri Province. Turk. J. Chem., 25, 113-121.

Yadav RK, Goyal B, Sharma RK, Dubey SK, Minhas PS(2002) Post-irrigation impact of domestic sewage effluent on composition of soils, crops and ground water – a case study. Environ. Int. 28, pp.481–486.

Chapter 7

Assessment of Seawater Water Intrusion Mapping Using Electrical Resistivity Tomography in Coastal Tracts of Pondicherry, India

Srinivasamoorthy[$] K., Balasubramanian M. Gopinath S. Saravanan K., Prakash R., Vinnarasi Antony Ravindran* A. and Sarma V.S.[#]

Department of Earth Sciences, Pondicherry University, Puducherry, India
* Department of Geology, Geophysical and Coastal Research Lab, Manonmaniam Sundaranar, University, V.O. Chidambaram College, Thoothukudi, India
[#]National Geophysical Research Institute, Uppal Road, Hyderabad

ABSTRACT

Coastal aquifers are the sources of freshwater which are being over extracted due to increase in population, agriculture and industrial activities. This over exploitation has resulted in decline in groundwater level and intrusion of sea water into the coastal aquifers making it unfit for human utilities. An attempt has been made in the coastal tracts of Pondicherry region using Electrical resistivity imaging. The survey has been attempted in six locations spread thought the study area and the interpretation has been attempted by RES2DINV software. The profile length attempted was between 150 to 300 m with electrode spacings between 5 to 10 m. The apparent resistivity ranges between 0.1 to 1183 & ohm. Sea water intrusion was prominent along the northern parts of the study area and lower resistivity isolated in kirumambakkam has been interpreted as salinity due to the presence of clay minerals. In nallavadu saline intrusion has been identified at shallow depth.

Key words: Pondicherry, Imaging, coastal aquifers, sea water intrusion, formation salinity.

1. Introduction

India has a long coastline (7500 km) with 25% of the population living in the coastal areas. Maximum extraction of groundwater along the alluvial tracts of rivers and coastal areas confining semi and unconsolidated sediments along the coastline helped mankind to go in for deeper groundwater exploration, resulting in problems like salinity hazard, salt water intrusion and land subsidence. Saline intrusion in coastal aquifers is of major concern (Satish et al., 2011, Gopinath et al. 2015 a,b; 2016) because it constitutes the commonest of all the pollutants in freshwater aquifers. Excessive withdrawal of groundwater coupled by significant decrease in recharge contributes to the problem. The extent of saline water intrusion is influenced by nature of geological settings, hydraulic gradient, rate of groundwater withdrawal and its recharge.

Seawater intrusion into coastal aquifers leads to impairment of the quality of the freshwater aquifers. Hence, understanding of saline intrusion is essential for the management of coastal water resources (Gopinath and Srinivasamoorthy, 2015b). The intrusion of seawater has been identified by many approaches such as isotope studies, geochemical and geophysical studies (Srinivasamoorthy et al. 2011a,b, Balasubramanian et al. 2014). In studying the thickness and geometry of depositional systems, a common procedure is to make use of information from geological research, drilling, and exploitation boreholes. However, these methods are expensive and time consuming, preventing their use on a large scale. In contrast, geophysical measurements can provide a less expensive way to improve the knowledge of a set of boreholes (Muktar et al., 2000). The resistivity technique has its origin in 1920 (Koefoed, 1979) which gained importance due to its non-invasive nature and non requirement of water sampling, relatively inexpensive, used for rapid and economical monitoring of large areas, assist in the optimization of the required number of monitoring wells. Electrical conductivity / resistivity are intrinsic properties of groundwater chemistry that are readily interpreted in terms of the degree of groundwater contamination (Todd, 1980). The presence of seawater causes groundwater to be considerably saline, hence the aquifer resistivity is reduced considerably, and the resistivity method can delineate the boundaries of the body of saline water. Due to the fact that sharp resistivity contrast exists at the interface between fresh and saline water, it is proved to be the useful method to detect the intrusion of saline water. It is helps in understanding spatial relations between fresh, brackish, and saline water, which commonly coexist in coastal aquifers.

The area of investigation is the Union territory of Pondicherry, with expanding population, tourist activities, agricultural and industrial activities has endangered the coastal groundwater. The total water availability for all utilities (domestic, irrigation and industrial purposes) estimated comes about 200 MCM per year which is likely to be utilised by 2020. The groundwater resource estimation committee, Central Groundwater Board (CGWB, 1997) indicates that the groundwater development in Pondicherry regions is very high (179%) which may pose a threat to groundwater quality and quantity. As the ground water in shallow aquifers along the coast shows sign of increasing salinity due to its hydraulic connection and over extraction makes the region more vulnerable to saline water ingress. Hence an attempt has been made in the proposed study area to demarcate saline water intrusion using electrical resistivity methods.

2. Study Area

The proposed study area Puducherry region is situated on the Coromandel coast between 11°45′ and 12°03′ N latitudes and 79°37′ and 79°53′ E longitudes with an area of 293 sq. km (Fig.1). The region is blessed with abundant water resources due to annual rainfall of about 1250 mm received during both southwest and northeast monsoons. Two river systems Gingee and Pennaiyar flow in the study area that configures Bay of Bengal. Geology of the area is underlined by semi-consolidated and unconsolidated sedimentary formations which age ranging from lower cretaceous to recent lying on archean basement. The sediments are mainly clay, claystone, silt, siltstone, marl, limestone, sand, sandstone, and gravel. Three major aquifers are isolated with unconsolidated, unconsolidated to semi-consolidated and semi-consolidated in nature. Ground water occurs in these formations both under water table as well as under confined conditions and is being developed by means of dug wells, dug-cum bore wells and tube wells. The transmissivity and storativity ranges between 100 to 2000 sqm/sec and between 2.93×10^{-5} 1.36×10^{-4} respectively. Due to excess groundwater withdrawal reversal of gradient are noted in certain locations like Kalapet, Muthialpet, Mudaliarpet, Kirumambakkam, and Panithittu.

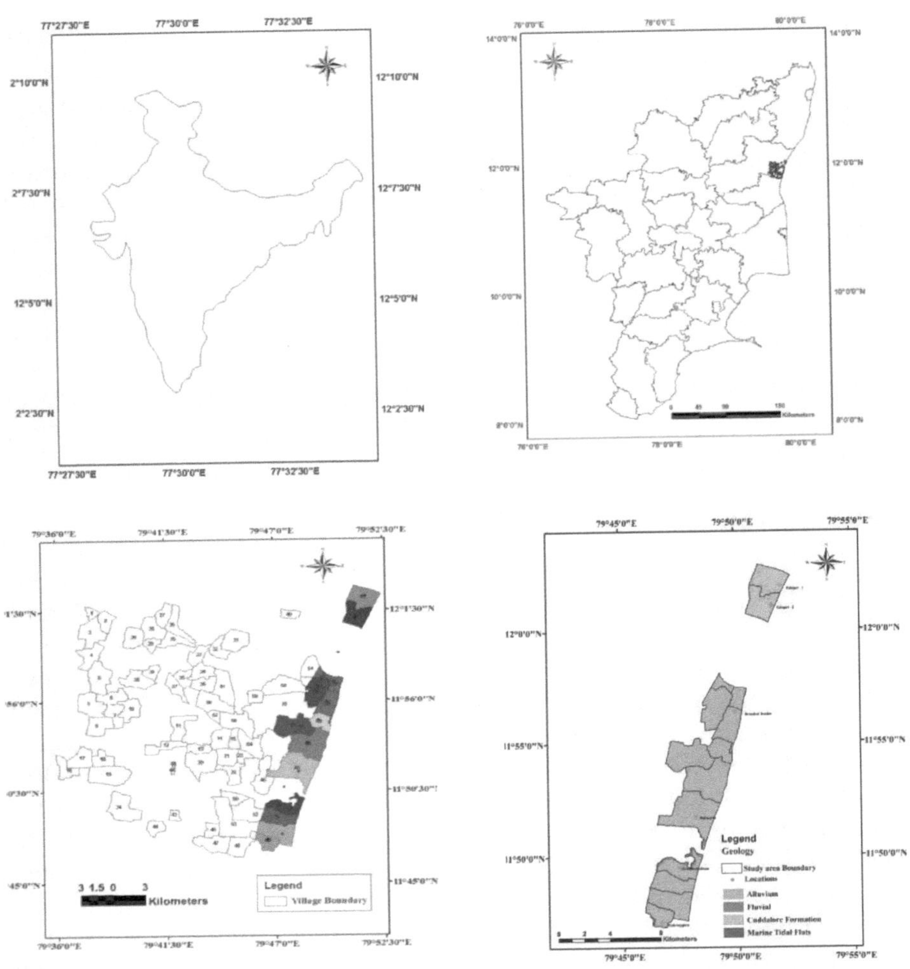

Fig. 1. Location, Geology and ERT soundings

3. Methodology

Electrical Resistivity Tomography (ERT) is one of the electrical geophysical techniques that are used to study the freshwater/seawater interface and soil salinization. In the present work BTSK WDDS-2/2B digital resistivity meter has been used to perform the survey. A total of 6 profiles were attempted perpendicular to the shore line within spacing of 5 m to 600 m away from the shore. Stainless steel electrodes were used for the survey that was pierced at depth between 30 to 40 cm. The Wenner - á protocol were applied with a number of electrodes varying from 20 to 34 with spacing of 5 m and/or 10m leading to an investigation depth ranging from 30 – 55.7 m. Salt water has been poured at the electrode points to ensure good contact with the earth and the surveys were attempted in flat surface. The electrodes are connected through multi-core cables to a switching panel which is placed in the middle of the profile. The current and potential terminals from the switch panel are connected to the respective terminals of the BTSK WDDS-2/2B Digital Resistivity Meter. The switching panel contains series of sockets connected to the electrodes through the multi-core cable system. The current terminal pin and the potential terminal pin which are connected with the current source and the resistivity measuring instrument can be inserted in the appropriate sockets for measuring the resistivity between any two electrodes without actually changing the electrodes along the profile. The multiple sounding along the selected profile registered the horizontal and vertical resistivity changes (Fig.2). These resistivity values are used to create a 2D Electrical Resistivity Images of the cross section of the profile. The pseudo section contouring method is used to plot the data collected through the field experiments (Loke, 1997a, Antony Ravindran 2010). The pseudo-section reflects the true resistivity distribution along the profile and therefore can be used as a base for qualitative interpretation. To minimize the differences between the measured and the calculated apparent resistivity values, the inversion method are applied (Antony Ravindran et al. 2012). The inversion method projects a 2D model of a subsurface by using the measured data by using RES2DINV software program (Loke, 1997 b).

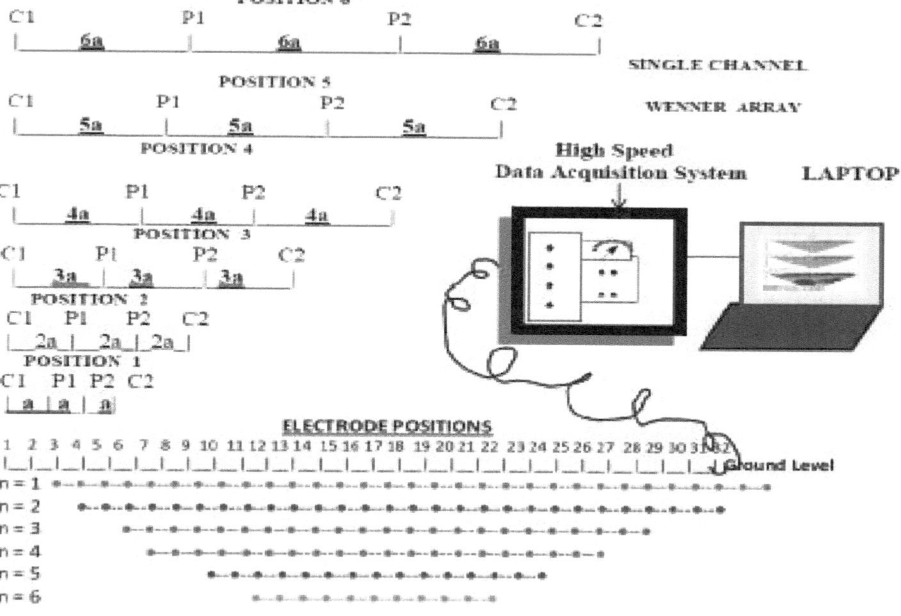

Fig. 2. Schematic diagram of Multi- Electrode system.

4. Results and Discussion

4.1. Kalapet 1

The first ERT, Profile performed near Kalapet bearing the Latitude E12°2′4" and Longitude 79°51′6" about 376.08 m orthogonal to the coast (Fig. 3). The subsurface layer resistivity obtained by the inversion process is controlled by the resistivity of the pore water and the resistivity of the host rock (Bear, 1997). The geoelectrical image shows a variation in resistivity distribution, with electrical resistivity ranging between 3.41-1484.0 Ωm. Lower electrical resistivity values (3.41 Ωm), observed at depth between 19.0 to 21.0 m irrespective of its orientation towards sea and the reduced values may be saturated strata for a subsurface seawater flow zone. Saline water has a resistivity below 1.0 Ωm, in particular seawater has an average resistivity of 0.2 ohm (Parasnis, 1977; Nowroozi et al., 1999), while a layer saturated by saline water and dissolved solids has resistivity in the range of 8 to 50 Ω m (Zohdy, 1993; Nowroozi et al. 1999). Therefore, based on these values of resistivity of layers saturated by saline water and dissolved solids, resistivity data obtained in this work highlight the presence of strata saturated with brackish to saline water. A gradual decrease in the resistivity value with depth indicates the wet nature of the subsurface formation. The dry formation with higher resistivity is confined to shallow depth and as depth increases the wetness of the formation increases with decrease in resistivity. The presence or absence of clay formations inter-bedding the sandstone formations influences the resistivity values where sediments without clay may vary from 1 to 100 ohm while the resistivity of wet clays alone may vary from 1 to 250 ohm. The presence of two patches indicates the incidence of wet clay formations at a depth of 13 to 17 m. Thus, a wide range of resistivity is often reported for a particular water saturated material. In order to get a brief idea and to infer the litho units involved the litho log were also taken for interpretation in which the 2D ERT was in good correlation. The zone influenced by saline water was the sandstone formation and the traces of saline water were noted below a depth of 19.0 m indicating the up coning of saline water into the sandstone formation. Starting from a distance of 19.0 m up to a distance of 21.5 m traces of saline water intrusion was observed and further migration towards inland was also noted indicating the contaminated nature of the sand stone formation. This fact was well in conformity with the higher EC (> 6000 μS/Cm) and Cl values observed from the nearby bore location.

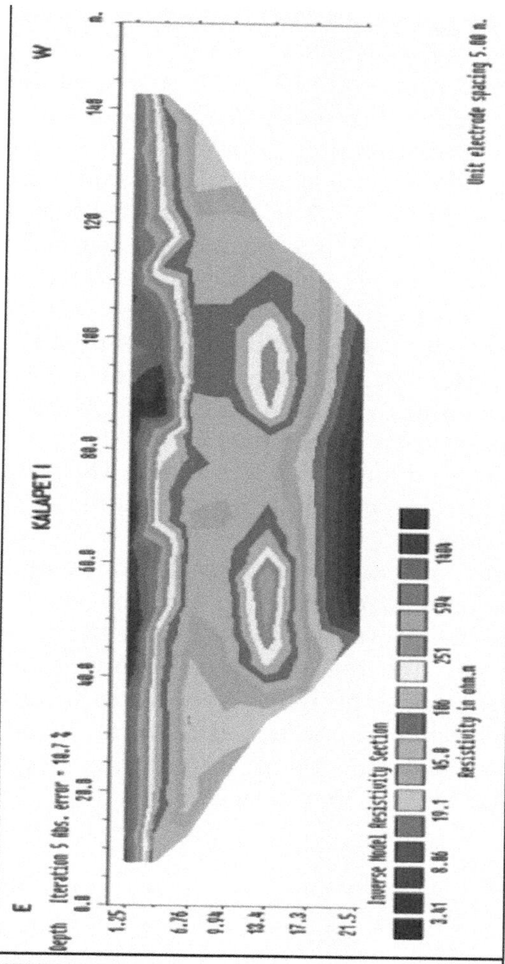

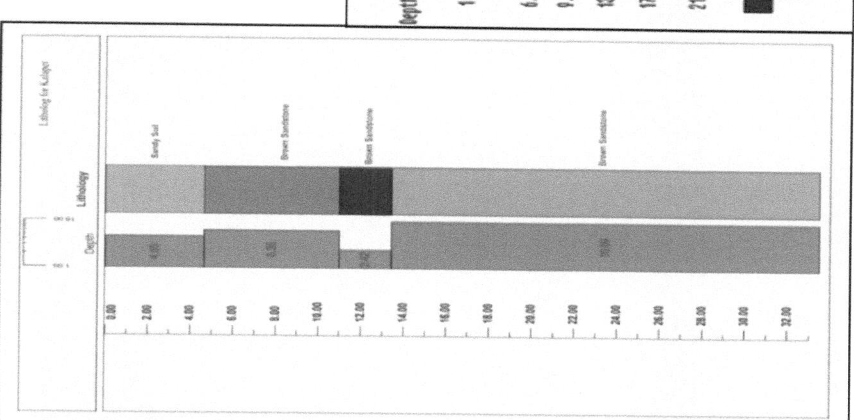

Fig. 3. Litho log and ERT profile of Kalapet - 1

4.2. Kalapet 2

The second ERT Profile (Fig. 4) was performed near Kalapet-1I bearing the Latitude E12°1′17" and Longitude N79°51'48" about 300 m orthogonal to the coast. The geoelectrical image shows a variation in resistivity distribution, with resistivity values ranging between 2.73-1183 Ωm. Lower electrical resistivity values (2.73 Ωm) were observed at a depth of 21.5 to 26.2 m irrespective of its orientation towards sea and the reduced values may be due to subsurface seawater flow zone. Gradual decrease in resistivity with depth indicates the wetted nature of the subsurface formation. The top layer with resistivity range of 1183 Ωm indicates dry sand formation without water content noted along eastern and western parts of the profile. Lower resistivity values (209 Ωm) are also noted in the central part of the profile indicating the soil formation admixed with water particles. An intermediate layer with lower resistivity values (15.5 Ωm to 209 Ωm) indicates the gradual wetting index increasing with depth also decreases the resistivity values. Thus, a wide range of resistivity is noted in the profile with reference to the presence or absence of water saturated material. Further the resistivity profile correlated with litho log from nearby bore hole location inferred that lower resistivity zones are the sandstone formations with varying grain size and the bottom most formations was interpreted as coarse grained sandstone which shows traces of saline water intrusion. From this survey it is confirmed that the saline migration into the coastal aquifers at shallow depth (<30m) from an orthogonal distance of 117.9 m to 257.9 m from the coast.

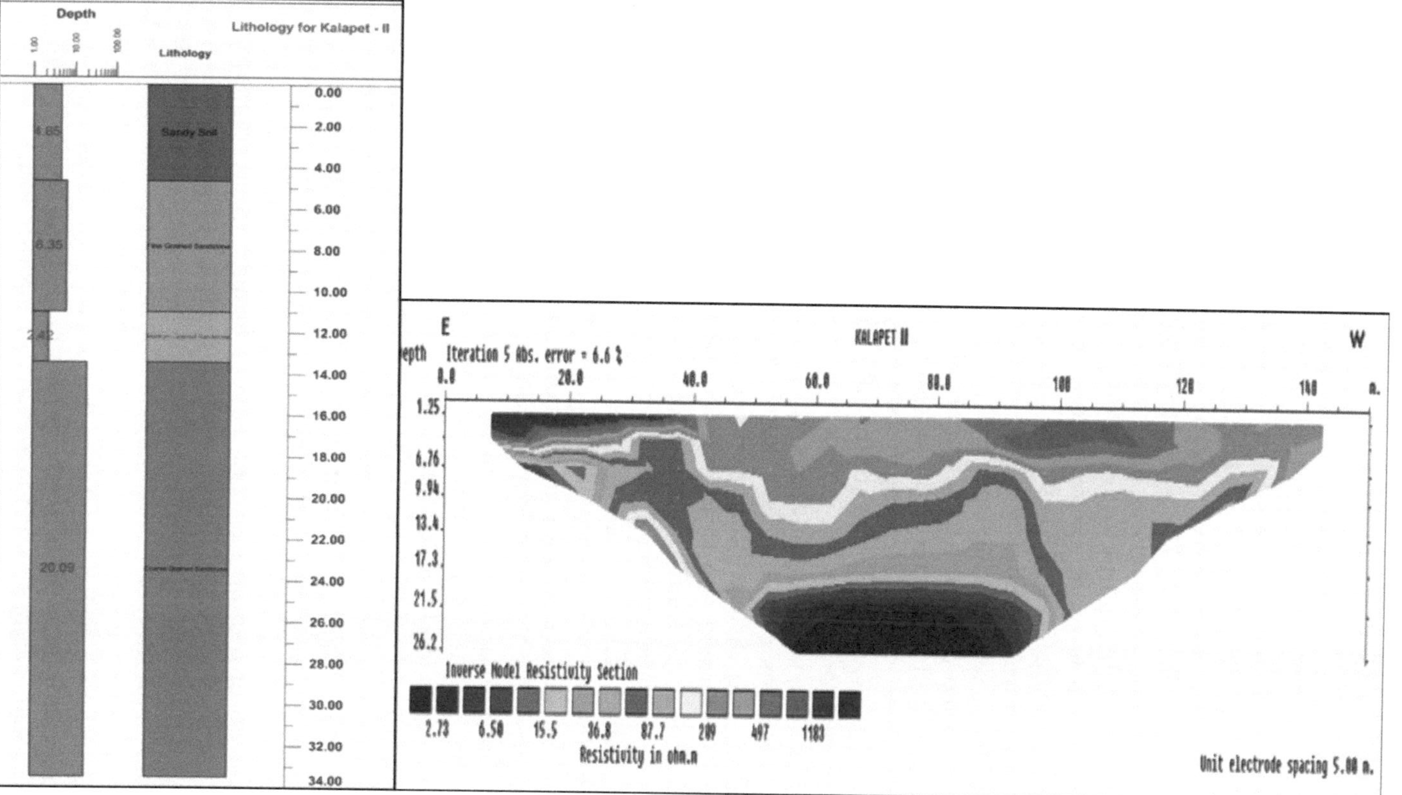

Fig. 4. Litho log and ERT profile of Kalapet - II

4.3. Botanical Garden

The third profile (Fig. 5) was performed at botanical garden bearing the Latitude E11°55′49" and Longitude N79°50′8" about 50 m orthogonal to the coast with an electrode spacing of 5m between the depths of 150m with a total modelled depth of about 26.2 m. The geoelectrical image shows variation in resistivity distribution, with resistivity ranging between 0.540 -166 ohm. The lower resistivity values were noted along the eastern part of the profile at depths between 6.76 to 17.3 m indicating the formation intruded by saline water. The extension of this zone is noted from 0 m to 60 m towards land. Along the western part of the profile after the distance of 60 m there was a gradual increase in resistivity values confining a geological divide at this particular depth. The first layer resistivity was ranging from 6.28 to 166 ohm where lower resistivity values are noted along the eastern part of the profile whereas along the western portion of the profile the thickness of the first layer showed an increasing trend with increasing resistivity value, might be the top soil. A gradual decrease in the resistivity value at a spread of 60 m is noted along the eastern part of the profile whereas along the western part of the profile there was a gradual decrease in resistivity with an increase in depth. From this profile it is evident that the lower resistivity value noted along the eastern portion of the profile might be interpreted as saline water intrusion up to an extent of 60 m towards inland. The litho log demarcated medium grained sandstone formation up to a depth of 4 m which is evident from the variation in the resistivity values. The second layer identified at a depth between 4m to 7m with a decrease in resistivity is mainly due to the presence of fine grained sandstone formation which is frequently wetted by the precipitation due to the higher porosity and permeability. The lower resistivity zone (0.548 ohm) along the eastern part of the profile direction confirms the role of saline water intrusion into coarse grained sand admixed with sticky clay. The presence or absence of clay might reduce the resistivity values but the presence of coarse grained sandstone formation with greater porosity and permeability should show an increased resistivity value even though found admixed with clay formation. As noted the geological divide observed at a spread distance of 60 m might be due to the variation in the thickness of the top soil, as noted along the eastern part of the profile the thickness of the top soil zone is very meager when compared with the thickness observed along the western portion of the profile. Frequent flushing of rain water due to the higher thickness of top soil has resulted in higher resistivity values, but along the eastern part the meager thickness of top soil and already saline water intruded sandstone formation has resulted in lower resistivity. From this survey it is confirmed that the saline migration into the coastal aquifers at shallow depth (<30m) from an orthogonal distance of 800 m to 860 m from the coast.

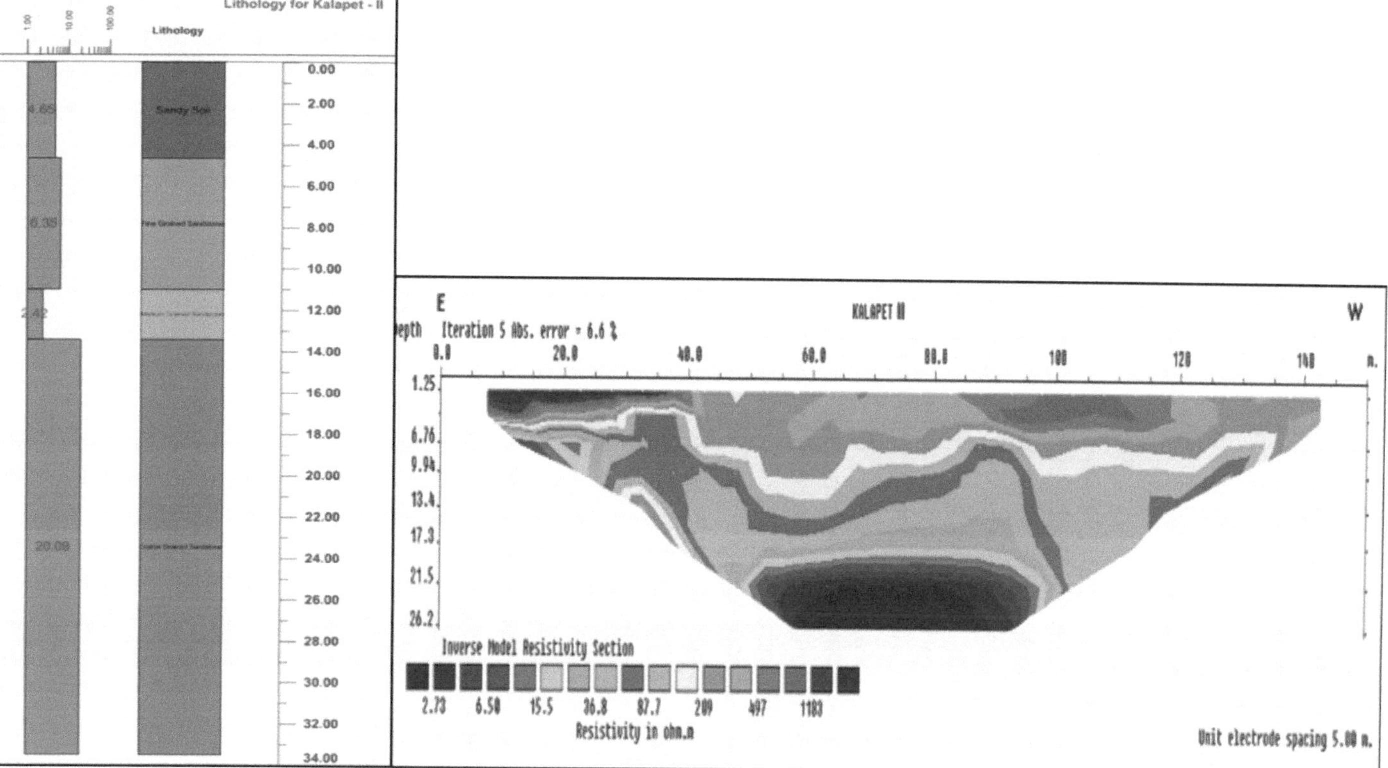

Fig. 5. Litho log and ERT profile of Botanical Garden

4.4. Nallavadu

This profile (Fig. 6) was performed at Nallavadu bearing the Latitude E11°51′44" and Longitude N79°48′14" about 1 Km orthogonal to the coast. The data acquired were inverted using RES2DINV to obtain depth ranging from 0 to 52.3 m using 10 m electrode spacing. The first layer, between the depths of 0 and 13.5 m has resistivity values of 13.8 Ùm. The second layer is between depth 13.5 m and 43.0 m with resistivity values between 0.78 Ùm and 3.29 ohm. The third layer has resistivity of 13.8 ohm between depth 43.0 and 52.3 m. The second and third layers have varying resistivity distributions. These layers appear to consist of top soil, clay, shale and sandstone. A prompt saline intruded zone is noted at a depth of 19.0 m to 34 m indicating the litho units saturated with saline water. At a depth of 43 to 52 m there appears to be a high resistivity zone with resistivity range of (22 ohm) evenly distributed throughout the profile. The litho log collected from a pumping well near by the survey location confirms the above statement with a top soil of 3.0 m thickness followed by medium grained, coarse grained sand along with clayey silt and clay formation starting from a depth of 3.0 m up to a depth of 43.0m indicating the saline intruded zone. It was doubted that the presence of clay formations might be the reason for lower resistivity values misinterpreted for saline water intruded zone. But when comparing with the sea water average resistivity value of 0.2 ohm and the resistivity of clay formations both wetted and dry as 1 ohm to 250 ohm the resistivity value observed in this particular zone (0.784 ohm) is in good conformity with the sea water resistivity values. Hence it is confirmed that, lower resistivity values are not mainly due to the presence of clay formations but due to the saline water intrusion into the aquifers. The groundwater sample collected from the observation bore hole used for the interpretation of the litho log confirms the saline nature of groundwater with higher EC (>4000 μS/Cm) and Cl (475 mg/L) ratios. From the profile it is confirmed that saline water intrusion is confined up to a distance of 1.280 Km inland.

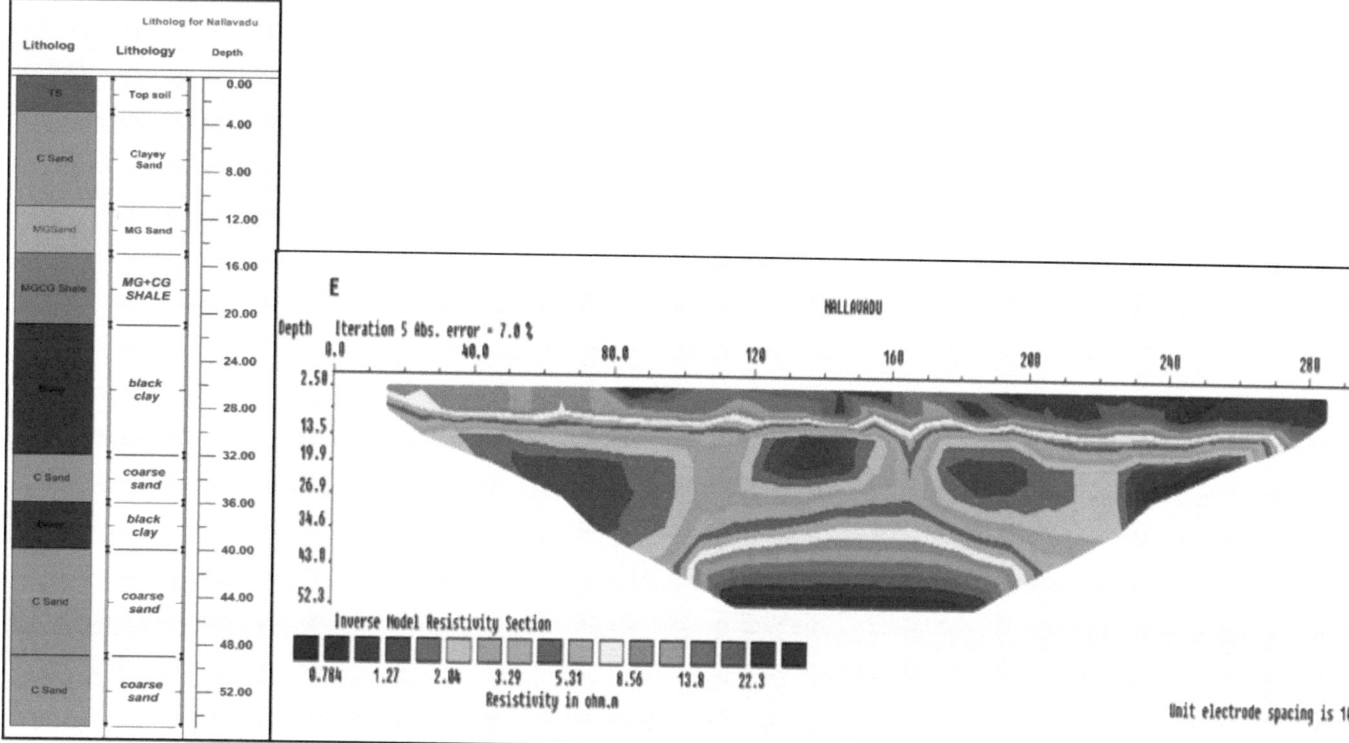

Fig. 6. Litho log and ERT profile of Nallavadu

4.5. Kirumambakkam

The profile (Fig. 7) was performed at Kirumambakkam bearing the Latitude E11°49′25" and Longitude N79°47′30" about 1.3 KM orthogonal to the coast with an electrode spacing of 5 m between the depths of 140 m with a total vertical depth of 26.2 m. The first layer resistivity was ranging from 13.3 to 35.5 ohm from depth ranging from 0.0 to 17.3 m. The second layer is between depths 17.3 m to 26.2 m with resistivity ranges between 6.37 to 10.4 ohm. From the first layer resistivity it is inferred that the formation might be made up of top soil due to its higher resistivity values. At the eastern part of the profile a low resistivity zone (6.37 ohm) identified might be the continuation of the second layer which is exposed at the eastern part of the profile direction. The first layer was found to be extending to greater depth along the eastern and western parts of the study area as patches of medium resistivity zones. For better information about the litho units the litho log profile conducted nearby bore hole location has been taken for interpretation and from that the first layer has been identified as the top soil followed by clayey sand and medium grained to coarse grained sandstone formations up to a depth of 18m this is in good conformity with the 2D resistivity soundings. The second layer is the clay with sand formation which has been recorded with lower resistivity values. The patch of second layer extension has been identified along the eastern part of the profile which is well visible from the resistivity values. From the profile it is inferred that no traces of saline water intrusion has been recorded but the low resistivity recorded is mainly due to the presence of clay sand formations.

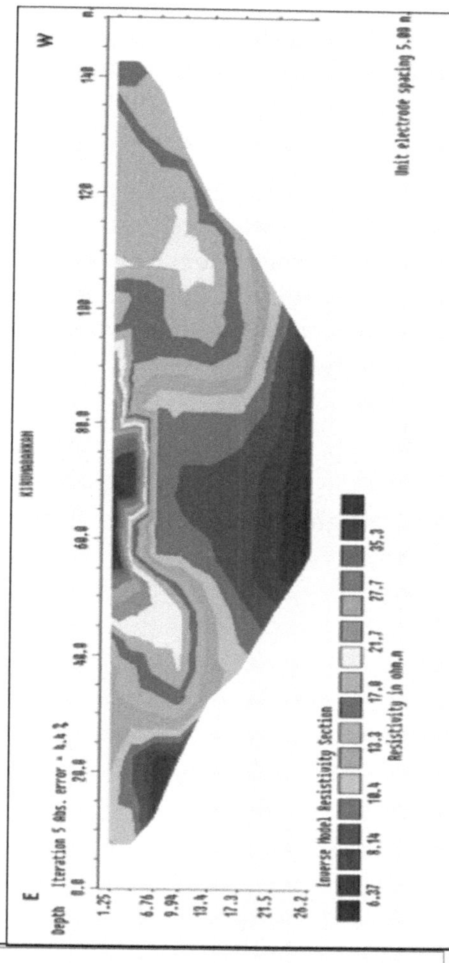

Fig. 7. Litho log and ERT profile of Kirumambakkam

4.6. Pudukuppam

The profile (Fig.8) was performed at Pudukuppam bearing the Latitude E11°47′50" and Longitude N79°47′28" about 622 m orthogonal to the coast with an electrode spacing of 5m between the depths of 140m, with a total depth of penetration of 21.5 m. The higher resistivity layer identified up to a depth of 3m may be the top soil zone with a resistivity range of 698 ohm. The second layer isolated with resistivity ranges between 11.9 ohm 91.3 Ùm at depths between 3.00 m to 17.3 m. The third layer with a resistivity range of 0.565 to 4.32 ohm between depths 17.63 to 26.2m has been demarcated as sea water intruded zone. A steeper decline in resistivity is noted from the top to bottom of the imaging indicating the inhomogeneities in the subsurface formation. From the litho log data the top soil up to a depth of 3 m is confirmed. The second layer with a low resistivity zone is interpreted as the medium grained sand formation. The third layer with a very low resistivity of (0.565 ohm) is the medium to coarse grained formation with traces of saline water intrusion. Hence from the plot it is inferred that saline water intrusion is high up due to the over extraction of groundwater.

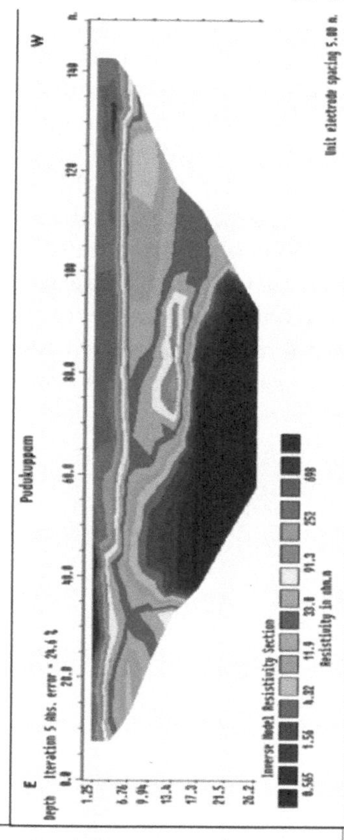

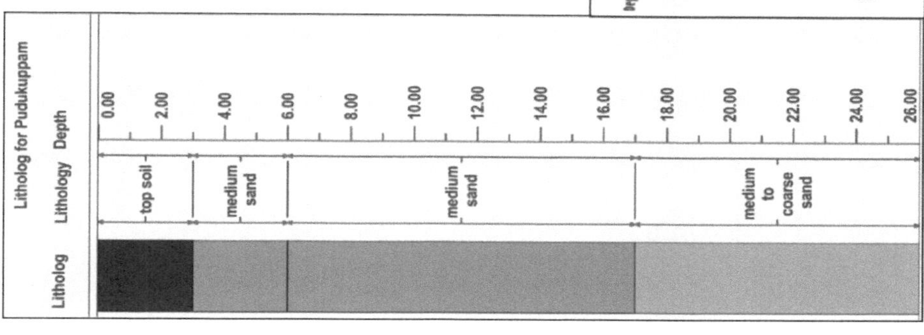

Fig. 8. Litho log and ERT profile of Pudukuppam

5. Isolation of Saline Sources

The traces of saline water intrusion found to be extending up to an extent of 1.5 KM inland in locations Kalapet I and II. Along the Southern parts of the study area in locations like Pudukuppam and Nallavadu saline water intrusion is found to extend up to a distance of 1 KM inland. Based on the resistivity values three different traces of saline water into the coastal aquifers are noted. The first is mainly due to the hydrodynamic connection between the coastal aquifers and the saline water, this type of intrusion is prominent along the northern part of the study area in locations like Kalapet I and II. In the central portion of the study area in locations like kirumampakkam and botanical garden influences of the litho salinity has a greater impact to determine the salinity in the coastal aquifers. The formations interbedded with clay admixtures recorded lower resistivity might not be taken as saline water intruded zone but the presence of clay mineral assemblages alters the quality of groundwater in those aquifers which is not suitable for human utilities. In the central part of the study area the presence of estuaries has a greater impact in increasing the salinity of groundwater. Hence the salinity in these locations is not due to direct sea water intrusion but might be due to the impact of estuaries with saline water.

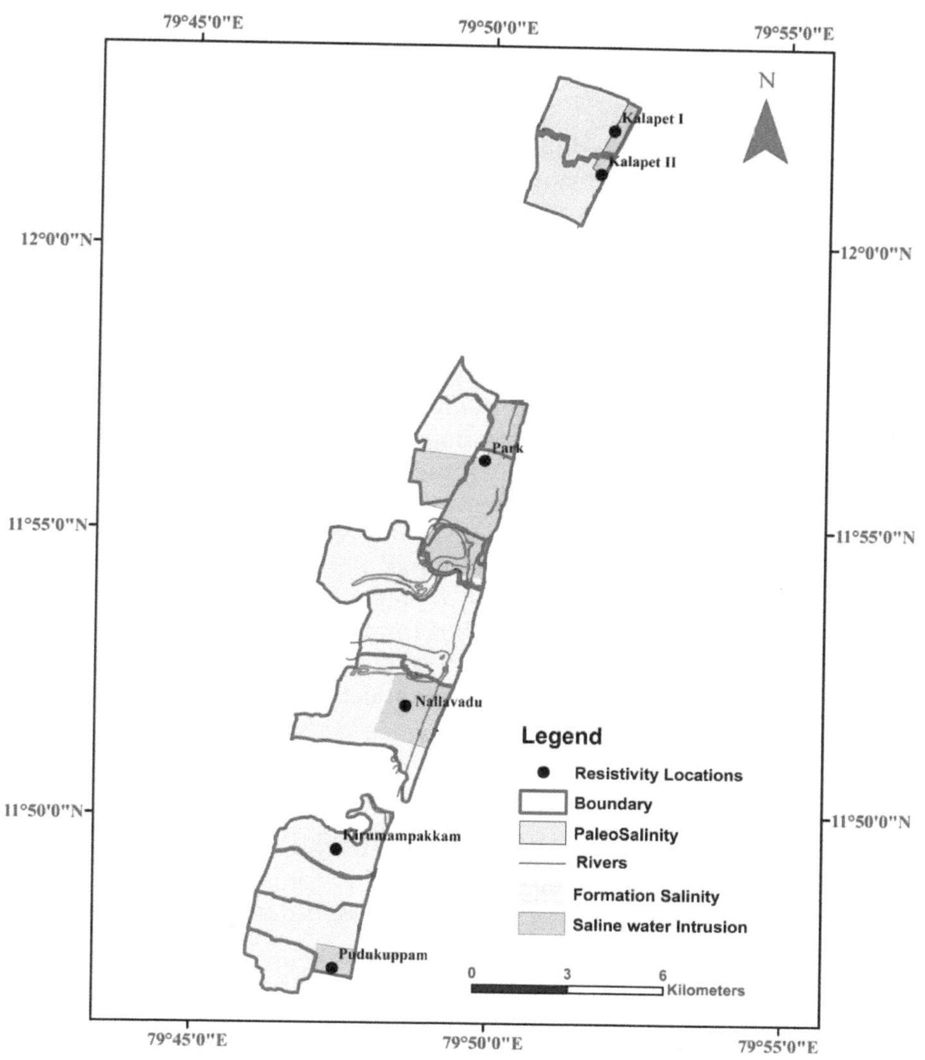

Fig. 9. Isolation of different saline sources in the study area

6. Conclusion

Geophysical resistivity investigations infer saline water prominent in locations like Kalapet 1, Kalapet 2 up to an extent of 1 km away from the coast. In botanical garden the top soil identifed as sand formation with prominent saline intrusion. In Nallavadu region salinity traces were observed in the shallower depth in medium to coarse grained sand stones. In Kirumambakkam region, no traces of saline water intrusion have been recorded but the low resistivity recorded is mainly due to the presence of clay sand formations. In Pudukuppam region the coarse grained sandstone formation have identified with traces of saline water intrusion. Isolation of saline traces using varying resistivity values have been attempted. From the resistivity values three different traces of saline water into the coastal aquifers are noted. The first is mainly due to the hydrodynamic connection between the coastal aquifers and the saline water, this type of intrusion is prominent along the northern part of the study area in locations like Kalapet. In the central portion of the study area the influences of the litho salinity has a greater impact to determine the salinity in the coastal aquifers. In the southern most tip traces of saline intrusion have been isolated but in majority part of the study area formation salinity due to the presence of clay formations seems to lower the resistivity values which is well identical when compared with the litho logs.

Acknowledgement

The corresponding author acknowledge the Dept of Science, Technology and Environment, Pondicherry, for the financial support (Sanction No.10/DSTE/GIA/JSA-I/2013/208, dtd.05.04.13.) to carry out this research work.

References

Antony Ravindran A (2010) Characterization of geology of subsurface shallow conglomerate using 2D Electrical Resistivity Imaging at Baragadi. Panna District, Madhya pradesh, India. Jour. App. Sci. and Env. Mgmt. 14(3): 33-36.

Balasubramanian M, Srinivasamoorthy K, Saravanan K, Sathish kumar G , Saravanan M, Prakask R (2011) Mapping of saline water intrusion along the coastal tracts of Pondicherry region using Electrical resistivity imaging methods, EnviroGeoChimica Acta (2014) 1(1): 46-52

Bear J, Sorek S, Ouazar D (1999) Seawater intrusion in coastal aquifers: Concepts, methods, and practices. Kluwer academic publishers.

Central Groundwater Board (1997) Groundwater Brochure, Pondicherry Region, U.T of Puducherry, Published report, pp27.

Gopinath S, Srinivasamoorthy K, Vasanthavigar M, Saravanan K, Prakash R, Suma CS, Senthilnathan D (2016) Hydrochemical characteristics and salinity of groundwater in parts of Nagapattinam district of Tamil Nadu and the Union Territory of Puducherry, India, Carbonates Evaporites, DOI 10.1007/s13146-016-0300-y.

Gopinath S, Srinivasamoorthy K (2015b) Application of Geophysical and hydrogeochemical tracers to investigate salinisation sources in Nagapatinam and Karaikal Coastal Aquifers, South India. Aquatic Procedia 4, 65–71.

Gopinath S, Srinivasamoorthy K, Prakash R (2014) Hydrochemical investigations for identification of groundwater salinization sources in Nagapattinam and Karaikal regions, Southern India. Journal of EnviroGeoChimica Acta 1(2), 153-160.

Gopinath S, Srinivasamoorthy K, Saravanan K, Prakash R, Suma CS, Faizal Khan, Senthilnathan D, Sarma VS, Padmavathi Devi (2015a) Hydrogeochemical characteristics of coastal groundwater in Nagapattinam and Karaikal aquifers implications for saline intrusion and agricultural suitability, 2(2), 1-11.

Koefoed O (1979) Geosounding principles 1, resistivity sounding measurements. Amsterdam: Elsevier.

Loke M H (1997a). Electrical imaging surveys for environmental and engineering studies - a practical guide to 2D and 3D survey, Penang, Malaysia, Universiti Sains Malaysia, unpublished short training course lecture notes.

Loke M H (1997b) RES2DINV ver. 3.3 for Windows 3.1, 95, and NT, Advanced Geosciences, Inc, pp. 66.

Mukhtar A L, Sulaiman W N, Ibrahim S, Latif P A, Hanafi M M (2000) Detection of Groudnwater pollution using resistivity imaging at SeriPetaling Landfill, Malaysia. J. Env. Hydrol. 8.

Nowroozi A A, Horrocks S B, Henderson P (1999) Saltwater intrusion into the freshwater aquifer in the eastern shore of Virginia: a reconnaissance electrical resistivity survey. J. App. Geophy. 42:1-22.

Parasnis D S (1997) Principles of Applied Geophysics. Chapman and Hall London. Prentice-Hall, p. 413-416.

Satish S, Elango L, Rajesh R, Sarma VS (2011) Assessment of seawater mixing in a coastal aquifer by electrical resistivity tomography, Int. J. Environ. Sci. Tec, 8(3), 483-492.

Srinivasamoorthy K, Vasanthavigar M, Chidambaram S, Anandhan P, Sarma VS (2011b) Characterisation of Groundwater Chemistry in an Eastern Coastal Area of Cuddalore District, Tamil Nadu. Journal of geological society of India. Vol.78, December 2011, pp.549-558

Srinivasamoorthy K, Vasanthavigar M, Vijayaraghavan K, Sarathidasan R, Gopinath S (2011a) Hydrochemistry of groundwater in a coastal region of Cuddalore district, Tamilnadu, India: implication for quality assessment. Arab J Geosci. doi:10.1007/s12517- 011-0351-2.

Todd D K (1980) Groundwater hydrology. Wiley, New York.

Zohdy A A R, Martin P, Bisdorf R J (1993) A study of seawater intrusion using direct-current soundings in the southeastern part of the Oxnard Plain, California". Open-File Report, 93-524. U.S. Geological Survey, 139 pp. In: Nowroozi, A.A., Horrocks, S.B , Henderson, P., 1999. "Saltwater intrusion into the freshwater aquifer in the eastern shore of Virginia: a reconnaissance electrical resistivity survey. J. App. Geophy. 42:1-22.

Chapter 8

Identification of Submarine Groundwarer Discharge in Coastal Aquifers Using Geo Chemical Parameters

Nepolian M., Chidambaram S., Sarathidasan J.
Noble Jacob[1] and Hemant Mohokar[1]

Department of Earth Sciences, Annamalai University, Annamalai Nagar, Isotope Application Division, BARC, Mumbai, India

ABSTRACT

A study on the Subsurface groundwater discharge (SGD) in the coastal Cuddalore region of south east India has been carried out. Measurement of surface water Radon, Water level, Electrical Conductivity (EC) and pH for a total of twenty four hours, by hourly interval the measurement has been calculated and promote correlated with tidal values calculated by WX Tide 32 software. The SGD measurements were made by using a modified seepage meter. The study reveals a match with water level variation and tide with minor variation due to influx of surface water. Saline discharges, fresh groundwater discharges and surface water mixing processes were identified along the coast. Lower SGD (37.24 – 79.16 cm/day) were observed during fresh groundwater discharge.

Key words: Coastal aquifers, subsurface groundwater discharge, radon, surface water

Introduction and Significance

Although not as obvious as river discharge, continental groundwaters also discharge directly into the sea. Like surface water, groundwater flows down-gradient. Therefore, groundwater flows directly into the ocean wherever a coastal aquifer is connected to the sea. Furthermore, artesian aquifers can extend for considerable distances from shore, underneath the continental shelf with discharge to the ocean at their points of outcrop. In some cases, these deeper aquifers may have fractures or other breaches in the overlying confining layers, allowing groundwater to flow into the sea (Fig. 1). While the magnitude of such discharge may be relatively minor in areas dominated by river flow, recent studies have indicated that groundwater may occasionally account for a significant fraction of the fresh water inflow (e.g., Valiela and D Elia, 1990; Moore, 1996). Because the composition of groundwater is normally different than receiving coastal waters, submarine groundwater discharge (SGD) may be important as a pathway for dissolved constituents such as nutrients. The concentrations of many parameters (e.g., nitrate) in groundwater is typically several times higher than seawater, even in pristine aquifers. In areas where groundwater contamination (e.g., organics, metals, radionuclides) has occurred, SGD may also be a pathway for anthropogenic material fluxes. Increasingly, groundwater is being recognized as a potentially significant, but still poorly quantified, source of nutrients and other dissolved elements to coastal ecosystems.

Study Area

The study area is located along south Cuddalore which is bounded by North latitudes 11°71′94″ East longitudes 79°77′44″ (Fig.2). The study area forms a part of coastal region composed of coastal alluvium located along the south eastern coast of India. The land is completely flat with large deposits of black and alluvial soil in inlands and coarse sand near the seashore.

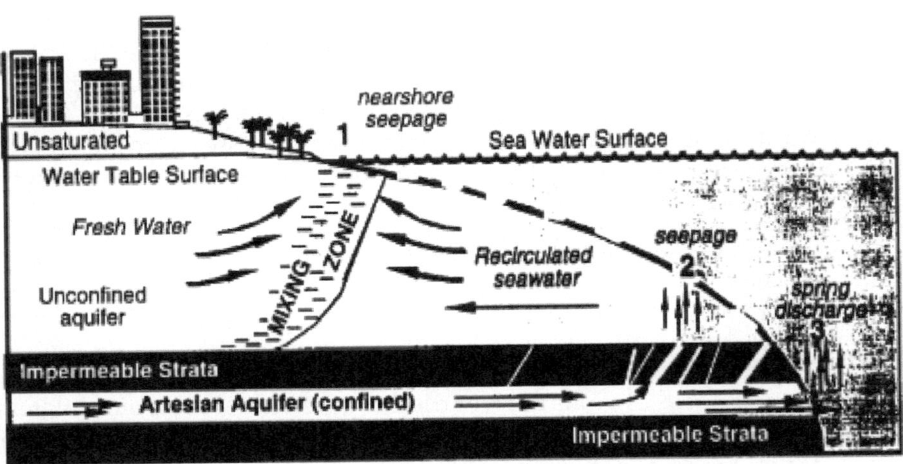

Fig. 1. Diagrammatic view of the relationships between coastal aquifers, seawater, and groundwater discharge. Three types of submarine groundwater discharge are illustrated: (1) nearshore seepage; (2) offshore seepage; and (3) submarine springs (Burnett et al. 2000a).

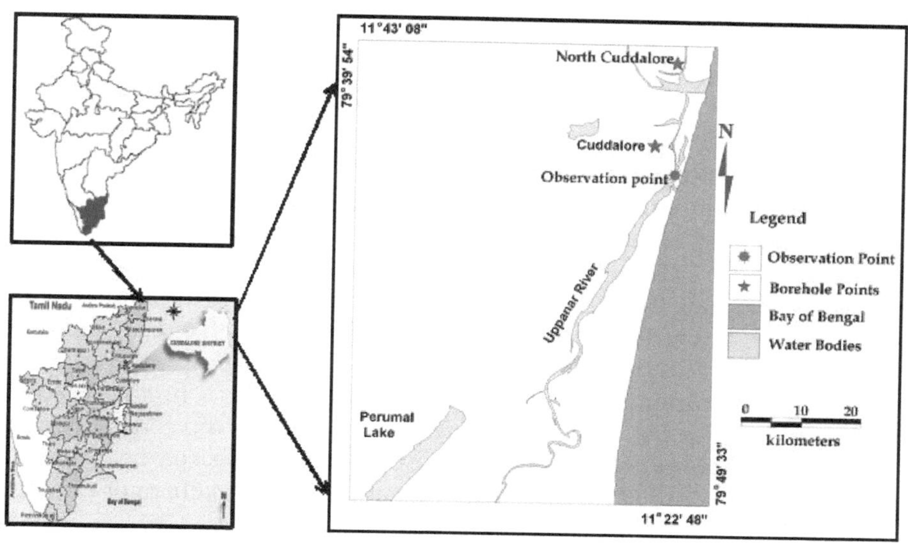

Fig. 2. Location map of the Study area

The study area has a peculiar climate that receives rainfall in almost all the seasons, the average annual rainfall is 1,400 mm, with major contribution from North-East monsoon. Uppanar River which originates from the southern part of Perumal Eri, flows NE and joins the sea along the south of Cuddalore. The location of observation site is on the Uppanar River mouth North latitudes 11°42′10″ East longitudes 79°46′42″ which falls North of Pichavaram mangroove. This river is ephemeral and carries floods during monsoon. It generally flows from west towards east and drains in Bay of Bengal. It has an elevation of about 3m amsl. The other major river Pennayar River runs north of the town, while Gadilam River runs across Cuddalore. The groundwater level ranges between 0.004 - 17.54 mbgl (CGWB 2009). The Storativity of this region is 7.72×10^{-5} - 9.5×10^{-3} and the Transmissivity is between 438 to 1900 m^2/day. The maximum temperature varies between 43.3°C and 39.60°C during 2009 and 2004, respectively (Srinivasamoorthy et al. 2011). The relative humidity recorded in Cuddalore ranges from 60 to 80% . Higher rates of humidity are observed during the month of North-East monsoon period. The relative humidity observed during the north east monsoon period is higher and low during summer.

The Wind prevailing in the region is generally moderate in strength with mean velocity variation from 6-14 Km/hr. Wind speed is higher in May and Lower in October. South westerly wind prevails over 33% of days in a year and Northeasterly wind over 32% of the days (Gowrishanker et al. 1980).

Methodology

Tide-affected Cuddalore region were selected for the field monitoring. The observations were carried out in 1h interval for 24 hrs (Indian Standard Time) IST, in the field. The study is mainly based on the insitu measurements and the estimation of SGD is based on the values of EC, WL, radon and tide. The time-wise measurements were made by using Hanna Instruments, Multiparameter water quality meter for Water level, pH; EC & Salinity were measured for every one hour during the month of August (South west monsoon period). Automated Radon monitoring was carried out by RAD7. The RAD7 was fixed on the river bank and water was pumped using a peristaltic pump (Fig 3a) to circulate water from river mouth.^{222}Rn values were measured as a function of tide levels and EC. For the in-situ measurement of ^{222}Rn in the coastal region, seawater is continuously pumped from 1m above the sea bed using a peristaltic pump and sprayed as a jet into an air-water exchanger (Fig.3b).

The circuit has a desiccant and purpose of the desiccant is to absorb moisture, since detection efficiency decreases at higher humidity. The equilibrium of ^{222}Rn between the liquid and gaseous phase is established within 30 min. The radon monitor (RAD-7; Durridge make) uses a high ^{218}Po (half life = 3.1 min; alpha energy = 6.00 MeV) and ^{214}Po+ (half life = 164 μs; alpha energy = 7.67 MeV), which are counted as a measure of ^{222}Rn concentration in air. An air filter at the inlet of the radon monitor prevents dust particles and charged ions from entering into the alpha detector. The ions are collected in energy specific windows which eliminate interference and maintain very low background. ^{222}Rn activities are expressed in Bq/m^3 (disintegration per second perm3) with 2s uncertainties. At room temperature, since the radon in air is about four times more than that in water at equilibrium, the measured radon concentrations in air are corrected accordingly. SGD was directly measured by the modified seepage meter. Seepage meter was placed on to the floor filled with 250ml of water. The EC of the water filled in was noted before and after the SGD measurements. The filled water is measured as liter/ hour to find the SGD. Water level was EC & salinity were measured using Standard IntellliCAL LDO Hach probes.

a

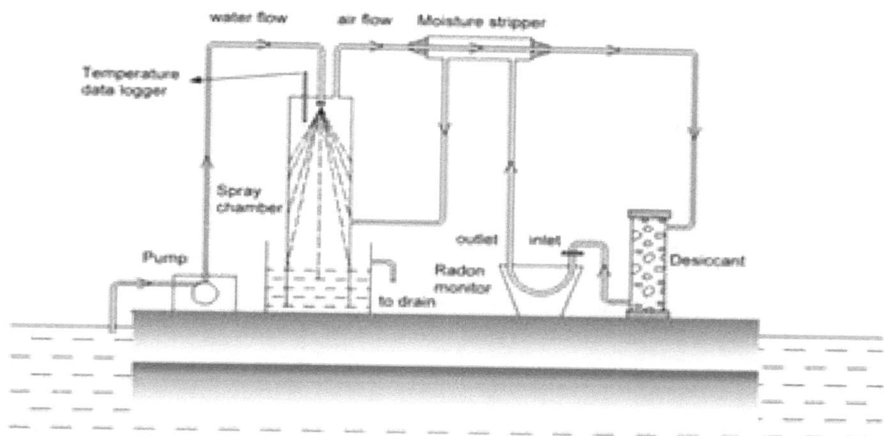

Fig. 3. The radon thus stripped out from the water is circulated through a closed air-loop via a desiccant tube into a ^{222}Rn counting system.

Results and Discussion

The maximum, minimum and average values for pH, EC, Radon Water level and SGD are shown in the Table 1.

pH, Radon, SGD

Continuous radon activity ranges from 125 Bq/m³ to 43.2 Bq/m³. The highest value was noted at 9.00 pm and that of the lowest was at 4.00 pm. The variation of radon value in the surface water may be due to the surfing of sea water and due to greater influence of fresh water influx or due to SGD in these regions (Nepolian 2013).

Table 1. Maximum, Minimum and Average table for Ec, pH, W.L, Rn, SGD

	EC(ms/cm)	pH	W.L(ft)	Rn (Bq/m³)	SGD(l/h)
Maximum	44.76	8.98	2.80	125	2370
Minimum	39.72	8.25	1.60	43.2	1910
Average	43.29	8.60	2.36	74.25	2057.14

EC varies from 39.72 to 44.76 ms/cm with a lowest at 4.00am and highest EC recorded during 10.00am.The pH ranges from 8.25 to 8.98 with highest values noted at 10.00am and the lowest observed at 5.00am. The maximum and minimum SGD is observed as 2.03 and 0.9 l/h at 12:00 and 20:00.The small rise in the WL proceeding both the Low tide condition may be due to the inflow of Uppanar river water (Nepolian 2013).

Radon as tracer for SGD

Radon in seawater ranges from 1.3-1.7 Bq l^{-1}(Ghose et al. 2000) shows only a few Bq/m³ in bay of Bengal water. But in Groundwater, concentrations reach up to 20 kBq/m³ and more (Schubert et al. 2014). The strong gradient at the groundwater/seawater interface results in an inverse correlation between seawater depth and radon concentration in the coastal sea, i.e., in high concentrations during low tides (due to intense SGD) (Fig 4) and vice versa as at 5-6 IST and 15-16 IST (Tsabaris 2010). The relation between the Radon and water level of the Cuddalore illustrate the inverse correlation between radon and water depth observed from 16:00 until the end of the monitoring period was observed. At the beginning of the day at 0:00, this relationship appears to be reversed. Similarly at 7-8 Low water depths are associated with low radon concentrations in the water column. The most likely reason for this unusual radon pattern is the high water turbulence (Schubert et al. 2014) or may be due to the influence of Uppanar River water input during the Low Tide (at 6.00 am). While radon get degassing from every open water surface, this gas evasion is strongly enforced if the water is very turbulent and contains many air bubbles. (The sea release radon due to intense water mixing between seawater and discharging groundwater). During low tide this pressure is released by a falling water table and the groundwater can migrate further seaward, inducing mixing within a zone of enhanced SGD that extends towards the shore more. The fluctuation of the radon with the WL observation shows that they are indirectly proportional to each other's. The increase of Rn during 3.00 am and from 13.00 to 16.00 hrs corresponds the decrease in the WL condition and similar situation is also reflected during 18.00-22.00 hrs.

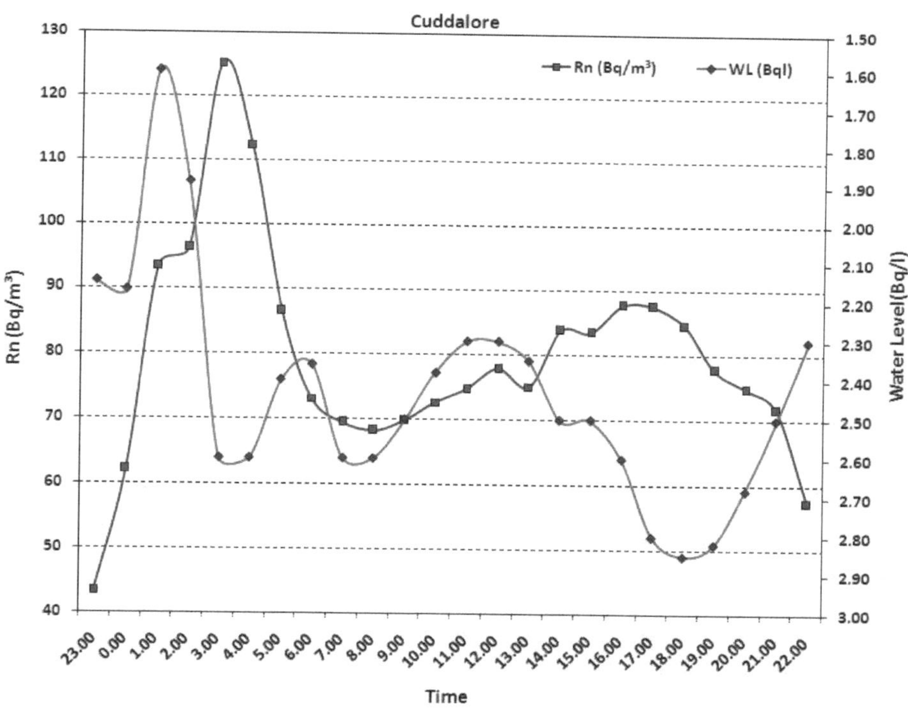

Fig. 4. Relationship of surface water level to Radon during the 24h observation.

EC and Radon

Radon is an excellent tracer of SGD (Burnett et al. 1996; Cable and Judge 1996; Corbett et al. 1997). The very large enrichment of ^{222}Rn concentration in groundwaters over surface waters (typically 1000-fold or greater), its unreactive nature, and short half-life (t1/2= 3.83 d) make

^{222}Rn an excellent tracer to identify areas of signicant groundwater discharge. Groundwater-surface water interaction has increased dramatically in the last few years; there is actually very little documentation available as of now. The knowledge on the origin, pattern and scale of radon migration in the groundwater environment are used to find the type of waters involving the presence of high radon concentrations.

Based on our field observations and analysis of data and literature review, we following assumptions were made to understand the different conditions and process based on the relationship between EC and Radon in ground water to determine the SGD (Fig. 5).

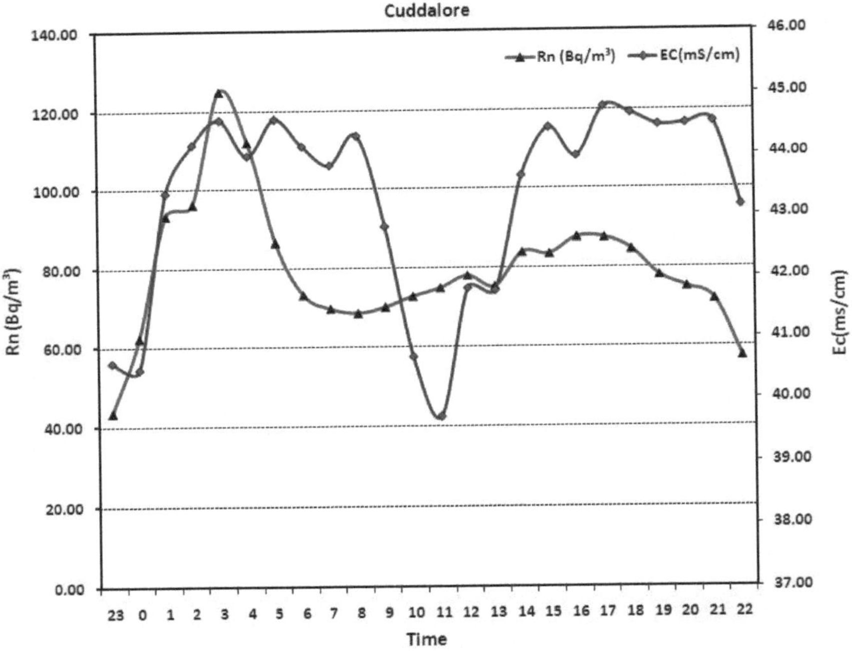

Fig. 5. Relationship of Electrical Conductivity to Radon

1. Increase of Radon and decrease of EC representing Fresh water SGD

From the continuous monitoring of ^{222}Rn in coastal waters, it is possible to quantify SGD (Kim 2003, Cable and Judge 1996 Corbett et al. 2000, Burnett and Dulaiova 2006).The lowering of EC is mainly either due to the increased freshwater input or decreased sea water input; in either case it represents SGD

2. Increase of Radon and increase of EC representing Saline SGD

At the second situation the increase of SGD is due to the ground water discharge with high salinity due to the expelling of sea water intruded into the aquifer during the early high tide period.

3. Decrease of Radon and decrease of EC representing Fresh Surface water

In the third condition the decrease of radon (ie) due to surface action of tide / surface waters. The value range of concentration activity of water-dissolved ^{222}Rn is very wide. The concentration of this gas in surface waters decreases very fast, as it is released into the atmosphere. Its concentration in the waters of streams and rivers oscillates from a fraction of a Becquerel to a few dozen Becquerel's in a litre. Moreover the lesser EC value indicates contribution from river water as freshwater flow.

4. Decrease of Radon and increase of EC representing Sea water

The fourth condition assumes that the decrease of Radon is due to the influence of surface water and the increase of EC adds to the fact that it's due to the influence of sea water.(confusing statements cross check the four statements and validate it with references).

Observed radon concentration in Cuddalore was inversely correlated with tide, indicating that groundwater was being added to the system when the hydraulic gradient was the greatest (low tide). Seepage rates are inversely proportional to tidal height. Studies have been conducted by several authors in Radon and tides (Przylibski 2000a, b, c, 2005, 2007; Przylibski and Zebrowski, 1996, 1999; Przylibski et al. 2001, 2002, 2004; Przylibski and Adamczyk-Lorenc, 2007) on the occurrence of radon, particularly ^{222}Rn isotope in groundwater environment.

Radon is SGD tracer and hence generally be stated that radon reveals robust data due to its distinct gradient at the groundwater/seawater interface, due to its inert behavior and its straight forward detectability on site(Mulligan and Charette 2006). However, in the case of rough seas intense radon degassing may limit its applicability as a tracer. Lower tide could result in increased seepage (and thus higher radon) representing the fresh groundwater at 17:00.

The shallow sandy aquifer has higher permeability. Generally the Permeability for Sand is , 10^{-5} to 10^{-3} cm/s Fetter (1994) it enhances the saline/fresh water movement along the pores during the HT and LT condition. The observation of the EC values show that higher EC was observed at 03.00hrs, which started to increase from 0.00 hrs. Further it get decreased till 10.00 hrs and later is shows an increasing trend with a small drop at 16.00 hrs and it decrease after 21.00 hrs.

Conclusion

The study broadly brings out the major geochemical process occurring in this coastal environment zone. The radon measurements indicate the difference in fresh water- groundwater interface in this region. The EC variation has been studied throughout the tidal cycle period and compared with parameters measured in the field like Radon, pH and Water level. The relationship of EC and Radon revealed four possible different conditions and process that may be existing and is used to determine the SGD status here. This study identified the processes that are dominant in the regions: saline SGD, Fresh water SGD, Saline surface water interference and the mixing of fresh Uppanar water with sea. The seepage studies conducted at the specific intervals also reveals the impact of the above said process and the rate of discharge was found to be lesser during the fresh water SGD. The maximum rate of discharge was observed during the Saline water discharge period.

Acknowledgement

The authors express their sincere thanks to Department Of Science and Technology, India, for providing the necessary nancial support to this research project wide ref No. DST/WAR-W/WSI/06/2010, dated 19.05.2011.

References

Burnett WC, Cable JE, Corbett DR, Chanton JP. (1996). Tracing groundwater flow into surface waters using natural ^{222}Rn. Proc. of Int. Symp. On Groundwater Discharge in the Coastal Zone. Land Ocean Interactions in the Coastal Zone (LOICZ), July 6–10, Moscow; 1996. p. 22–8.

Burnett WC, Dulaiova H, (2006). Radon as a tracer of submarine groundwater discharge into a boat basin in Donnalucata, Sicily. Cont Shelf Res; 26: 862–73.

Cable D M, Judge TA (1996) Person–organization fit, job choice decisions, and organizational entry. Organizational Behavior and Human Decision Processes, 67, 294–311.

Central Ground Water Board (2009).Report of the Group for Suggesting New and Alternate Methods of Ground Water Resources Assessment. Ministry of Water Resources, Government of India.Faridabad.mcgwb.gov.in/ GroundWater/GW_assessment.htm

Corbett D R, Burnett W C, Cable P H, Clark S B (1997) Radon tracing of groundwater input into Par Pond, Savannah River Site. Journal of Hydrology 203, 209–227.

Corbett D R, Dillon K, Burnett W C, Chanton J (2000) Estimating the groundwater contribution into Florida Bay via natural tracers 222Rn and CH4. Limnol.Oceanogr, 2000, 45, 1546–1557.

Ghose S, Alam M N, Islam M N (2000) Concentrations of ^{222}Rn, ^{226}Ra and ^{228}Ra in surface water of Bay of the Bengal, J. Environ. Radioactivity, 47, 291-300.

Gowrishanker S, Meyyappan M, Radhakrishanan M (1980) Lignite deposits and groundwater development on Neyveli area. Bull ONGC 17:1–16

Kim K (2003). Long-term disturbance of ground water chemistry following well installation. Ground Water 41 : 780–789

Schubert M, Jan Scholten, Axel Schmidt, Jean François, Comanducci, Mai Khanh, Pham, Ulf Mallast, Kay Knoeller (2014) Submarine Groundwater Discharge at a Single Spot Location: Evaluation of Different Detection Approaches Water 2014, 6, 584-601; doi:10.3390/w6030584.

Moore W S (1996). Large groundwater inputs to coastal waters revealed by Ra-226 enrichments. Nature 380, 612–614.

Mulligan A E, Charette M A (2006) Intercomparison of submarine groundwater discharge estimates from a sandy unconfined aquifer. J. Hydrol., 327, 411–425.

Nepolian (2013) Hydrogeochemical Study In The Coastal Waters Of Cuddalore Region With Special Reference To Tides, M.Phil Dissertation, Annamalai University.

Przylibski T A, Zebrowski A (1996). Origin of radon in medicinal waters of ´SwieradówZdrój. Nukleonika, 41, 4, 109–116.

Przylibski T A, Zebrowsk A (1999).Origin of radon in medicinal waters of La²dekZdrój (Sudety Mountains, SW Poland), J. Environ. Radioactive, 46, 121–129.

Przylibski T A, Koz³owska B, Dorda J, Kie³czawa B.: (2002). Radon-222 and 226Ra concentrations in mineralized groundwaters of Gorzanów (K³odzko Basin, Sudeten Mountains, SW Poland), J. Radioanal. Nucl. Chem., 253, 1, 11–19.

Przylibski T A, Mamont-Ciesla K, Kusyk M, Dorda J, Koz³owska B (2004).Radon concentrations in groundwaters of the Polish part of the Sudety Mountains (SW Poland), J. Environ. Radioactiv., 75, 2, 193–209.

Przylibski T A, Mroczkowski K, Zebrowski A, Filbier P (2001) Radon-222 in medicinal groundwaters of SzczawnoZdrój (Sudety Mountains, SW Poland), Environ.Geol., 40, 4/5, 429- 439.

Przylibski T A (2007).Radon as a natural radioactive tracer for studying crystalline rock aquifers - a few usage concepts, ActaUniversitatisWratislaviensis, 3041, Hydrogeologia, 125–142.

Przylibski T A (2000a) Size estimation and protection of the areas supplying radon to groundwater intakes, Archiv. Environ. Prot., 26, 1, 55–71.

Przylibski T A (2000b).222Rn concentration changes in medicinal groundwaters of La²dekZdrój (Sudety Mountains, SW Poland), J. Environ. Radioactive, 48, 3, 327–347.

Przylibski T A (2000c).Estimating the radon emanation coefficient from crystalline rocks into groundwater, Appl. Radiat. Isotopes, 53, 3, 473–479.

Przylibski T A (2005).Radon, Specific component of medicinal waters in the Sudety Mountains, OficynaWydawniczaPolitechnikiWroc³awskiej, Wroc³aw, (in Polish).

Srinivasamoorthy K, Vasanthavigar V, Chidambaram S, Anandhan P and Sarma V S (2011), Characterisation of groundwater chemistry in an eastern coastal area of cuddalore district, Tamil Nadu" J Geol Soc India 78, p. 549–558.

Tsabaris C, Scholten J, Karageorgis AP, Comanducci JF, Georgopoulos D, Wee Kwong LL, Patiris DL, Papathanassiou E (2010) Underwater in situ measurements of radionuclides in selected submarine groundwater springs, Mediterranean Sea. Radiat.Protect. Dosim. 142, 273–281.

Valiela I (1990) Transport of groundwater-borne nutrients from watersheds and their effects on coastal waters.Biogeochemistry 10 (3), 177– 197.

Chapter 9

Understanding the Influence of Colloidal Concentration on Groundwater Chemistry

Pradeep K., Chidambaram S., Manivannan R. and Seshadhri[1]

Department of Earth Sciences, Annamalai University, Annamalai Nagar
[1]SRI, IGCAR, Kalpakkam, India

ABSTRACT

A study on colloids in groundwater and its relationship with the geochemical parameters is reasonably essential because of its facilitating nature to transport radionuclides in groundwater. Colloid sampling has been carried out in the coastal aquifers of southeast coast of India. Two different bore wells were selected based on the distance from the coast to collect groundwater samples. On the basis of the result achieved from this study, a reliable representation is emerging. The values of pH and EC are governed by the different size fraction of colloids. In this study, the relation between different size fractions of groundwater colloids and its pH, TDS and EC were carried out. These parameters were analyzed in the lab and its influence with four fractions of colloid like 1.2, 0.8, 0.45 and 0.22 μm have been discovered by smooth scatter plot. The accumulation of particles weighting by normal weighing instrument. There is an higher accumulation of particles found in Kalpakkam than Anupuram in all the size fractions. For separation of particles high frequency pressure was used by ultrasonic cleaner had been carried out. This method is very simple and more reasonable than any other methods.

Key words: Colloids, Groundwater, Coastal aquifers, Size fractions, High frequency pressure.

Introduction

The task of separating solids from liquids is important in nearly every field of industrial production: waste water treatment and environmental protection, mineral processing industry, coal and ore, basic chemicals and synthetic fertilizer, colour and pigment chemistry, biotechnology, biomedicine, food industry, and drinking water treatment, just to name a few (Iritani and Ohashi 1992; Weber and Stahl, 2002). Colloids or nanoparticles are naturally occurring solids comprised of inorganic or organic material found in all groundwater and defined in size between 1 and 1000nm (Stumm, 1992). The presence of colloids cannot justify that there would be transport of radionuclide. But when an inorganic or organic colloid attached with a radionuclide or any other contaminant is often called pseudo or intrinsic or eigen colloids. Numerous studies have demonstrated that the presence of colloidal material in groundwater may facilitate the transport of organic and inorganic contaminants (Sheppard et al. 1979; Means and Wijayaratne 1982; Takayanagi and Wong 1983; Chiou and Malcolm 1986). Following well purging, there is a cursory evaluation of water quality stability, usually pH, temperature, and specific conductivity (Puls and Barcelona 1996). Many studies have been demonstrated contaminant transport by colloidal mobility (Gschwend and Reynolds, 1987; Eichholz et.al.1982; Enfield and Bengtsson,1988; Robertson, 1984). Colloidal particles generated in batch experiments by Sheppard et al (1979) and puls et.al (1989) were shown to retain substantial proportion of radionuclides. Further work by Sheppard et al (1980) concluded that the transport of radionuclides by colloidal clay particles should be considered in contaminant transport models. The major feature of the low-flow purge method is the velocity with which water enters the pump intake and that is imparted to the formation pore water in the immediate vicinity of the well screen (Puls and Barcelona 1996). Various techniques have been developed to reduce or prevent polarization and fouling which is one of the major bottleneck in the application of the filtration process for separation of solid-liquid system. Indeed, various ingenious techniques have been developed for reducing the amount of cake forming; including cross flow filtration (Iritani, Sumi 1991), dynamic filtration with rotating cylindrical membrane (Murase et al. 1989), upward and inclined filtration (Iritani et al.1991; Iritani and Watanabe 1992; Iritani and Mukai 2000). The filter cake (filter cake refer to colloidal particles and gel layer to macromolecules) increases the flow resistance and considerably reduce the specific permeate rate. It could be shown that fine particles are deposited at the membrane even for high crossflow velocities (Altmann and Ripperger 1997). The study aims to identify the amount of colloids in groundwater from different size fractions and to understand the relationship between colloids and pH, EC with respect to space.

Study Area

Two townships namely Kalpakkam (12p 30.313'N, 80p 09.388'E) and Anupuram (12p 33.923'N, 80p 07.755'E) were selected from Thirukalikundram taluk, Kanchipuram district for the proposed study. Lithologically, Anupuram township is from charnockite rock of Archaean age and Kalpakkam township is belongs to clayey sand of Quaternary alluvium deposite. Geomorphologically Kalpakkam is from Older coastal plain and Anupuram is from shallow weathered/shallow buried pediplain. These two locations separated by Backingham canal but Kalpakkam is near to it.The average annual rainfall in the study area is 1,237 mm. The water table in Kalpakkam is shallow and that of the Anupuram township is deeper. The water level shows that there is a wide range in depth of water table from 1 to 17 m below the ground level (mbgl).

Methodology

The water level of both wells was measured by water level indicator. Groundwater was purged the stagnant prior by peristaltic pump (Solinst & model 410) which is connected with 12v battery. Then the groundwater sample was collected with the adequate pumping rate of 200-300 ml/min to reduce excessive drawdown and stirring up sediments. pH and EC were analysed insitu while collecting of 50 lit water samples from bore wells for the colloidal studies. The collected water samples were filtered in the laboratory by polycarbonate membrane filter with different size such as 1.2μm, 0.8μm, 0.45μm and 0.22μm. For all the work Milli-Q water was used. All filter membranes were packed safely and put it in the ultrasonic cleaner (Leelasonic-60) to detach the colloidal particles free from membrane and it was swab by 5-10ml of milli-q water. Then the milli-q water collected to centrifuge for separate the liquid phase. Finally the particles were collected and weighting in the normal weighting instrument.

Result and Discussion

The effect of colloidal material on contaminants in the saturated zone depends on the nature of relative interactions between the contaminants and the colloids, the groundwater, and the soil matrix. As a general rule, metals tend to attach onto negatively charged colloids.

Table 1. pH and EC values of groundwater in every size fractions

Locations	Filter Size	Northeast Monsoon	
		pH	EC in µs/cm
Anupuram	Before Filter	6.95	756
	After 1.2 µm	7.17	729
	After 0.8 µm	7.24	708
	After 0.45 µm	7.36	701
	After 0.22 µm	7.45	683
Kalpakkam	Before Filter	6.77	682
	After 1.2 µm	7.04	680
	After 0.8 µm	7.28	680
	After 0.45 µm	7.52	680
	After 0.22 µm	7.92	680

The table-1 shows the pH and EC variations of groundwater samples in different size fractions. pH and EC value was observed as 6.95, 756 µs/cm in Anupuram and 6.77, 682 µs/cm in Kalpakkam during the collection of water sample. After the filtration in 1.2 µm size fraction colloids with suspended particles and macromolecules had separated from the groundwater sample pH value increased in both locations as 7.17 and 7.04. Inversely EC decreased in both locations as 729µs/cm and 680µ/cm. These fluctuation continuous and end with the pH value of 7.45 in Anupuram and 7.92 in Kalpakkam. In the same manner EC value ends as 683µs/cm but Kalpakkam shows remaining the same.

pH value of the samples from both locations had acidic in nature in before filtering (Fig. 1). After filtration from all the size fractions it started increase towards finer fractions. In electrical conductivity values of both locations were getting decrease towards the finer fractions.

The fluctuation of EC in kalpakkam shows that after filtering it maintain the constant value. But in Anupuram fluctuation of EC getting decline and after 0.8µm fraction it gets little stable then again decline continuous up to the end (Fig. 2).

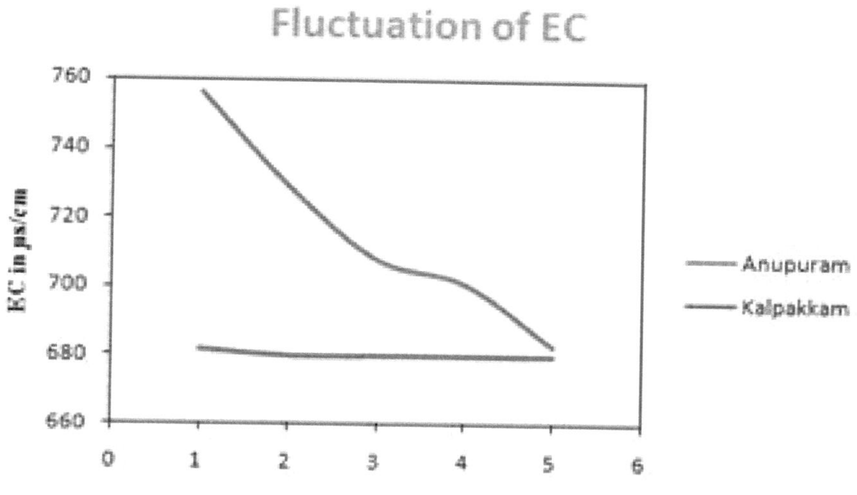

Fig. 1. Fluctuation of EC in different fractions

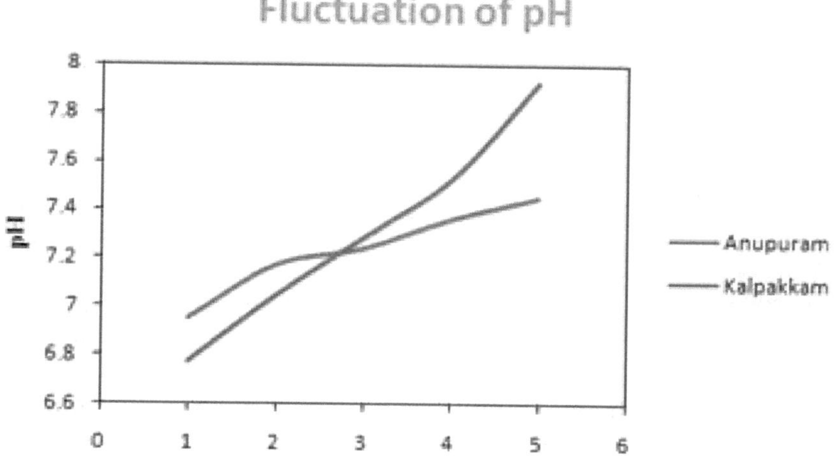

Fig. 2. Fluctuation of pH in different fractions

Even though both two locations were exhibits increasing trend of pH towards finer fractions but Kalpakkam has higher peaks. There was a pH value of 6.77 in kalpakkam location before filtering and it reached 7.92 after 0.22µm fraction reveals that 1.15 has increased which means 17% of pH value has increased from groundwater to after 0.22µm fraction. In kalpakkam location there is a drastic increase of pH after 0.45µm fraction filtration. However Anupuram also having increasing of pH value but there is a stabilisation pH observed after 1.2μm and 0.8μm fraction filtering (Table 2). After these fractions pH value increases as usual.

Table 2. Amount of colloids in different fraction

Location	Weight of colloids (1.2 μm) in grams	Weight of colloids (0.8 μm) in grams	Weight of colloids (0.45 μm) in grams	Weight of colloids (0.22 μm) in grams	Totalin grams
Anupuram	0.75	0.14	0.44	0.51	1.84
Kalpakkam	1.60	1.37	2.65	3.19	8.81

Table shows the weight of accumulation of colloids in different size fraction of filter circles. In all the size fractions kalpakkam got the higher amount of accumulation in the filtrates. But fraction wise 1.2μm fraction has more accumulation of 0.75g in a Anupuram and 0.22μm fraction has more accumulation of 3.19g in Kalpakkam. The important feature observed in this table is the location Kalpakkam having the accumulation of 1.37g of colloids in the 0.8μm fraction which is more than the higher accumulation fraction of 1.2μm in Anupuram.

Conclusion

The variations of pH and EC due to the presence of colloids in different size fractions have been investigated. The fluctuations of those parameters is characterised by smooth scatter plot. Also the amount of colloids determined by weighting before and after filtration of groundwater in the polycarbonate filters. All these experiment Kalpakkam having higher amount of accumulation of colloids than Anupuram. Moreover 0.22μm fraction has higher amount of colloids in Kalpakkam and 1.2μm fraction has higher accumulation in Anupuram which concludes that attachment of colloids with suspended particles leads the higher accumulation in 1.2μm fraction in Anupuram and Kalpakkam has higher amount of colloids in finer fractions. The fluctuation graphs were explained the relationship of pH and EC with the colloid concentration. It exhibits that there is an increasing trend of pH towards finer fraction because accumulation of colloids increase the stability of the colloids and it increases pH. Additionally it reflects the decreasing trend in EC due to removal of all the constituents in water affects the electrical conductivity.

Acknowledgement

The authors express their sincere thanks to Atomic energy Regulatory Board (AERB), India, for providing necessary financial support to this research project wide ref. No. (AERB/CSRP/Proj.No.60/11/2015).

References

Altmann J, Ripperger S (1997). "Particle deposition and layer formation at the crossflow microfiltration." Journal of Membrane Science 124: 119-128.

Chiou C T, Malcolm R L (1986) Water solubility enhancement of some organic pollutants and pesticides by dissolved humic and fulvic-acids. Environmental Science & Technology 20(5): 502-508.

Eichholz G G, Wahlig B.G, Powell G F, Craft T.F (1982) Subsurface Migration of Radioactive Waste Materials by Particulate Transport. Nuclear Technology 58:511.

Enfield C G, Bengtsson G (1988) Macromolecular Transport of Hydrophobic Contaminants in Aqueous Environments. Groundwater 26(1):64.

Gschwend P M, Reynolds M D (1987) Monodisperse Ferrous Phosphate colloids in an Anoxic Grounwater Plume. J. of Contaminant Hydrol. 1:309.

Iritani E, Sumi H (1991) Analysis of filtration mechanism of cross-flow upward and downward ultrafiltration. Journal of Chemical Engineering of Japan 24(1): 39-44.

Iritani E and Ohashi K (1992) "Analysis of filtration mechanism of dead-end electroultrafiltration for proteinaceous solution." Journal of Chemical Engineering of Japan 25(4): 383-388.

Iritani E, Watanabe T (1991). "Upward and inclined ultrafiltration under constant pressure by use of dead-end filter." Kagaku Kogaku Ronbunshu 17(1): 206-209.

Iritani E and Watanabe T (1992) "Effects of pH and solvent density on dead-end upward ultrafiltration." Journal of Membrane Science 69(1-2): 87-97.

Iritani E Y and Mukai (2000). "Effects of electric field on dynamic behaviors of deadend inclined and downward ltrafiltration of protein solutions." Journal of Membrane Science 164: 51-75.

Means J C, Wijayaratne R (1982). "Role of natural colloids in the transport of hydrophobic pollutants." Science 215(4535): 968-970.

Murase T E, Iritani et al. (1989). "Periodic dynamic filtration by use of a creamic filter with rotating cylinder." Kagaku Kogaku Ronbunshu 15(6): 1179-1186.

Puls R W, Barcelona M. J (1996). Low-flow (minimal drawdown) groundwater sampling procedures." U.S. Environmental Protection Agency, Groundwater Issue: EPA/540/S-95/504, April 1996.

Puls RW, Barcelona MJ (1989). Groundwater Sampling for Metals Analyses.EPA/540/4-89/001.

Robertson W D (1984) Contamination of an Unconfined Sand Aquifer by Waste Pulp Liquor: A Case study. Groundwater 22(2):191.

Sheppard J C, Campbell M. J. et al. (1979). "Retention of neptunium, americim, and curium by diffusible soil particles." Environmental Science & Technology 13(6): 680-684.

Sheppard J C, Campbell MJ, Cheng T, Kittrick J A (1980). Retention of Radionuclides by Mobile Humic Compounds. Environ. Sci. Technol. 14(11), 1349-1353.

Sheppard JC, Campbell M J, Kittrick J A (1979) Retention of Neptunium, Americium and Curium by Diffusible Soil Particles. Environ. Sci. Technol. 13(6). 680-684.

Stumm W (1992) Chemistry of the Solid-Water Interface. John Wiley & Sons, New York, 428 pp.

Takayanagi K, Wong G T F (1983). Organic and colloidal selenium in sothern chesapeake bay and adjacent waters." Marine Chemistry 14(2): 141-148.

Weber K, Stahl W (2002). "Improvement of filtration kinetics by pressure electrofiltration." Separation and Purification Technology 26: 69-80.

Chapter **10**

Hydrogeochemical Characteristics of Groundwater at Various Lithological Conditions with Space and Time

Prasanna M.V., Chidambaram S.[1]

Department of Applied Geology, Faculty of Engineering and Science
Curtin University Sarawak, CDT 250, 98009 Miri, Sarawak, Malaysia
[1] Department of Earth Sciences, Annamalai University, Annamalai Nagar

ABSTRACT

In Gadilam river basin, groundwater is found in both hard rock and sedimentary terrain and is the most important source of rural/urban drinking water supply. Gadilam river basin has gained its importance due to the presence of Neyveli Lignite open cast mine (NLC), group of industrial estate (SIPCOT), extensive depressurization of Cuddalore aquifer and boreholes of New Veeranam Scheme (NVS) in the downstream of the basin. A total of 178 water samples were collected in four different seasons at various lithologies. The collected samples were analyzed for various physico-chemical parameters using standard procedures. Geochemical indicators of groundwater were used to identify the chemical processes that control hydrogeochemsitry. Interpretation of hydrogeochemical data suggests that leaching of ions followed by weathering and anthropogenic impact controls the chemistry of the groundwater. Isotopic study reveals that recharge from meteoric source in sedimentary terrain and rock water interaction with significant evaporation prevails in hard rock region. This thorough understanding helps us to make judicial usage and to maintain the quality of this commodity.

Key words: Groundwater, hydrogeochemistry, water-rock interaction, isotope, Gadilam river basin.

Introduction

Groundwater is a natural resource, which is being renewed by different process. The balance between the annual recharge and development has to be maintained for its sustainability. Geochemical processes occurring within the groundwater and reactions with aquifer minerals have a profound effect on water quality. Hydrogeochemical composition of groundwater can also be indicative of its origin and history of passage through underground materials which water has been in contact.

In Tamilnadu 7% of population of the country is endowed with only 3% of water resources of India. The state's water resources are dependent on monsoon rainfall. The total water potential of the state including groundwater is 46,540 MCM (1643 TMC). The total surface water potential of the state is 24,160 MCM (853 TMC) including the contribution (7391 MCM or 261 TMC) from the neighbouring States, viz., Kerala, Karnataka and Andhra. Water is a serious limiting factor as the State has harnessed the available surface water potential. Of the total annual water potential of 46,540 MCM in the state, surface flow accounts for nearly half. Among the two major resources of irrigation water in the state namely, surface and ground water, the surface water potential of about 2.4 million hectares (24,160 MCM) has almost been fully (more than 95%) tapped since the late sixties. Ground water resource is, therefore, the only alternative source for further development. The net area irrigated rose from 21.11 lakh ha. in 1950s to 27.75 lakh ha. in the 1990s and came down to 21.48 lakh ha during 2004-05. The state's per capita irrigation potential in per capita terms is 0.08 ha, compared to the all India average of 0.15 ha. The State is dependent on the ground water resources for providing irrigation to additional acreage and to stabilize. The three main sources of irrigation in the state are rivers, tanks and wells. There are 39,366 tanks, 2295 irrigation main canals and 18,16,434 irrigation wells in the state (Neelakantan et al. 2007). The rivers play a major role in replenishing the aquifers apart from tanks.

In Tamilnadu, there are about 34 river basins, grouped into 17 major river basins, all flowing eastwards from the Western Ghats and Deccan Plateau. Among the 17 river basins, many basins are water deficient and some have surplus water. Even though the annual average rainfall is in the order of 925 mm, there are a number of rain shadow regions receiving less than or around 400 mm rainfall making them drought prone. Gadilam is one such river located in the northern part of Tamilnadu. In Gadilam river basin, groundwater is found in both hard rock and sedimentary terrain and is the most important source of rural/urban drinking water supply. In the recent years due to the advent of industrial growth, large-scale application of synthetic fertilizers for agriculture production and use of pesticides and insecticides for production has caused serious concern regarding susceptibility of groundwater contamination. In addition to this, multi-various human activities particularly those involved in land disposal of industrial effluents and sewage effluents, sludge and solid waste, septic tank effluents, urban run-off as well as various unknown human activities contaminate groundwater sources.

Gadilam river originates in the hard rock region and flows through the sedimentary terrain. The important large scale extraction sites in this basin are Neyveli Lignite Corporation and the boreholes of New Veeranam Scheme (NVS). The open cast mining of Lignite requires heavy pumping at the rate of 9,000 - 10,000m^3hr^{-1} as water table condition has to be brought down below the level of mining (Anandhan 2005). Water from the Veeranam Lake is supplied through transmission line to Chennai city. To augment the supply during summer season, 48 deep bore wells were drilled and operations are done alternatively to pump the groundwater from deep aquifer and the pumped water is connected to the New Veeranam Scheme (NVS) pipelines. Apart from these large scale extraction features, an industrial estate SIPCOT (Small Industries Promotion Corporation of Tamilnadu) with groups of industries, which generate multi facet chemicals and raw materials are distributed along the downstream of the river Gadilam, near the coast at Cuddalore (Prasanna 2008).Aravindhan et al. (2004) has prepared the groundwater prospective zone map using remote sensing and GIS in the hard rock region of this basin. He has also discussed about the nitrate pollution, presence of antimony and mercury in the groundwater in the hard rock region of this basin. Apart from anthropogenic impact, natural processes also contribute to the groundwater system. In recent days, the geochemical study of the groundwater has attained its momentum. Hence, this study focuses unraveling the geochemical processes involved in the groundwater of this basin.

The vagaries in monsoon and insufficient surface water flow conditions have led to increase of agricultural and domestic extraction (Aravindhan et al. 2004). The increase of human population and the economic activities in this coastal region has grown in scale; the demand for large scale supplies of freshwater from various competing end users has increased. In this scenario deep aquifers have been bored in the coastal region, increasing the fragility of the system, aiding the movement of the seawater into these fresh water aquifers (Prasanna et al. 2007). The development of the industrial estates in the coastal environments creates more sensitivity to the system. Declining quality and quantity of water supply in this region can be attributed to the over exploitation of water and improper management of the existing resource (Prasanna 2008). In this existing scenario, an attempt has been made to understand the complexity of hydrogeochemical characters in the study area. The entire Gadilam river basin has been selected to have a holistic picture.

Study Area

Gadilam river basin is located in parts of Cuddalore and Villupuram districts of Tamilnadu, India. It lies in between 79° 0′ E to 79° 47′ E longitudes and 11° 30′ N to 11° 55′ N latitudes (Fig. 1). It occurs within the Survey of India toposheets of 58M/1, 2, 5, 6, 9, 13, 10 and 14, covering a total area of about 1,394 sq.km. The Gadilam river basin is bounded by Ponnaiyar river basin in the North and the Vellar river basin in the south. The total length of the river is about 112 km with a catchment area extending about 900 sq.km. The average annual rainfall of the basin is about 1643mm. The water level ranges from 3.10mbgl to 98.85mbgl (below ground level) with an average of 62.37bgl. The river basin has different rock types along its flow path. The lithological set up of the basin shows that hard basement rocks are exposed in western part of the study area and sedimentary formation in the east with a faulted contact between both(Aravindan et al. 2004). River originates in the hard rock region and flows through the sedimentary terrain. The basin covers different stratigraphic units viz. Archaean, Cretaceous and Tertiary to Recent alluvium (Fig. 2). The charnockites, granites, syenites and gneiss of archaean complex constitute the upstream area. The younger cretaceous and tertiary formations are found in the midstream and recent alluvium in the downstream of the Gadilam river basin. Lithologically, hard rocks constitute the western part of the basin. In the sedimentary region, northern part essentially comprises of recent alluvium and the rest by sandstone (Prasanna et al. 2010).

Materials and Methods

The water samples were collected during March 2005, July 2005, November 2006, and January 2006 to broadly cover all seasons (Fig. 2). A total of 178 water samples were collected from 156 bore wells, river (14) and tank (8) in four different seasons (November 2006 represents NE monsoon; January 2006 represents Post monsoon; March 2005 represents Summer; July 2005 represents SW monsoon). In summer (SUM) and south west monsoon (SWM), twelve groundwater samples were collected from Alluvium, ten samples from Tertiary, fourteen samples from Archaean and two surface water from tank. In north east monsoon

(NEM) and post monsoon (POM), twelve groundwater samples were collected from Alluvium, twelve samples from Tertiary, seventeen groundwater samples from Archaean, one groundwater sample from Cretaceous, seven surface water from river and two surface water from tank. Cretaceous formation covers less than 5% of the study area. In certain location during summer and southwest monsoon periods the wells were dry. River Gadilam is an ephemeral river with water flows only during NEM and POM periods. One litre of water samples was collected in polyethylene bottle four times a year to broadly cover seasonal variation with respect to lithology. Then it was sealed and brought to laboratory for analysis and stored at 4°C before analysis.

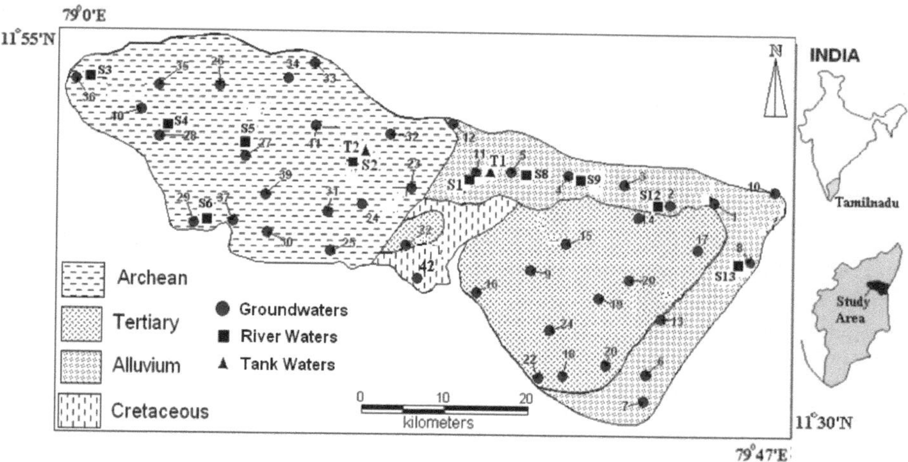

Fig. 1. Location, geology and sample location map (*modified after Prasanna et al. (2010)*

The samples collected were analyzed for major cations like, Ca and Mg by Titrimetry, Na and K by Flame photometer (CL 378); anions, Cl and HCO_3 by Titrimetry, SO_4, PO_4, and H_4SiO_4 by Spectrophotometer (SL 171 minispec). EC and pH were determined in the field using electrode (Eutech). The analyses were done by adopting standard procedures (APHA 1995).

Stable Isotopic studies ($\delta^{18}O$ and δD) were carried out for 60 samples from different locations of Gadilam river basin during south west monsoon (2005) and post monsoon (2006) following standard procedures of International Atomic Energy Agency (IAEA). Isotopic studies were done for groundwater, surface water and rain water samples. Oxygen ($\delta^{18}O$) and Deuterium (δD) present in samples were analyzed by using Isotopic Ratio Mass Spectrophotometer (Finnigan Delta^{plus}Xp, Thermo

Electron Corporation, Bermen, Germany, the standard deviation of our measurements is ±1.72% for Oxygen and ±0.8% for Hydrogen). All the measurements were carried out against laboratory substandard that were periodically calibrated against the international isotope water standards recommended by the IAEA (V-SMOW). The isotope results obtained are reported in terms of ä units (permil deviation of the isotope ratio from the international standard V-SMOW) δ being defined by

$$\delta = (R_{sample} - R_{SMOW} / R_{SMOW}) \times 10^3$$

where, R = D/H or $^{18}O/^{16}O$

Results and Discussion

General water Chemistry

The average, mean, and standard deviation values of the water samples collected during four different seasons in different formations are given in Table 1. In Alluvium formation, Cl is the dominant anion followed by HCO_3, SO_4 and PO_4 during SUM, NEM and POM seasons. But in SWM, HCO_3 is the dominant anion followed by Cl, SO_4 and PO_4. Higher concentration of chloride in this region is may be due to sea water intrusion along the coast (Chidambaram et al. 2007). In POM, Na is the dominant cation followed by Ca, K and Mg. But in SUM, SWM and NEM, Na is the dominant cation followed by Ca, Mg and K. Na is found to be the dominant cation it may be due to the weathering of Alkali feldspar in rocks or due to cationic contribution from sea water.

In Tertiary Formation, Cl is the dominant anion followed by SO_4, HCO_3 and PO_4 during SUM. In POM, Cl is the dominant anion followed by HCO_3, SO_4 and PO_4. In SWM, HCO_3 is the dominant anion followed by SO_4, Cl and PO_4. In NEM, HCO_3 is the dominant anion followed by Cl, SO_4 and PO_4. The contribution of HCO_3^- is may be due to chemical weathering of silicate and carbonate minerals (Srinivasamoorthy 2004). In SUM, NEM and POM, Na is the dominant cation followed by Ca, K and Mg. But in SWM, Na is the dominant cation followed by Ca, Mg and K.

In Archaean formation, HCO_3 is the dominant anion followed by Cl, SO_4 and PO_4 during SUM, SWM and NEM seasons. But in POM, Cl is the dominant anion followed by HCO_3, $SO_4$and PO_4. The contribution of chloride may be due to the leaching of secondary salts during weathering processes. In SUM, SWM and POM, Na is the dominant cation followed by Ca, K and Mg. But in NEM, Na is the dominant cation followed by Ca, Mg and K.

Table 1. Maximum, Minimum, Average and Standard deviation for chemical composition of groundwater (in mg/l) except EC(in ìs/cm) and pH.

Alluvium

	Summer				South West Monsoon				North East Monsoon				Post Monsoon				
	Max	Min	Avg	Std	Max	Min	Avg	Std	Max	Min	Avg	Std	Max	Min	Avg	Std	F.Avg
pH	7.30	6.60	6.99	0.21	8.65	6.40	7.29	0.65	7.83	6.01	7.17	0.61	9.20	6.83	8.15	0.91	7.40
EC	6337.31	420.18	1424.31	1605.68	6278.93	387.00	1543.82	1538.59	6728.00	468.00	1829.58	1645.38	6005.00	345.00	1652.59	1621.56	1612.58
Cl	2384.01	62.03	347.77	646.29	1994.06	26.59	357.67	526.91	2782.82	96.42	498.17	734.87	2375.15	51.00	393.60	639.42	399.30
HCO_3	677.10	48.79	289.84	241.36	1293.20	91.50	397.77	334.67	634.39	54.30	270.75	151.37	1500.59	12.20	271.44	399.28	307.45
SO_4	341.00	23.00	96.28	84.30	189.00	27.00	90.23	39.78	368.00	1.00	69.62	106.03	312.20	5.00	107.21	85.42	90.83
PO_4	10.10	0.25	6.08	3.76	6.80	0.10	4.20	2.55	6.60	0.01	1.25	1.92	6.80	0.00	0.90	1.97	3.11
H_4SiO_4	34.00	8.10	19.02	9.15	50.00	1.60	30.06	14.15	147.50	8.80	58.60	36.30	160.00	27.00	73.92	40.62	45.40
Ca	261.33	11.00	64.11	69.41	264.00	16.00	103.14	76.12	352.00	6.00	64.33	93.63	168.00	10.00	50.16	41.09	70.44
Mg	91.20	1.00	14.47	25.00	62.39	0.00	17.09	17.51	134.40	1.00	27.51	35.31	124.80	4.79	19.50	33.78	19.64
Na	1210.03	81.00	279.94	310.56	1400.20	49.30	267.35	365.38	1091.95	74.71	317.34	274.98	1149.43	52.00	293.71	307.29	289.59
K	42.00	3.00	11.71	11.11	48.71	3.00	13.61	12.21	36.40	1.00	15.14	10.46	108.40	1.00	25.21	32.74	16.42
TDS	4436.44	267.81	994.45	1122.35	4395.25	271.00	1080.07	1077.38	4710.00	328.00	1291.44	1149.91	4204.08	242.00	1157.33	1134.97	1130.82

Tertiary

	Summer				South West Monsoon				North East Monsoon				Post Monsoon				
	Max	Min	Avg	Std	Max	Min	Avg	Std	Max	Min	Avg	Std	Max	Min	Avg	Std	F.Avg
pH	7.30	6.10	6.69	0.35	8.52	6.00	7.36	0.87	7.50	5.56	6.68	0.57	8.60	6.21	7.76	0.72	7.12
EC	867.00	250.00	492.93	214.50	1280.00	197.00	543.72	330.01	822.00	165.00	448.75	220.10	981.00	286.21	476.58	198.16	490.49
Cl	241.05	26.59	84.19	64.26	239.29	8.86	89.69	73.97	194.97	8.00	81.67	52.13	255.23	44.00	95.32	61.99	87.71
HCO_3	274.50	18.30	72.51	74.60	323.30	36.60	100.65	81.60	195.19	12.20	84.36	55.25	195.10	24.40	72.52	46.72	82.51
SO_4	171.00	28.00	89.10	52.89	247.00	18.00	98.60	66.33	196.00	0.01	50.29	63.45	141.00	7.00	46.89	38.38	71.22
PO_4	10.10	0.07	3.92	4.69	6.40	0.15	2.36	2.40	7.45	0.09	1.83	2.57	1.40	0.00	0.30	0.42	2.10
H_4SiO_4	69.00	18.10	31.62	17.51	54.00	7.30	27.16	15.29	72.40	8.00	39.37	18.93	120.00	18.00	68.38	27.02	41.63
Ca	55.99	8.00	24.90	14.18	88.00	10.00	36.00	27.36	40.00	6.00	23.16	11.83	34.00	10.00	22.08	7.78	26.54
Mg	9.00	0.00	3.40	3.13	33.60	1.00	8.48	10.03	14.40	0.00	6.51	4.35	14.40	1.00	6.40	4.20	6.20
Na	195.40	45.98	91.53	51.13	272.00	1.80	82.49	83.15	191.00	6.00	69.36	61.11	183.91	22.99	72.30	45.89	78.92
K	14.00	0.00	4.60	4.70	21.90	1.00	7.01	6.71	14.60	1.00	7.68	3.89	23.90	1.00	7.41	5.71	6.68
TDS	646.27	175.00	349.06	158.05	897.22	138.76	380.88	231.05	633.30	122.60	319.83	162.71	687.00	200.35	329.56	140.48	344.83

Archaean

	Summer				South West Monsoon				North East Monsoon				Post Monsoon				
	Max	Min	Avg	Std	Max	Min	Avg	Std	Max	Min	Avg	Std	Max	Min	Avg	Std	F.Avg
pH	8.10	6.73	7.14	0.33	8.00	6.80	7.14	0.32	8.60	6.78	7.46	0.41	9.20	8.10	8.52	0.32	7.56
EC	2610.32	657.13	1500.32	580.51	2725.89	618.29	1498.53	642.83	2810.00	617.00	1364.24	591.00	2574.00	490.37	1129.36	583.25	1373.11
Cl	771.03	59.03	342.01	225.96	771.03	53.17	361.46	248.08	638.10	35.44	233.55	152.64	620.37	35.45	229.38	168.76	291.60
HCO_3	524.60	284.50	425.76	90.61	579.50	274.50	416.11	92.70	414.79	183.00	278.44	84.44	268.39	109.80	185.15	48.27	326.36
SO_4	124.60	43.00	70.97	21.56	158.20	53.40	87.03	28.66	336.22	1.50	67.27	98.55	288.18	4.00	116.98	82.67	85.56
PO_4	15.00	4.80	8.76	2.75	6.40	3.80	5.36	0.81	3.60	0.20	0.75	1.01	1.60	0.00	0.18	0.40	3.76
H_4SiO_4	63.00	5.80	27.10	21.81	53.00	21.20	39.63	10.19	94.00	6.60	50.08	28.69	120.00	22.00	72.82	28.03	47.41
Ca	225.99	35.99	108.14	50.11	240.00	55.99	143.00	61.72	111.99	23.99	56.47	27.78	96.00	18.00	49.05	24.64	89.16
Mg	40.80	4.79	20.50	13.82	52.80	4.79	24.25	18.19	33.59	4.80	16.35	7.99	24.00	0.00	12.98	6.95	18.52
Na	405.00	55.20	231.60	90.11	491.00	50.30	188.58	122.44	581.20	69.80	192.60	129.67	551.72	54.90	188.04	128.68	200.21
K	208.00	1.00	56.11	66.48	115.30	3.30	31.24	34.95	17.40	9.40	10.12	1.93	78.00	6.60	14.95	17.78	28.11
TDS	1827.23	459.99	1050.22	406.35	1908.12	432.80	1048.97	449.98	1967.00	436.00	972.94	403.76	1802.47	343.26	795.56	409.11	966.92

Spatial distribution of EC of samples was done for all four major seasons (Fig. 3). The study reveals that in SUM and SWM seasons, higher EC was noted in the regions covered by the Archaean formation in the upstream side and its also noted along the downstream side of the coastal areas. In NEM and POM seasons, higher concentration was observed in the upstream side, along the lithological boundaries and near the coastal area. This indicates the leaching of secondary salts in the upstream side and anthropogenic activities in the downstream side.

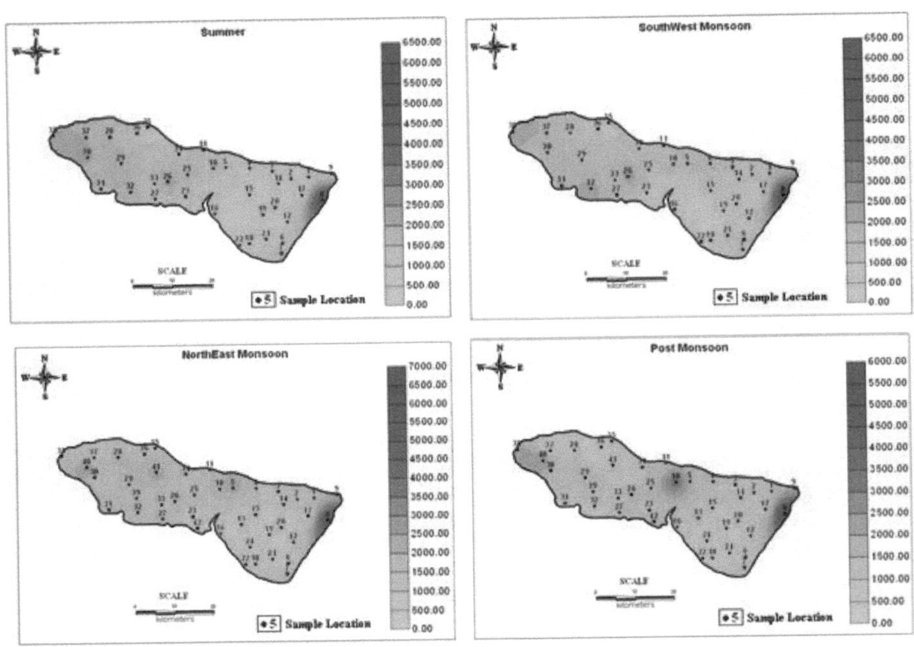

Fig. 3. Spatial distribution of EC in groundwater.

Hydrogeochemical Classification

Hydrogeochemical facies interpretation is a useful tool for determining the flow pattern, origin of chemical histories of groundwater. The diamond field of Piper (1944) for Alluvium formation (Fig. 4) shows variation with seasons. The plot mainly falls in fields 2, 3, 4 and 1. In NEM and POM, the samples are concentrated in Na-Cl type indicating saline nature in the groundwater. In SUM and SWM, cluster of samples fall in Na-Cl and mixed Ca-Na-HCO_3, Ca-Mg-Cl facies. The data plot of Piper Trilinear diagram (Fig. 4) indicates the change of $CaHCO_3$facies in summer to NaClfacies in winter, as calcium is replaced by sodium during North East monsoon (Aravindan et al. 2004). In Tertiary formation

(Fig. 4), the plot falls in fields 2, 3, 4 and 5. The dominant fields are 2, 3 and 4. SUM and SWM seasons are well represented in plot 2 and 3, indicating the dominance of alkaline and strong acids. In NEM and POM, cluster of samples fall in Na-Cl type and some samples also represent mixed Ca-Mg-Cl and Ca-Cl facies.

In Archaean (Fig. 4), most of the plots fall in field 2 with less representation in 3 and 4. In NEM and POM, alkaline and strong acid are predominant indicating the intensive weathering action of silicate bearing minerals and secondary leaching of ions (Srinivasamoorthy 2004). In SUM and SWM, majority of samples fall in Na-Cl and mixed Ca-Mg-Cl with minor representation from mixed Ca-Na-HCO_3 and Ca-HCO_3facies. In Cretaceous formation (Fig. 4), NEM samples fall in Ca-HCO_3facies indicating the dominant of alkali earth and week acids. But in POM, plot fall in Na-Cl facies indicating the predominant of alkaline and strong acid.

Gibbs (1970) proposed a method for identifying the relationship between water composition and mechanism controlling the chemistry. There are two plots for cations and anions. In Alluvium (Fig. 5), the plot for cations and anions fall in rock dominant zone and minor representation is noted in evaporation zone irrespective of season. In Archaean formation (Fig. 5), the plot for cations fall within rock dominance zone and some good representation of SUM and SWM seasons is noted in the evaporation zone. Same trend is observed in anions but some good representation of SUM season is also noted in evaporation zone. In Tertiary and Cretaceous the plot for both cations and anions fall within the rock dominance zone. In general the Gibbs plot reveals that rock water interaction is also a chief factor which controls the chemistry of groundwater in the study area. Minor representations in evaporation zone may indicate contribution from secondary leachates.

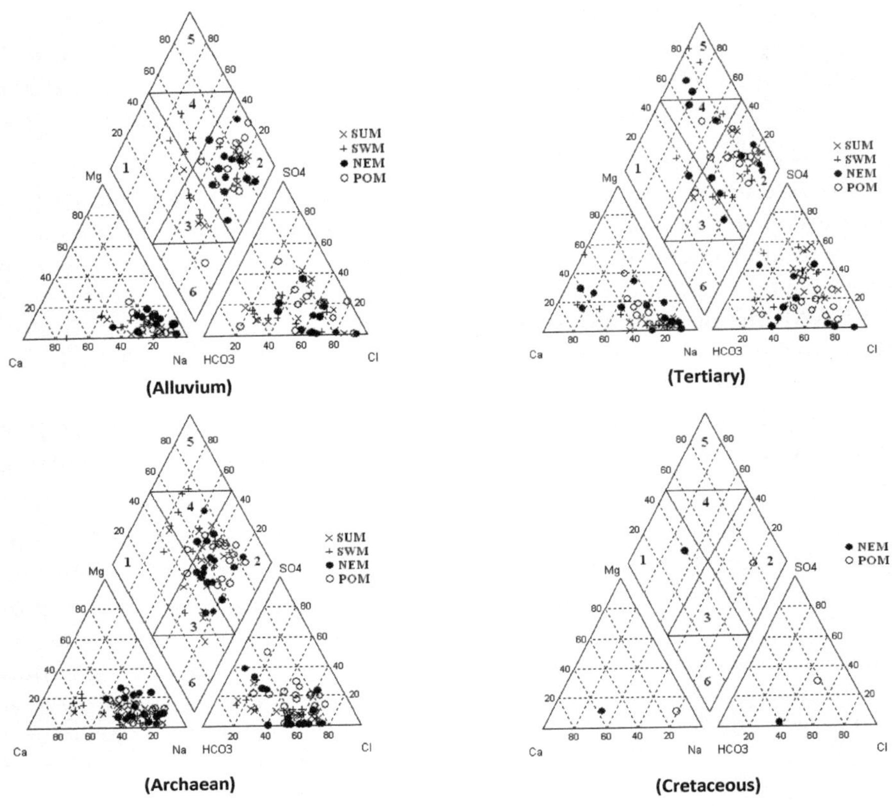

Fig. 4. Hill Piper Plot

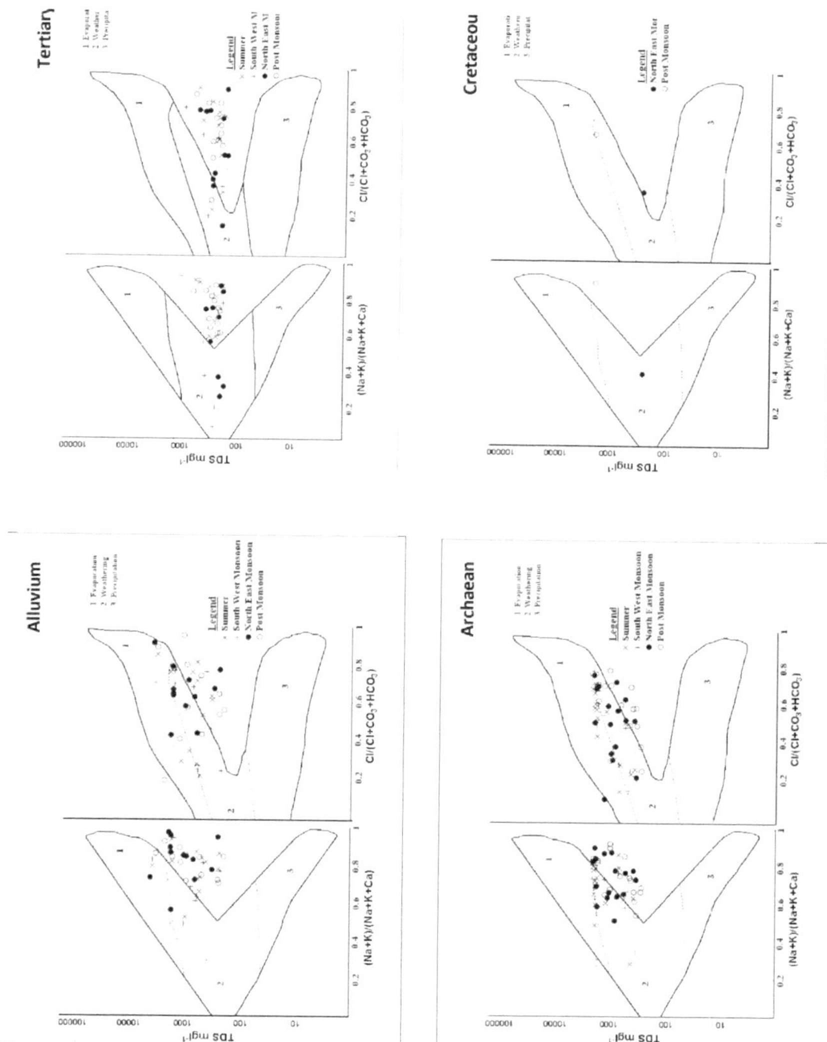

Fig. 5. Gibbs plot for the groundwater (after Gibbs, 1970)

Ionic Strength

Ionic strength is a measure of total concentration of ions which emphasizes increased contribution of species with charges greater than one to solution non-ideality (Domenico and Schwartz 1990).

$$I= 0.5 \Sigma m_1 z_1^2$$

Where m_1 is the atomic/molecular weight and z_1 is the valance of the respective ion. Ionic strength of fresh water is less than 0.005(Hem 1959). In the study area, higher ionic strength was noted in Alluvium and Archaean formation irrespective of seasons, indicating lesser inflow of fresh water into the system (Fig. 6). Low ionic strength was noted in Tertiary formation indicating higher inflow of fresh water/ recharge (Prasanna et al. 2006).

Partial pressure of Carbon dioxide (pCO_2)

In SUM, the Log pCO_2 values ranges from -2.5 to -1. The samples from Tertiary formation and certain samples in Alluvial terrain show nearly the atmospheric value (-3.5) (Fig. 7). Samples from Archaean formation show higher Pco_2 value ranging from -2.39 to -1.19 suggesting that the additional CO_2 has been acquired from the soils during the process of infiltration towards the zone of saturation (Prasanna et al. 2006). In SWM, certain samples in Alluvium and Tertiary formations show nearly similar or slightly lesser than the atmospheric log pCO_2 values. It may be interpreted that their residence time in the aquifer matrix is considerably lesser. Higher Log pCO_2 values are noted in Archaean formation. In NEM, all the samples irrespective of terrains, show higher Log pCO_2value except few representations of near atmospheric log pCO_2. In POM, majority of samples irrespective of terrains fall around the atmospheric Log pCO_2 value indicating recent recharge waters by the monsoon rainfall (Prasanna et al. 2010).

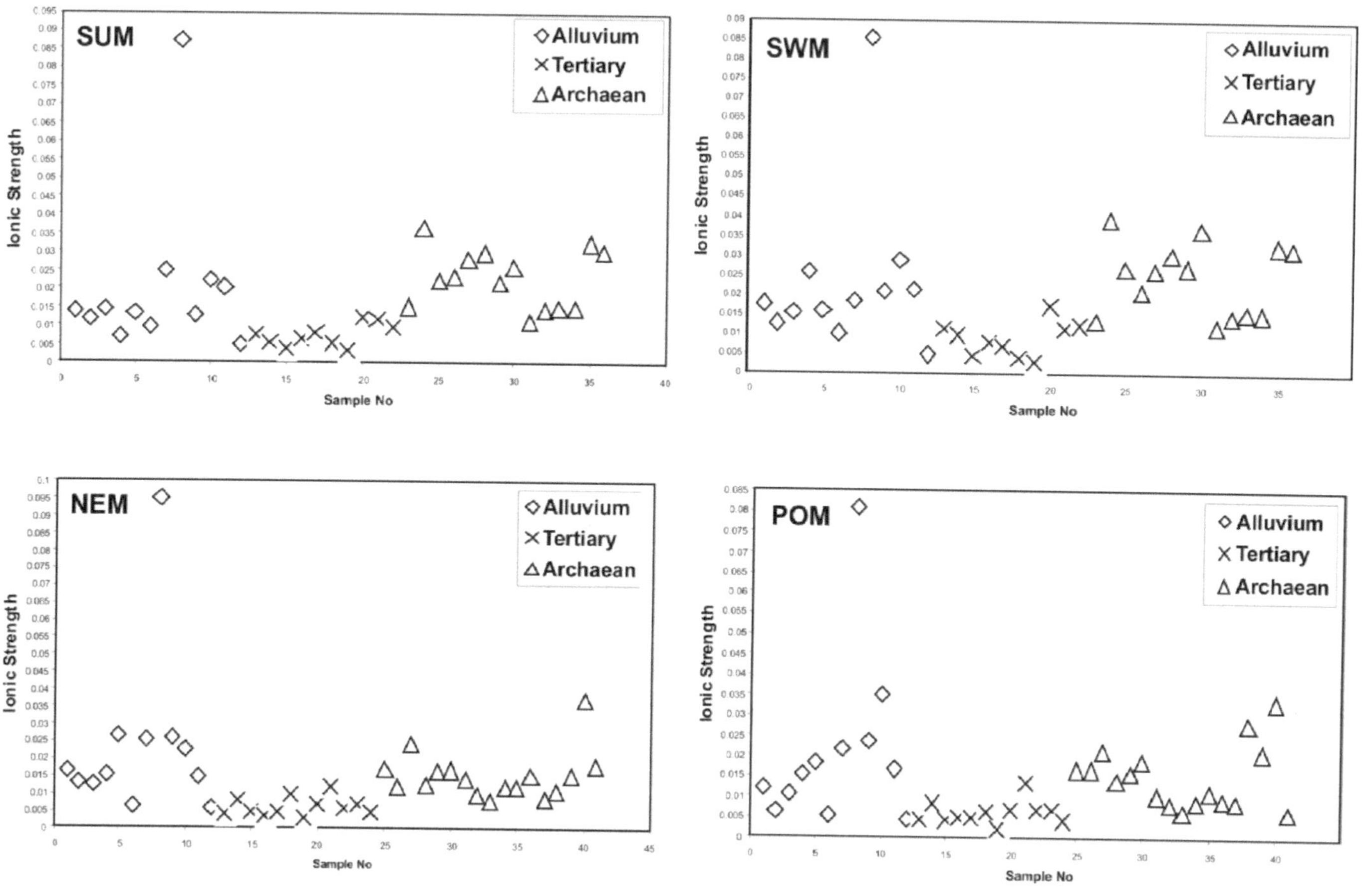

Fig. 6. Ionic Strength for the groundwater.

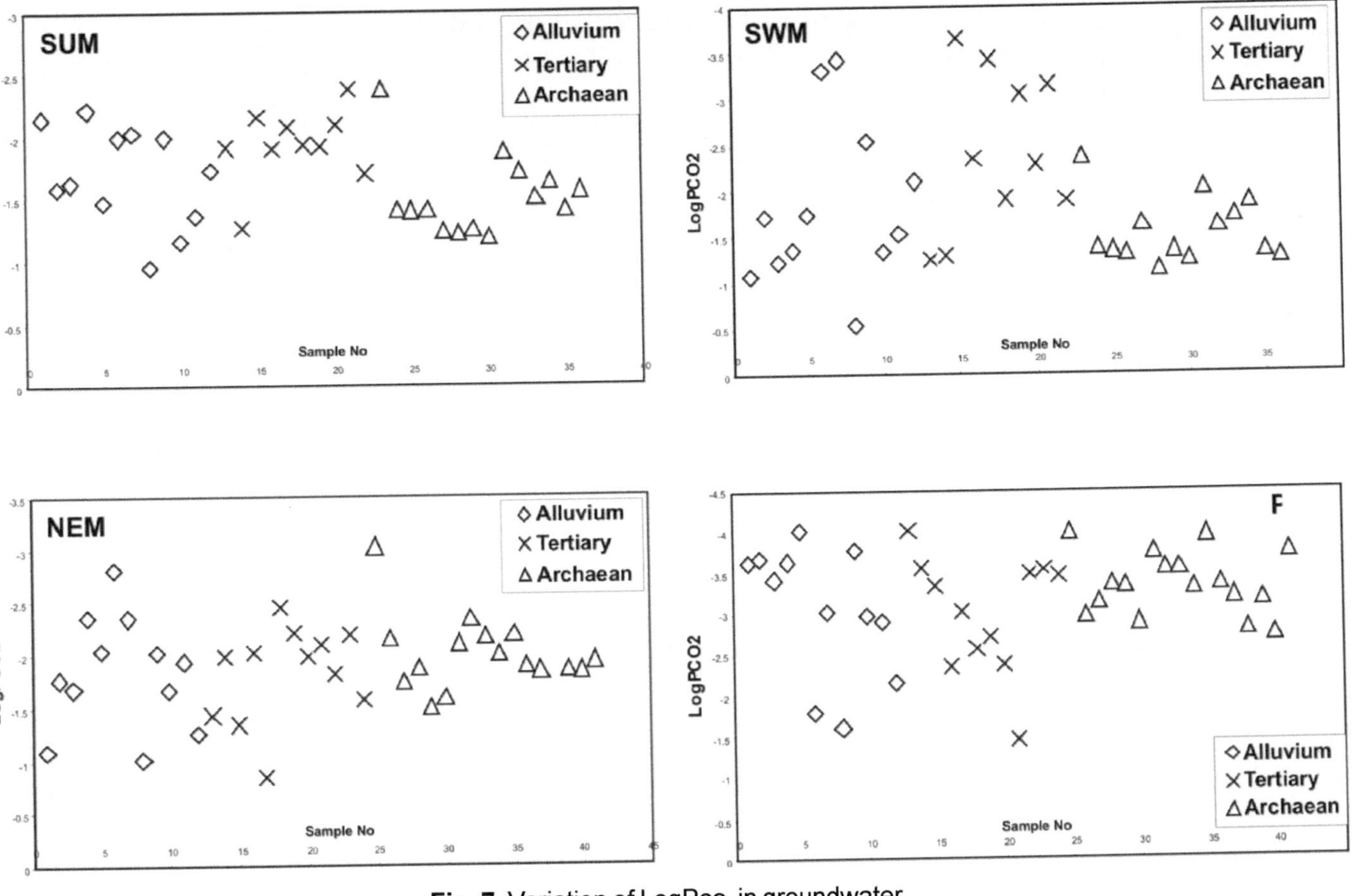

Fig. 7. Variation of $LogPco_2$ in groundwater.

Isotopic signatures in groundwater

During SWM, the collected groundwater samples of Tertiary formation have a range from -7.95% to -6.25% for $\delta^{18}O$ and from -52.85% to -45.93% for δD. The mean isotope values of groundwater are -7.33% and -50.90% for $\delta^{18}O$ and δD respectively. The δD and $\delta^{18}O$ values of the groundwater samples fall nearer to Local meteoric water line (LMWL) indicating recharge from meteoric source (Prasanna et al. 2010) (Fig. 8). The groundwater samples belong to Alluvium formation have a range from -7.19% to -6.17% for $\delta^{18}O$ and from -50.49% to -44.63% for δD. The mean isotope values of groundwater are -6.85% and -48.33% for $\delta^{18}O$ and δD respectively. Groundwater samples of this formation fall parallel to the LMWL or GMWL indicating recharge has taken place by the local precipitation (Chidambaram et al. 2007). Allison et al. (1984) developed a concept that under certain conditions of direct infiltration often has an ^{18}O-^{2}H composition that plots parallel to the local meteoritic water line. It is because of the mixing that occurs between the evaporated soil moisture and subsequent rain that infiltrates and displaces the residual soil water downwards, this mixture ultimately reaches the water table. Since the Alluvial formation are porous and permeable, easy infiltration of rain water is depicted in the isotopic ratios.

The groundwater samples belonging to Archaean formation have a range from -5.7% to -4.12% for $\delta^{18}O$ and from -43.32% to -38.39% for δD. The mean isotope values of groundwater are -4.96% and -41.66% for $\delta^{18}O$ and δD respectively. In Archaean formation, enrichment of heavier isotopes is noted may be due to two factors. 1) The samples are displaced from the LMWL, this may be because of the extensive evaporation form the unsaturated zone. 2) Kinetics effect by vapour diffusion is greater than those associated with evaporation from open surfaces (Ian Clark and Peter Fritz 1997). The slope of ä^{18}O-ä^{2}H relation was only ~2, much lower than the range for the evaporation from the open water surfaces. Hence, either one of the above two factors or in combinations may be the responsible for low slope. This displacement may also be due to recharge of the evaporated surface waters. Since few surface water tanks are located in the hard rock terrains of the study area, these tanks may significantly contribute water to the ground system. The evaporation taking place in the fractures is also well witnessed by the higher saturation index (S.I) of Calcite, precipitation of Calcite is also witnessed in the fractures along the well cuttings of the study area (Chidambaram et al. 2007). Though few rock water interaction mechanisms are also noted in this terrain. δD and $\delta^{18}O$ values of the surface water samples fall away from the GMWL line indicating that the waters are mostly affected by evaporation processes (Prasanna 2008).

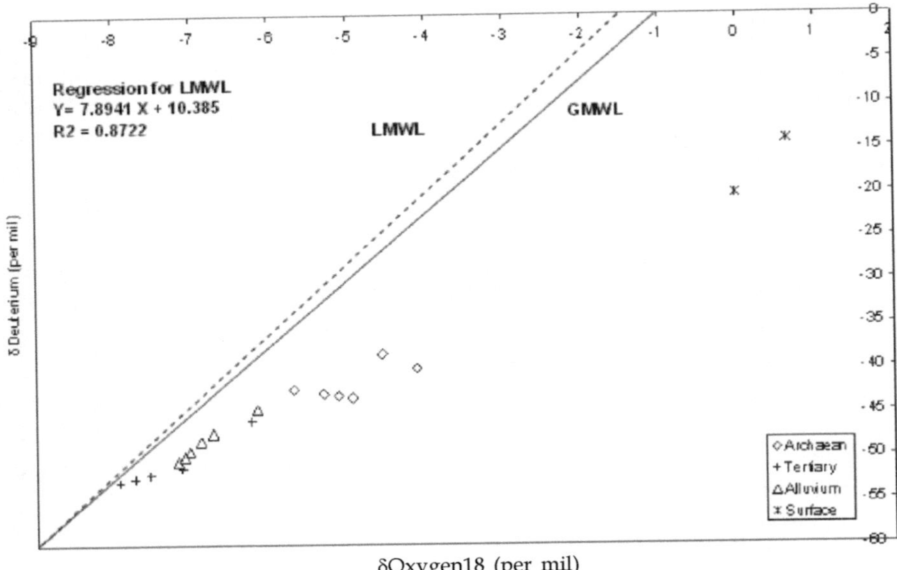

δOxygen18 (per mil)

Fig. 8. Relationship between $\delta^{18}O$ and δD in SWM.

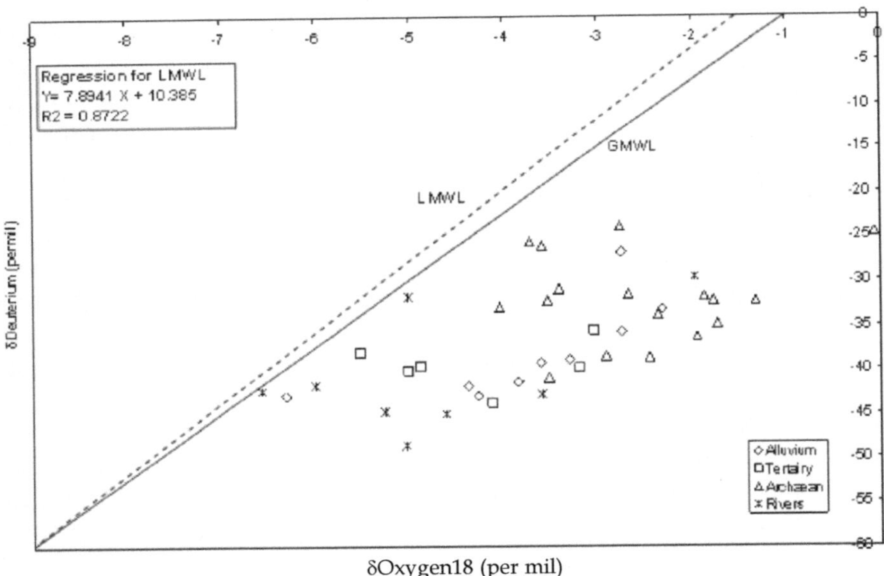

δOxygen18 (per mil)

Fig. 9. Relationship between $\delta^{18}O$ and δD in POM.

During POM, river water samples plot to the right of the LMWL along a shallow-slopping trend, suggesting a typical evaporation pattern (Fig. 9). The groundwaters in Tertiary formation fall within a cluster either close to or to the right of the LMWL, indicating that these samples have undergone some evaporation prior to infiltration. In Alluvium formation, samples fall parallel to the LMWL indicating recharged by the local precipitation. The majority of Archaean terrain groundwaters plot along a shallow-slopping trend to the right of the LMWL, indicating that evaporation concentration is a significant process for these samples.

The δD and $\delta^{18}O$ composition for the groundwater in the downstream of the river basin are different from the isotopic composition of the groundwater in the upstream of the basin. Majority of groundwater samples in the downstream have a relatively depleted isotopic signature (high negative values) that is similar to certain river water samples. This indicates that infiltrating river water or evaporated river recharge is the source for the groundwater in the downstream during this season. Under low or average flow condition, river water tends to be isotopically enriched relative to rainfall because of surface water evaporation.

Conclusion

The evolution of ions along its flow path, reaction of the medium with the aquifer matrix, variation of water quality with space and time, migration of plume, identification of recharge etc can be determined by this type of studies. Groundwater is generally near acidic to alkaline in nature. The study indicates that the groundwater of the alluvium and the Archaean formations have higher ion concentration irrespective of season. The variation in the water chemistry with season indicates that the secondary leaching of salts, weathering and anthropogenic activities play a major role. Water-rock interaction is also identified as one of the factor contributing the variation in water chemistry. Isotopic study reveals that recharge from meteoric source in sedimentary terrain and water-rock interaction with significant evaporation prevails in hard rock region. The impact of monsoon is well witnessed in the sedimentary formations than the Archaean. Based on the findings of this study, it is essential to make a judicial usage and to maintain the quality of the commodity with lesser anthropogenic impact.

Acknowledgement

The authors wish to express thanks to Department of Science and Technology, India for providing the necessary financial support to carry out this study(SR/S3/ENGF-01/2002). They are also thankful to the Centre for Water Resource, Development and Management (CWRDM), Kozhikode, India for helping in the stable isotope analysis of the water samples.

References

Anandhan P (2005) Hydrogeochemical studies in and around Neyveli mining region, Tamilnadu, India. Ph.D Thesis, Department of Earth Sciences, Annamalai University, 189p.

APHA (1995) Standard methods for the examination of water and waste water, 19th edition, APHA, Washington DC, USASS.

Aravindan S, Manivel M, Chandrasekar S.V.N (2004) Groundwater quality in the hard rock area of the Gadilam river basin, Tamilnadu. Journal of Geological Society of India, 63: 625-635.

Chidambaram S, Prasanna M. V, Vasu K, ShahulHameed A, UnnikrishnaWarrier C, Srinivasamoorthy K (2007) Study on the stable isotope signatures in groundwater of Gadilam river basin, Tamilnadu, India. Indian Journal of Geochemistry, 22(2): 209–221.

Domenico P A, Schwartz W (1990) Physical and chemical hydrogeology. John Wiley, New york, 824p.

Gibbs RJ (1970) Mechanisms controlling world's water chemistry, Science, 170: 1088-1090.

Hem JD (1959) Geochemistry of water calculation and use of ion activity. USGS water supply, 1935 C, 17p.

Ian D Clark, Peter Fritz (1997) Environmental Isotopes in Hydrology, Lewis Publishers, New York.

Neelakantan K S, Reddy K.S.S.V.P, Thomson Jacob C, Raghuram K.P (2007) Water Resources in Tamilnadu. ENVIS Newsletter, v.4(1).

Piper AM (1944) A graphic procedure in the geochemical interpretation of water analysis. Trans AmerGeophys union, 25: 914 - 923.

Prasanna M V (2008). Hydrogeochemical studies in the Gadilam river basin, Tamilnadu. Unpublished Ph.D. thesis, Annamalai University, 300p.

Prasanna M V, Chidambaram S, ShahulHameed A, Srinivasamoorthy K (2010) Study of evaluation of groundwater in Gadilam basin using hydrogeochemical and isotope data. Environmental Monitoring and Assessment. 168: 63-90.

Prasanna M V, Chidambaram S, Srinivasamoorthy K, Anandhan P, John Peter A (2006) A study on hydrogeochemistry along the ground water flow path in different litho units in Gadilam river basin, Tamilnadu (India). International journal of Chemical sciences, 2(2): 157-172.

Prasanna MV, Chidambaram S, Srinivasamoorthy K, John Peter A, Anandhan P (2007). Hydrogeochemical Characterization of Groundwater in Gadilam River Basin, through Statistical Analysis.International journal of Environment and social sciences. An International Quarterly Journal of Environment and Social Sciences, 2(1): 21-26

Srinivasamoorthy K (2004) Hydrogeochemistry of groundwater in Salem District, Tamilnadu, India.Unpublished Ph.D Thesis, Annamalai University.

Chapter 11

Understanding the Seawater Intrusion Along the Coastal Aquifers Using Geochemical Techniques

Chidambaram S., Sarathidasan J., Nepolian M.
Noble Jacob[#] and Hemant Mohokar[#]

Department of Earth Sciences, Annamalai University, Annamalai Nagar Tamil Nadu, India

[#] Isotope Application division, BARC, Mumbai, India

ABSTRACT

The development and management of fresh groundwater resources in coastal aquifers are seriously constrained by the impact of seawater intrusion. The coastal aquifers are fragile and this situation becomes more intense like that of the coastal Cuddalore region. The geochemical signature of groundwater of Cuddalore coast was studied for Pre Monsoon, Southwest, North east and Post Monsoon seasons by collecting total of 64 groundwater samples in each season. The collected samples from were analysed for physical parameters, such as electrical conductivity (EC), pH and major ion concentrations, such as Ca, Mg, Na, K, Cl, HCO_3, PO_4 and SO_4. Interpretation of analytical data shows the predominance of Na–Cl type in PRM and Ca-Mg-Cl type is the dominant facies during POM. The Chaddas evaluation diagram shows that the hydrogeochemical evaluation plot for PRM samples fall in Na-Cl type. The ratios of Na/Cl and Cl/HCO_3 is also enumerated the fact that the most of the samples from PRM and few samples from other seasons also indicates dominance of salt water intrusion.

Key words: Coastal groundwater, Geochemical signature, Piper, Chadda, Seawater intrusion.

Introduction

Coastal regions face many hydrological problems like flooding due to cyclones, wave surge and drinking freshwater scarcity due to problem of saltwater intrusion. The natural balance between freshwater and saltwater in coastal aquifers is disturbed by groundwater withdrawals and other human activities that lower groundwater levels reduce fresh groundwater flow to coastal waters and ultimately causing saltwater to intrude coastal aquifers (Barlow, 2003). A coastal aquifer is defined as water bearing geological formation hydraulically connected to the sea (Johannes, 1980). The movement of water in coastal aquifers is generally towards the sea/estuary. This is as a result of the positive hydraulic gradient set up by the balance between recharge inland and discharge towards the sea/estuary. When dealing with the exploitation, restoration and management of fresh groundwater in coastal aquifers, the key issue that arises is that of saltwater intrusion. The extent of saltwater intrusion into a coastal aquifer depends on several factors, including the rate of groundwater that is withdrawn compared to the total freshwater recharge to the aquifer, the distance of the wells and drainage canals from the source or sources of saltwater, the geological structure and distribution of hydraulic properties of the aquifer (Barlow, 2003). Saltwater can intrude into coastal freshwater aquifers through several pathways, including lateral flow from the ocean, an upward intrusion from deeper and more saline zones of the groundwater system as well as by downward intrusion from coastal waters. Saltwater intrusion reduces freshwater storage in coastal aquifers and can result in the abandonment of freshwater supply wells when concentrations of dissolved ions exceed drinking-water standards. Hence it is important to understand the groundwater quality evolution and seawater intrusion mechanisms for groundwater resource management in coastal regions. Hence a study had been carried out to in characterization of the spatial and temporal characteristics of saline water in Cuddalore coastal aquifers by using geochemical techniques and this study elucidates the discrimination of fresh groundwater.

The study area forms a coastal region of Cuddalore district, Tamilnadu, India, lying between north latitudes 11° 44′ 34″ and 11°29′ 59"and east longitudes 79°35′ 06"and 78°46′ 11" (Fig. 1). The area is entirely underlined by sedimentary formations of Tertiary and Quaternary age. The average annual rainfall recorded in the study area for the period 2003–2013 is 138.70 mm. The study area is drained by Ponnaiyar River in the north, Uppanar River in the south. Vellar River controls the drainage of the southern regain along with its tributary Manimuthanathi. Uppanar River starts from the southern part of the Perumal eri. The drainage pattern of this area is subdentritic. The major occupation is agriculture; more than 60% of people are engaged in agricultural. All the industries of SIPCOT are situated near the NE part of coastal region in the study area. The geology of the study is ranging from Cretaceous to recent. The most common outcrops in the present study area are Alluvium, Sandstone,

Clays and Limestone. These are covered in places by Lateritic, Kankar and Alluvium. The general geological succession of the formation encountered in the district is presented in Table 1.

Study Area

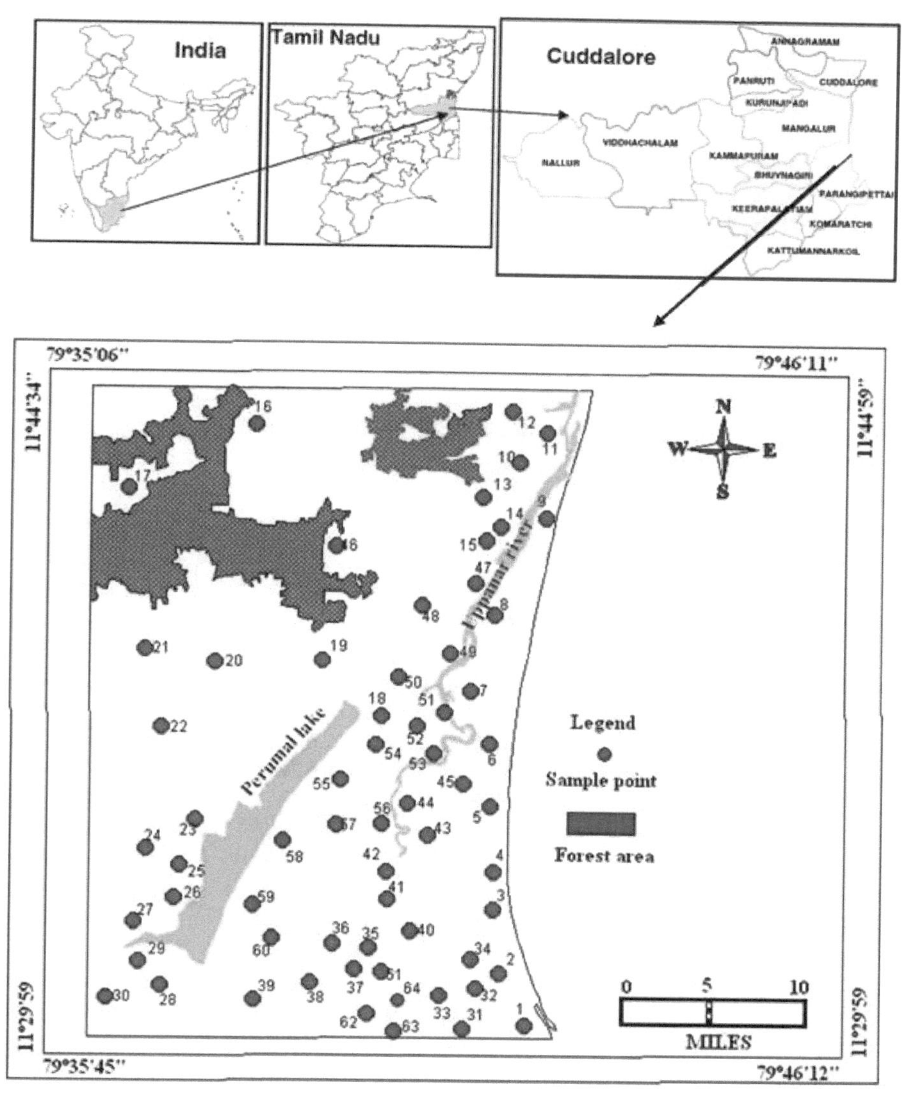

Fig. 1. Location Map of the Study area

Table 1. Geology of the Study area

Era	Age	Formation	Lithology
Quaternary	Recent to Sub Recent	Alluvium and Laterite	Soils Alluvium and Coastal sand, Clays Kankar and Laterite
Unconformity			
Tertiary	Mio Pliocene	Cuddalore Sandstone	Sandstone Argillaceous and Pebble bearing grits, Clays (variegated) With lignite seams and Pebble beds.

Methodology

A total of 64 groundwater samples were collected from the study area during Pre Monsoon(PRM), Southwest(SWM), North east(NEM) and Post Monsoon(POM) seasons. The sample bottles were labeled, sealed and transported to the laboratory and preserved by adopting standard preservation methods. The major anionic and cationic concentrations were determined in the laboratory using the standard analytical procedures (Table 2). The accuracy of complete chemical analysis of a groundwater sample was checked by computing the cation–anion balance.

Cation-Anion Balance = [TCC – TCA)/ (TCC + TCA] X 100

The reaction error (*E*) of all the groundwater samples was less than the accepted limit of ±10%, an added proof of the precision of the data.

Table 2. Instrumental and Volumetric Methods Used for Chemical Analysis of Groundwater in the Study Area

Chemical Parameter	Method and Instrument
pH	pH meter (Systroics)
EC	EC meter (Systronics)
TDS	TDS meter (Hanna)
Ca^{2+}, Mg^{2+}	EDTA titrimetric
Na^{+}, K^{+}	Flame photometer (Elico)
HCO_3, Cl^-	Titrimetric
SO_4^{2-}, H_4SiO_4, PO_{-4}	UV- Vis Double Beam Spectrophotometer (Elico)
NO_{-3}	Spectrophotometer (Systronics)

Results and Discussion

The maximum, minimum, average values in mgl^{-1} for different seasons given in (Table 3).

The dominance of anion shows that; $Cl^- > HCO_3^- > SO_4^{2-} > NO_3^- > PO_4^{3-} > F^-$ and that of cation is $Ca^{2+} > Na^+ > Mg^{2+} > K^+$

Electrical Conductivity (EC)

Electrical conductivity (EC) is a measure of the total salt content of water based on the flow of electrical current through the sample. Higher EC in the water samples reflects leaching or dissolution of the aquifer material or mixing of saline water. The Electrical Conductivity (EC) of the study area it is ranging from 125 to 13975 µs/cm. The Highest EC value noted during PRM when compared with other seasons located along the coastal region. This may due to the saline water intrusion (Chidambaram et al. 2007; Karmegam et al. 2010) or due to anthropogenic activities (Thilagavathi et al. 2012).

Spatial Distribution of EC

The attribute values of the specific locations were used to create a spatial data, in order to obtain an overall behavior of the electrical conductivity in the study area. In PRM, EC ranges from 928 - 13975 µs/cm (Fig 2). Higher concentration was observed in central and SW part of the study area. In SWM (Fig 2), EC ranges from 125 - 2630 µs/cm. Higher concentration was observed in Northern part of the study area. In PRM there will be absence of rainfall and in that period the groundwater is mainly influenced greater groundwater extraction resulting in saline water intrusion into the aquifers. The spatial distribution in the study reveals that in NEM (Fig 2), EC ranges from 820 - 7792.50 µs/cm. Higher concentration of EC observed in the central and SW part of the study area. In POM, EC ranges from 588.33 to 5731.67 µs/cm (Fig 2). In general lower concentration of EC in northwestern and higher concentration of EC in central part of the study area, indicating.

Table 3. Maximum, Minimum and Average of the Chemical constituents in groundwater representing all four sampling seasons (All Values in mgl^{-1} except EC inµscm^{-1}, Temperature in ^{0}C and pH) saline water intrusion from the nearby coastal area. It is also interesting to note that the greater percentages of salinity indicate the dissolution and leaching of secondary salts during monsoon and sea water ingression during PRM (Chidambaram et al. 2010).

	Temp	pH	EC	TDS	Ca^{2+}	Mg^{2+}	Na^{+}	K^{+}	Cl^{-}	HCO^{-}_{3}	PO^{-}_{4}	SO^{-2}_{4}	H_4SiO_4	F^{-}	NO^{-}_{3}
PRM (Premonsoon)															
Average	30.00	7.4	4487.36	2833.24	71.32	30.88	75.09	17.58	245.91	137.58	0.36	34.04	104.74	0.35	23.63
Maximum	32.20	9.80	13975.00	8949.00	220.00	121.20	178.00	105.00	575.00	439.20	2.30	68.40	232.00	0.83	86.40
Minimum	27.60	6.20	928.00	587.00	20.00	8.00	7.00	1.00	70.89	35.00	0.00	2.50	20.00	0.03	1.30
SWM (South west monsoon)															
Average	32.42	7.09	813.00	568.76	57.04	39.18	46.45	10.20	141.63	144.52	6.10	5.41	16.47	0.90	7.69
Maximum	35.90	7.80	2630.00	1841.00	145.56	98.58	142.43	25.25	312.25	356.79	28.02	23.10	67.26	1.32	35.10
Minimum	30.10	6.40	125.00	87.50	2.25	1.69	11.20	1.28	65.25	41.98	0.08	0.34	1.46	0.56	1.50
NEM (Nor th east monsoon)															
Average	31.21	7.18	2650.18	1701.00	64.18	35.03	60.77	13.89	193.77	141.05	3.23	19.72	60.61	0.62	15.66
Maximum	34.05	8.40	7792.50	5038.00	135.36	83.79	131.63	60.88	389.71	351.70	14.01	37.47	130.82	1.05	45.80
Minimum	29.45	6.2	820.00	492.75	18.37	9.85	12.85	1.69	71.73	44.66	0.04	5.74	15.86	0.37	3.25
POM (Post monsoon)															
Average	30.36	7.19	2039.12	1325.24	63.02	36.85	57.70	18.65	177.24	147.43	4.73	15.24	45.63	0.67	16.36
Maximum	32.23	8.07	5731.67	3723.60	131.02	90.65	132.09	48.13	323.71	316.23	15.84	28.43	106.87	1.13	50.38
Minimum	29.23	6.23	588.33	370.73	19.37	9.05	12.23	3.88	67.77	49.81	0.06	3.93	11.97	0.35	3.47

Total Dissolved Solids (TDS)

TDS in the water samples of the study area reveal that it varies between 87.50 to 8949.00 ppm. TDS was noted to be higher in PRM followed by NEM, POM and SWM reflecting the nature of EC. Subsequently, four classes of water were proposed based on the procedures adopted from Carroll (1962) (Table 4).

Table 4. Water Quality Classification Based on TDS Content (Carroll, 1962)

TDS in ppm	Water quality	No. of Samples			
		PRM	SWM	NEM	POM
<1000	Fresh water	6	47	16	18
1000- 10,000	Brackish water	44	3	34	32
10,000- 100,000	Salty water	-	-	-	-
>100,000	Brine	-	-	-	-

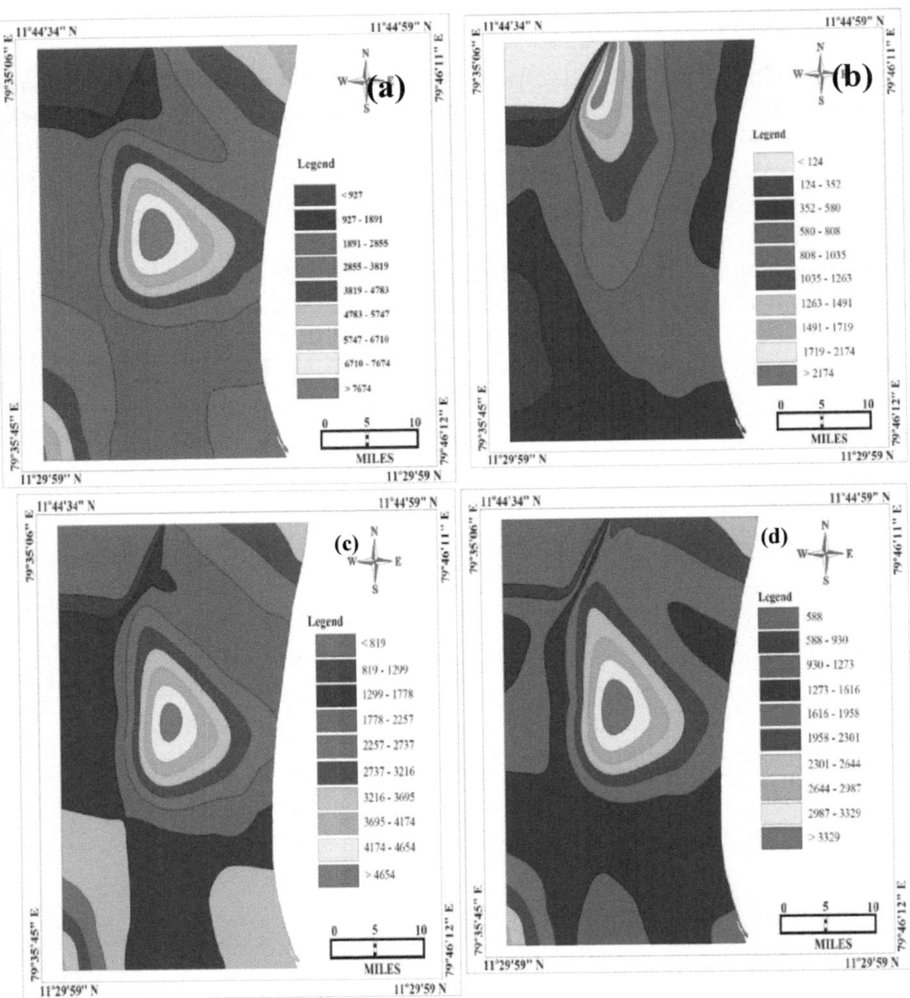

Fig. 2. Spatial distribution of EC (µs/cm) for a) PRM and b) SWM c) NEM and d) POM

The Table 4 shows that nearly 44 samples of the PRM samples are brackish and further decreased percentage of samples during SWM which may be due to dilution process. Further there is an increase of TDS in NEM and subsequently it decreases in POM due to the leaching and dilution process, this influences the monsoonal changes in the study area.

Salinity Hazard

Based on Electrical Conductivity, waters are classified into four major classes (Richards, 1954). The boundary values between the individual classes are 250, 750 and 2250 ìscm^{-1}. These limit values were determined on the basis of the relationship between the electrical conductivity of waters and the electrical conductivity of saturated soil extracts.

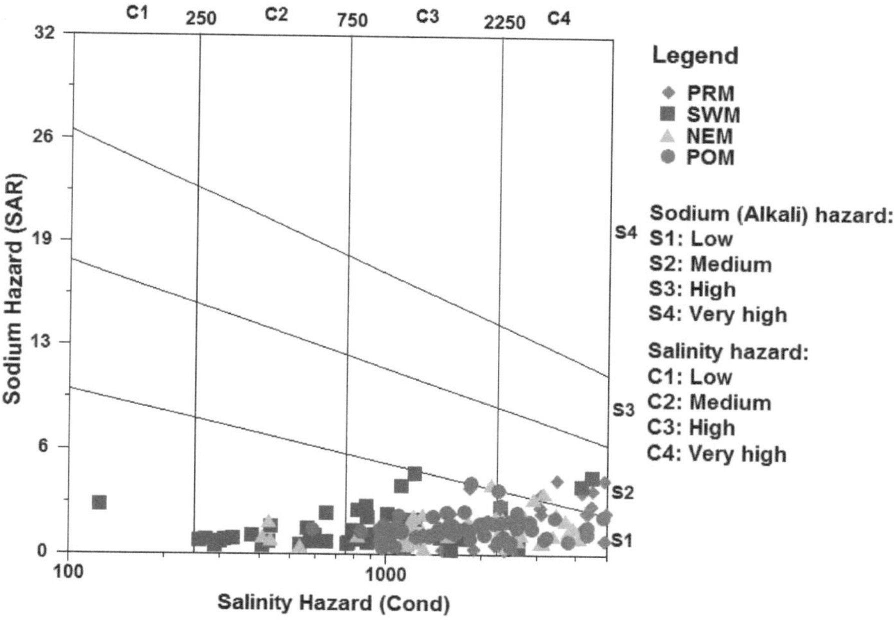

Fig. 3. Wilcox diagram representing irrigation quality of groundwater samples for the four seasons.

- Low-Salinity Water (EC < 250 μscm^{-1}) can be used for irrigation
- Medium-Salinity Water (250 < EC < 750 μscm^{-1}) can be used if the soil is washed with a moderate amount of water.
- High-Salinity Water (750 < EC < 2250 μscm^{-1}) cannot be used on soils with restricted drainage.
- Very High-Salinity Water (EC > 2250 μscm^{-1}) is not suitable for irrigation under ordinary conditions, but it can be used under very special circumstances.

The Wilcox classification of ground water for irrigation purpose (Fig. 3) applied by United State Salinity Laboratory for obtaining the classification of irrigation water. This diagram shows that during the

PRM and NEM the samples fall in C3 category (High Salinity Hazard) and the samples of POM, SWM range from C2-C4 (medium to very high salinity hazard) of which the PRM and POM shows good representation of very high salinity hazards. Few samples of PRM and POM fall beyond the plot preview representing higher values of conductivity.

Hydrogeochemical Facies

Geochemical evolution can be understood from the Piper plot (Fig. 4), which has been divided into six sub-fields, viz. 1 (Ca–HCO_3 type); 2 (Na–Cl type); 3 (mixed Ca–Na–HCO_3 type); 4 (mixed Ca–Mg–Cl type); 5 (Ca–Cl type) and 6 (Na–HCO_3 type). In study area, samples are clustered in the fields of 1, 2, 3 and 4, & 5 the majority of the samples are concentrated in the Mixed Ca - Mg - Cl type (Fig. 4).Ca and Mg are major cations and Cl and HCO_3 is the major anion in this groundwater during PRM. Majority of the samples fall on Na-Cl types during PRM it is clearly indicating the dominance of seawater intrusion (Prasanna et al., 2008). It is also observed from the Piper plot that certain groundwater samples also show that alkaline earths (Ca^{2+} and Mg^{2+}) exceed the alkalis (Na^{+}& K^{+}). Few samples represented the field 2 and 3.In general, the total hydrochemistry of the study area is dominated by alkali and strong acids. SWM samples falls in recharging zone of zone 1(Udayalaxmi et al., 2010), most of the POM samples falls in zone 4, whereas NEM samples falls in zone 4 and 5.

Most of the samples from PRM fall in field 2, with high sodium and chloride are due to the process of removal of other ions from the system either by adsorption or by precipitation (Chidambaram et al., 2007) due to the saturation, they can also be understood as the discharge region (Thilagavathi et al., 2012) and migration of the samples from transition zone of mixed Ca- Mg-Cl and Na-Cl facies towards the sea water composition, these samples may also indicate the long residence time of shallow groundwater (Prasanna et al., 2010) where Na exceeds Ca and Mg, Cl exceeds HCO_3 and SO_4 (Prasanna et al., 2008). Na+Cl water type in discharge zone shows the saline nature (Prasanna et al., 2009).

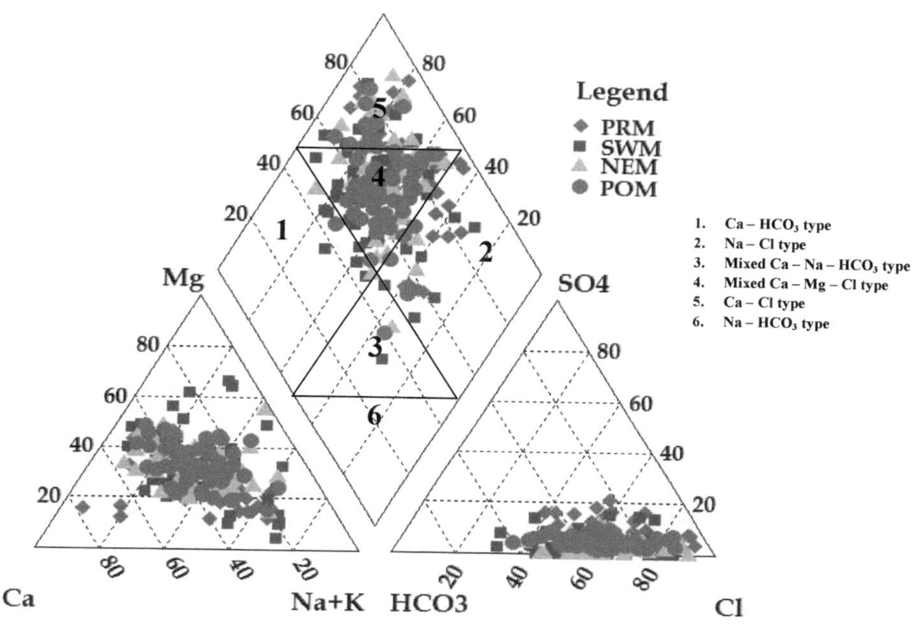

Fig. 4. Piper plot exhibiting the chemical facies of groundwater samples for different seasons

Chadda's Plot

A hydrochemical diagram proposed by Chadha (1999) has been applied in the study to understand the representation of the factors in the four fields responsible for groundwater geochemistry. Data were converted to percentage reaction values (milliequivalent percentages) and expressed as the difference between alkaline earths (Ca + Mg) and alkali metals (Na + K) for cations, and the difference between weak acidic anions (HCO_3 + CO_3) and strong acidic anions (Cl + SO_4). The hydrochemical processes suggested by Chadha (1999) are indicated in each of the four quadrants of the graph. These are broadly summarized as: The four fields of the Chadda plot explains

Field 5—Ca^{2+}-HCO''_3 type recharging waters

Field 6—Ca^{2+}-Mg^{2+}-Cl'' type reverses ion exchange waters

Field 7—Na^+-Cl'' type end-member waters (seawater)

Field 8—Na^+-HCO''_3 type base ion exchange waters

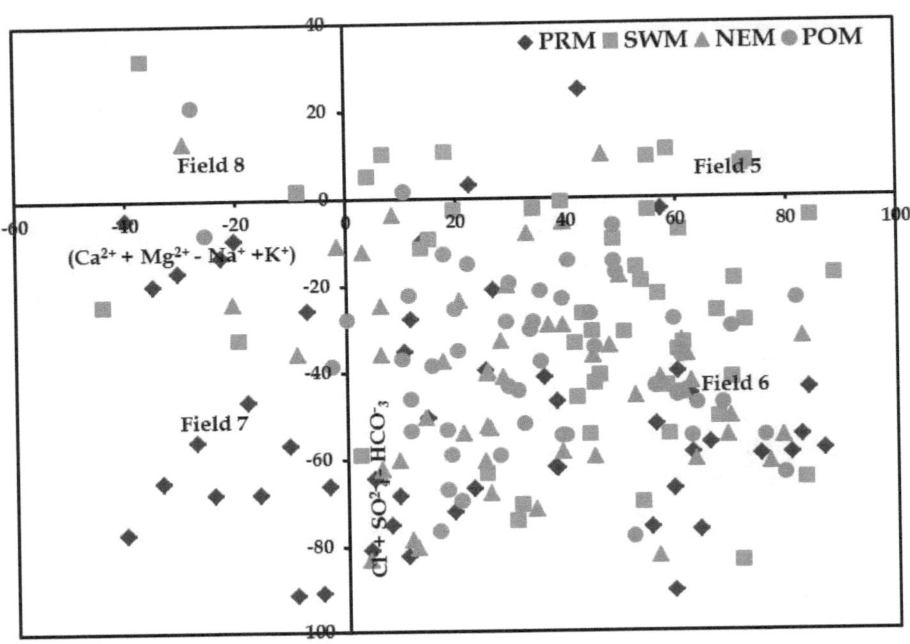

Fig. 5. Chadha's geochemical process evolution plot

The plot show that majority of samples of study area fall in Field -6 (Reverse ion-exchange type) during irrespective of seasons with few representation in Field -5 and 8 (Fig. 5), indicating Recharging water and Base Exchange respectively. Most of the PRM samples fall on Field 7 (Na^+- Cl^-) type along the coastal region. It is interesting to note that greater impact of sea water intrusion is noted in PRM sample compared to other season.

Cl^-/HCO^-_3 Ratio

Effect of salinization could be classified using the Cl^-/HCO^-_3 ratios (Todd, 1959). The relationship between Cl^-/HCO^-_3 ratios and Cl^- also shows mixing of fresh groundwaters and seawater, and the samples with lower ratios can be characterized as fresh waters. Simpson classification includes five classes: <0.5 for good quality water, 0.5 to 1.3 for slightly contaminated water, 1.3 to 2.8 moderately for contaminated water, 2.8 to 6.6 injuriously contaminated and >15.5 for highly contaminated water. Considering the threshold value of Cl^- concentration (63 mg/l) and the ratio of Cl^-/HCO^-_3, it was found that about 30% of the groundwater was strongly affected by the saline water, 67% were slightly or moderately affected and 3% of samples were not affected by sea water. The most of the ground waters show the strong contamination with sea water in PRM.

The ratios of Cl^-/HCO^-_3 ranged between 0.48-20.65, 0.48-11.35, 0.68-10.07 and 0.58-7.85 during PRM, SWM, NEM and POM season respectively (Fig. 6). This linear relationship indicates simple mixing of fresh groundwaters with saline waters. It's also interesting to note that there is slight deviation in the linear relationship at the lower range of the mole ratios this is mainly due to the variation in the HCO^-_3 values. The decrease of this HCO^-_3 ion represents an the increase of the Cl^-/HCO^-_3 ratio even at the lower values of the Cl^- represented in the 'X' axis. This decrease of HCO^-_3 value or increase of Cl^- relatives to HCO^-_3 ions, is mainly may be due to land use patter (industrial effluents) along the coast.

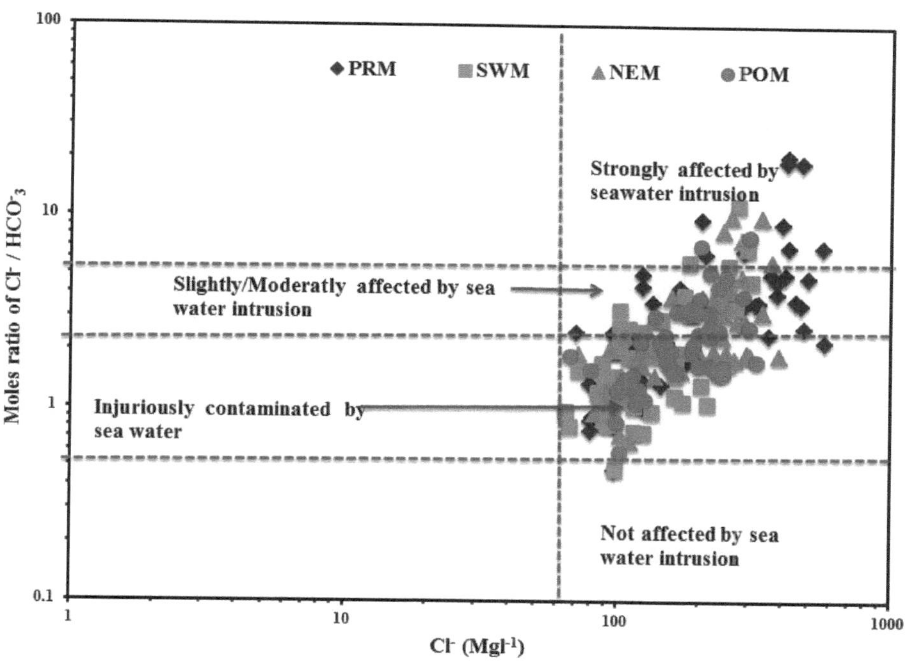

Fig. 6. Plot between Cl and mole ratio of Cl/ HCO_3

Majority of SWM and POM season samples fall on slightly/moderately affected by sea water intrusion region compared to other season. Few of the samples fall on injuriously contaminated by sea water during irrespective of the seasons. Few PRM samples fall on strongly affected by seawater intrusion region along coastal compared to other season.

Na^+/Cl^- Mole Ratio

Effects of seawater encroachment have been evaluated by studying a series of ionic ratios (Petalas and Diamantis 1999; Sanchez Martos et al. 1999; Vengosh et al. 2002; El Moujabber et al. 2006). Figure 7 shows molar ratios of Na^+/Cl^- versus Cl^- concentrations. Conservative seawater–fresh water mixing is expected to show a linear increase in Na^+ and Cl^- (Sanchez Martos et al. 1999). Ratio values of groundwater falling near the seawater ratio (0.86) indicate that fresh groundwaters were contaminated with the saline waters. The Na^+/Cl^- ratios ranged from 0.07 to 1.80, 0.08 to 2.21, 0.11 to 1.72 and 0.11 to 1.97 during PRM, SWM, NEM and POM season respectively most of the samples fall near the contamination region. The ratio of Cl^- to Na^+ and Mg^{2+} indicate a predominant seawater source for these ions.

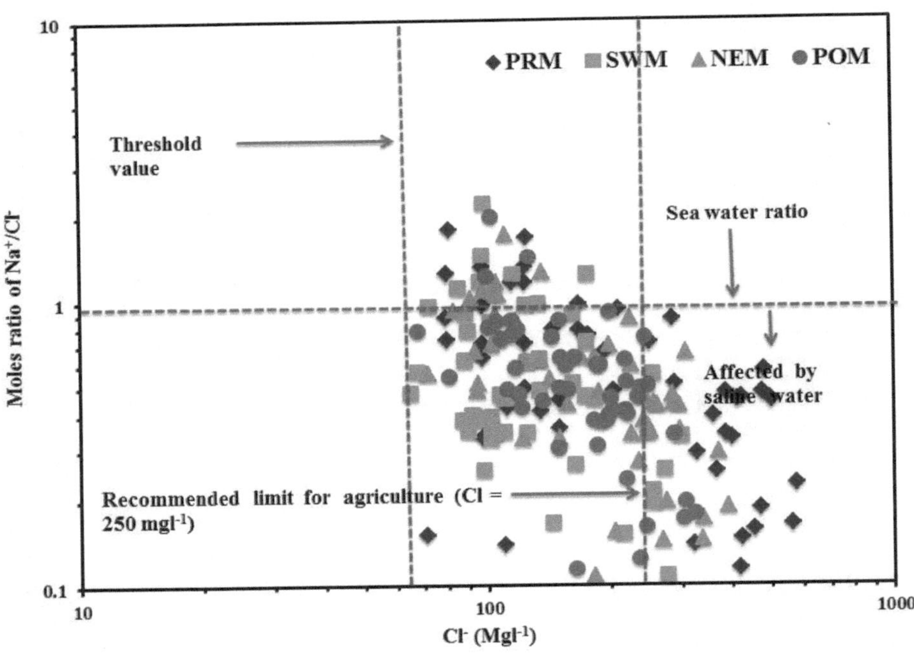

Fig. 7. Plot between Cl and moles ratio of Na/Cl

The Na^+/Cl^- ratio show that 74%, were near the seawater ratio, while the other 26% exceeded the seawater value (Fig. 7), indicating a large proportion of groundwater was affected by seawater intrusion to certain degrees during PRM. 24% of the samples exceed the seawater values during SWM and NEM season respectively. 14% of the samples exceed the seawater value during POM. It is also noted that most of the PRM sample affected by seawater intrusion compared to other season. Very high Na^+/Cl^- ratios may be indicative of anthropogenic contamination, like industries and fertilizers (Jones et al. 1999).

Vengosh Explanation

Seawater has distinct ionic ratios such as Na^+/Cl^- = 0.86, SO^{2-}_4/Cl^- = 0.05; $Ca^{2+}/(HCO^-_3 + SO^{2-}_4)$ and Mg^{2+}/Ca^{2+} =5.2 (Vengosh et al. 2002; Vengosh et al. 1994). The average values of these ratios in the samples of the study area are given in (Table. 5). The Na^+/Cl^- ratio is greater than unity in about 24% of the samples of the study area during PRM. 22% of the samples fall on exceeds the seawater ionic ratio during SWM and NEM season respectively. 8% of the samples exceed the seawater ratio during POM. The Na^+/Cl^- molar ratio could reach unity due to the mixing of seawater and freshwater (Vengosh and Rosenthal, 1994). Relative to seawater, the groundwater has relatively low Na^+/Cl^- ratios, high Mg^{2+}/Ca^{2+} ratios and high $Ca^{2+}/(HCO^-_3 + SO^{2-}_4)$ ratios (greater than unity) (Vengosh et al.,2002, Mondal et al.,2010).

Each probable source of salinization is characterized by a distinguishable chemistry and well known ionic ratios. For example, if seawater intrusion is expected to be the only source of salinization then the Mg^{2+}/Ca^{2+} ionic ratio will be greater than 5. The Mg^{2+}/Ca^{2+} ionic ratio greater than 5 is a direct indicator of seawater contamination (Metcafe and Eddy 2000), as noted in within seawater ionic ratio during PRM, NEM and POM respectively. 10% of the samples exceed the seawater ionic ratio during SWM season. Likewise, seawater and seawater diluted with freshwater have distinguished geochemical characteristics (Metcafe and Eddy 2000). The value of these ratios in the study area is greater than unity and it may be due to deep saline upconing or ion exchange process. Modification of the geochemical characteristics of these saline waters is caused by water rock interaction with three possible mechanisms which may be involved: (1) base-exchange reactions with clay minerals (Vengosh et al. 1994); (2) adsorption onto clay minerals; and (3) carbonate dissolution-precipitation (Vengosh et al. 1994; Ghabayen et al. 2006).

The chemical signature of water polluted with urban wastewater or from the reuse of treated wastewater is very distinguishable. Sewage effluent has a relatively high Na^+/Cl^- ratio (greater than unity) and comparatively with high SO^{2-}_4/Cl^- ratio (0.09) (Ghabayen et al. 2006). These ratios are attributed to applications of Na-Cl salts and domestic waste water characteristics. This range is also reflected within range of the samples of the study area. Agriculture return flows have a distinctive chemical composition relative to other salinization sources. This water is characterized by high SO^{2-}_4/Cl^- ratios (much greater than 0.05) (Vengosh et al. 2002). A high SO_4^{2-}/Cl- ratio is attributed to the application of gypsum fertilizers (Vengosh et al. 2002). Lesser values of K^+/Cl^- and SO^{2-}_4/Cl^- indicate aquifers with relatively lesser anthropogenic impact. K^+ seems to be also contributed from anthropogenic sources, this finding get strengthen by the comparison of K^+/Cl^- ratio that found quite higher and equivalent to waste water infiltration. Weathering of K-feldspar is much slower than weathering of plagioclase, and less K^+ than Ca is released into solution (White et al., 2001). The predominantly lower than seawater K^+:Cl^- ratios indicate that K^+ has been removed from inflow waters. Depletion of K^+ in pan inflow waters is relatively common and has been attributed to uptake by plants, and adsorption onto exchange sites on soil particles (Drever, 1997), as the large fraction of exchangeable K in the soils.

$Ca^{2+}/(HCO^-_3 + SO^{2-}_4)$ ratio depicts Sea water intrusion process in most of the samples and but still the ratios may also result from several complex reactions in the aquifer (Kumar et al. 2006).When an aquifer has an underlying layer of saline water and is pumped by a well penetrating only the upper freshwater portion of the aquifer, a local rise of the interface below the well occurs. This phenomenon is known as upcoming (Singaraja et al, 2013). The interface is near horizontal at the start of pumping. With continued pumping, the interface rises to progressively higher levels until eventually it reaches the well. This generally necessitates the well to be shut down because of the degrading influence of the saline water. When pumping is stopped, the denser saline water tends to settle downward and to return to its former position. Observations of all these ratios (Table.5) indicate that there is also a possibility of saline upcoming conditions in the study area.

Table 5. The comparative ionic ratio of potential salinization sources and study area groundwater parameter for seawater intrusion for all seasons *a) Vengosh and Rosenthal (1994) b) Vengosh and Ben-Zvi (1994) c) Vengosh et al. (1994) d) Mercado (1985) e) Vengosh et al. (2002)*

Ratio	Seawater intrusion	Deep saline up coning	Agricultu re return flows	Wastewater infiltration	Study area			
					PRM	SWM	NEM	POM
Na^{+}/Cl^{-}	0.86–1[a]	<0.8[b]	–	1.1[c]	0.07-1.80	0.08-2.21	0.11-1.72	0.11-1.97
SO_4^{2-}/Cl^{-}	0.05a,c	*0.05c,d	>>0.05e	0.09d	0.01-0.41	0.0-0.09	0.02-0.20	0.02-0.17
K^{+}/Cl^{-}	0.019	<0.019	–	>>0.02	0.0-0.72	0.01-0.91	0.01-0.45	0.02-0.36
Mg^{2+}/Ca^{2+}	> 5b	>1a	–	–	0.18-1.58	0.19-31.79	0.44-2.88	0.47-3.52
$Ca^{2+}/(HCO_3^{-}+SO_4^{2-})$	0.35–<1	<1a	–	–	0.30-3.08	0.06-4.28	0.27-2.59	0.20-2.75

Conclusion

The dominance of anion shows that; $Cl^- > HCO_3^- > SO_4^{2-} > NO_3^- > PO_4^{3-} > F^-$ and that of cation is $Ca^{2+} > Na^+ > Mg^{2+} > K^+$.Lower concentration of EC in North Western and higher concentration of EC in central part of the study area, indicating saline water intrusion from the nearby coastal area. It is also interesting to note that the greater percentages of salinity indicate the dissolution and leaching of secondary salts during monsoon and sea water ingression during PRM.

The cluster of samples irrespective of season fall in C3S1 and C4S1 category this may be due to dissolution of dissolved salts, dispersivity of ions, agricultural practices etc. In POM season samples representing C3S1 category. The majority of the samples in all the seasons are represented in Class II and Class III and then few samples are represented in Class IV. In PRM and SWM most of the samples observed in Class III which is fairly suitable for irrigation and in POM and NEM samples are observed in Class II suitable for irrigation purpose. Ca and Mg are major cations and Cl and HCO_3 is the major anion in this groundwater during PRM. Majority of the samples fall on Na-Cl types during PRM it is clearly indicating the dominance of seawater intrusion. It is also observed from the Piper plot that certain groundwater samples also show that alkaline earths (Ca^{2+} and Mg^{2+}) exceed the alkalis (Na^+& K^+). Few samples represented are represented in the Na–Cl type and mixed Ca–Na–HCO_3 type. In general, the total hydrochemistry of the study area is dominated by alkali and strong acids. SWM samples falls in recharging zone of zone (Ca–HCO_3 type), most of the POM samples falls in zone (mixed Ca–Mg–Cl type), whereas NEM samples falls in zone (mixed Ca–Mg–Cl type) and Ca–Cl type. The Chaddha plot show that majority of samples of study area fall in Field -6 (Reverse ion-exchange type) during irrespective of seasons with few representation in Field -5 and 8, indicating Recharging water and Base Exchange respectively. Most of the PRM samples fall on Field 7 (Na^+- Cl^-) type along the coastal region. It is interesting to note that greater impact of sea water intrusion is noted in PRM sample compared to other season. The ratios of Cl^-/HCO_3^- ranged between 0.48 - 20.65, 0.48 – 11.35, 0.68 – 10.07 and 0.58 -7.85 during PRM, SWM, NEM and POM season respectively. Majority of SWM and POM season samples fall on slightly/moderately affected by sea water intrusion region compared to other season. Few of the samples fall on injuriously contaminated by sea water during irrespective of the seasons of which PRM is predominant along coast.

Acknowledgement

The authors express their sincere thanks to Department of Science and Technology, India, for providing the necessary nancial support to this research project wide ref .No DST/WAR-W/WSI/06/2010, dated 19.05.2011.

References

Barlow P M (2003) Ground water in fresh water-salt water environments of the Atlantic Coast.U.S. Geological Survey Circular 1262.

Carroll D (1962) Rainwater as a chemical agent of geologic processes: a review, USGS Water Supply Paper. pp.1535.

Chadha DK (1999) A proposed new diagram for geochemical classification of natural waters and interpretation of chemical data. Hydrogeology Journal,v. 7, pp.431–439.

Chidambaram S, Ramanathan AL, Prasanna MV, Karmegam U, Dheivanayagi V, Ramesh, R., Johnsonbabu, G., Premchander, B. and Manikandan, S. (2010) Study on the hydrogeochemical characteristics in groundwater, post- and pre-tsunami scenario, from Portnova to Pumpuhar, southeast coast of India Environ Monit Assess v.169, pp. 553–568.

Chidambaram S, Ramanathan AL, Prasanna MV, Anandhan P, Srinivasamoorthy K, Vasudevan S (2007) Identification of hydrogeochemically active regimes in ground waters of Erode district, Tamilnadu-A statistical approach. Asian Journal of Water, Environment and Pollution v.5 (3), pp. 93–102.

Drever IJ (1997) The Geochemistry of Natural Waters, 2nd Edition; Englewood Cliffs NJ (Eds) Prentice Hall, pp.388.

El Moujabber M, BouSamra B, Darwish T, Atallah T (2006) Comparison of different indicators for groundwater contamination by seawater intrusion on the Lebanese coast. Water Resour Manage v.20, pp.161–180.

Ghabayen S, McKee M, Kemblowski M (2006) Ionic and isotopic ratios for identification of salinity sources and missing data in the Gaza aquifer. Journal of Hydrology v. 318,pp.360–373.

Johannes RE (1980) The ecological significance of the sub marine discharge of groundwater.Mar Ecol, ProgSer; 3:365–73.

Jones BF, Vengosh A, Rosenthal E, Yechieli Y (1999) Geochemical investigations. In: Seawater intrusion in coastal aquifers: concepts, methods, and practices. Kluwer, Dordrecht, the Netherlands, pp 51–71.

Karmegam U, Chidambaram S, Prasanna MV, Sasidhar P, Manikandan S, Johnsonbabu G, Dheivanayaki V, Paramaguru P, Manivannan R,

Srinivasamoorthy K, Anandhan P (2010) A study on the mixing proportion in groundwater samples by using Piper diagram and Phreeqc model, Chin.J.Geochem v. 30, pp. 490–495.

Kumar M, Ramanathan AL, Rao MS, Kumar B (2006) Identiûcation and evaluation of hydrogeochemical process in the groundwater environment of Delhi, India. Environ Geol v. 50, pp. 1025–1039

Metcafe Eddy (2000) Integrated aquifer management plan: nal report. Gaza Coastal Aquifer Management Program, USAID Contract No. 294-C-00-99-00038-00

Mercado A (1985) The use of hydrogeochemical patterns in carbonate sand and sandstone aquifers to identify intrusion and flushing of saline waters. Groundwater v. 23 (5),pp.635–645.

Mondal NC, Singh VP, Singh VS, Saxena VK (2010) Determining the interaction between groundwater and saline water through groundwater major ions chemistry. Journal of Hydrology v. 388, pp. 100–111.

Petalas CP, Diamantis IB (1999) Origin and distribution of saline groundwaters in the upper Miocene aquifer system, coastal Rhodope area, northeastern Greece. Hydrogeol J 7:pp.305–316.

Prasanna MV, Chidambaram S, Pethaperumal S, Srinivasamoorthy K, John Peter A, Anandhan P, Vasanthavigar M (2008) Integrated geophysical and chemical study in the lower sub basin of Gadilam River, Tamilnadu, India. Environmental Geosciences, v. 15(4), pp. 145–152.

Prasanna MV, Chidambaram S, ShahulHameed A, Srinivasamoorthy K (2009) Study of evaluation of groundwater in Gadilam Basin using hydrogeochemical and isotope data. Environmental monitoring and assessment, DOI 10.1007/s 10661-009-1092-5.

Prasanna MV, Chidambaram S, ShahulHameed A, Srinivasamoorthy K (2010) Study of evaluation of groundwater in Gadilam basin using hydrogeochemical and isotope data. Environ Monit Assess v.168, pp. 63–90.

Richard LA (1954) Diagnosis and improvement of saline and alkali soils. USDA Handbook, v.60, pp.160.

Sanchez Martos F, Pulido Bosch A, Calaforra JM (1999) Hydrogeochemical processes in an arid region of Europe (Almeria, SE Spain). ApplGeochem v.14,pp. 735–745.

Singaraja C, Chidambaram S, Anandhan P, Prasanna MV, Thivya C, Thilagavathi R, Sarathidasan J(2013) Hydrochemistry of groundwater in a coastal region and its repercussion on quality, a case study- Thoothukudi district, tamilnadu, India. Arab J Geosci. DOI 10.1007/s12517-012-0794-0

Thilagavathi R, Chidambaram S, Prasanna MV, Thivya C, Singaraja C (2012) A study on groundwater geochemistry and water quality in layered aquifers system of Pondicherry region, southeast India. Appl Water Sci. doi:10.1007/s13201-012-0045-2.

Todd D K (1959) Ground water Hydrology, New York: John Wiley & sons.

Udayalaxmi G, Himabindu D, G Ramadass (2010) Geochemical evaluation of groundwater quality in selected areas of Hyderabad, A.P., India Indian Journal of Science and Technology Vol. 3, pp. 5.

Vengosh A, Gill J, Davisson ML, Hudson GB (2002) A multiisotope (B, Sr, O, H, and C) and age dating study of groundwater from Salinas Valley, California: hydrochemistry, dynamics, and contamination process. Water Resour Res v. 38(1): 1–17.

Vengosh A, Heumann KG, Juraski S, Kasher R (1994) Boron isotope application for tracing sources of contamination in groundwater. Environ Sci Technol v.28 (11): 1968–1974.

Vengosh A, Rosenthal E (1994) Saline groundwater in Israel: its bearing on the water crisis in the country. J Hydrol v.156,pp. 389–430.

White DE, Hem JD, Warming GA (2001) Chemical composition of subsurface waters. US Geol. Surv. Prof. Pap. 440-F: 67. United states Geological survey, Washington, D.C.

Chapter **12**

Use of Geochemical Methods to Evaluate the Hydraulic Connection of Aquifers: A Case Study from Southeast Coast of Tamil Nadu, India

Singaraja C.*, Chidambaram S.[1], Anandhan P.[1]
Thivya C.[1] and Thilagavthi R.[1]

* Department of Geology, Presidency College, Chennai, Tamil Nadu
[1]Department of Earth Sciences Annamalai University, Annamalai Nagar Tamil Naud

ABSTRACT

Aquifers are usually composed of permeable layers that can ways a considerable amount of groundwater. Usually, aquifer units are correlated through the concept of lithostratigraphy. For low-permeable aquifers, it is difficult to define the spatial distribution of hydrogeological units, and this study attempts to use geochemical methods to identify the groundwater flow paths in an area of Southeast coast of Tamil Nadu, India. Multiple geochemical analyses, including groundwater chemistry; stable isotopic compositions of $\delta^{18}O$ and δD contents were performed. Spatial distribution of $\delta^{18}O$, δD, Cl^-/HCO_3^- show that northeastern part of the study area seriously affected by seawater intrusion. The highly permeable aquifers at the east coastal region are characterized by high hydraulic conductivity as compared with other regions of the study area. Hydrogeochemical characteristics based on bivariate diagrams of major and minor ions show that changes in the chemical composition of groundwater are mainly controlled by the salinization process followed by cation-exchange reactions.

Key Words: isotopic; seawater intrusion; hydraulic conductivity; India.

Introduction

Coastal aquifers serve as most significant sources for freshwater supply in many countries around the world, particularly in waterless zones. The fact that coastal zones contain some of the heavily populated areas in the world makes the need for freshwater even more discriminating. The intensive extraction of groundwater from coastal aquifers reduces freshwater outflow to the sea and creates local water aquifer depression, causing seawater migration inland and rising toward the land. As seawater intrusion progresses, existing pumping wells become saline and have to be neglected (Prasanna et al., 2010; Warner et al., 2013; Singaraja et al., 2015).

Using multiple sources of chemical and isotope data combined with hydrogeological information may significantly improve the current understanding of seawater intrusion processes. Studies on the problem of salinization in groundwater of inland areas using stable isotopes are reported in a more of literatures (Datta et al., 1994; Caron et al., 2008; Prasanna et al., 2010; Erik et al., 2014; Rautio and Korkka-Niemi, 2015). In general, isotopic compositions are very different for seawater, fresh water and their mixtures. Therefore, the distribution, origin and age of different water can be determined by analysing their isotopic compositions and studying their variation. It's useful to study the mixing and origin of saline water by analysing stable isotopes such as $\delta^{18}O$ and δD based on geochemical support (Luoma et al., 2015; Rautio et al., 2015).

Many researchers had worked on the groundwater of the present study area on various concepts like; the land use and land cover pattern along the metal pollution in groundwater to highlight affect of the industrial (SIPCOT) effluents (Puthiyasekar et al., 2010), hydrological influences on the water quality (Ravichandran, 2003; Solai et al., 2012) and hydrochemial characteristic of coastal aquifer (Mondal et al., 2011; Selvam et al., 2015). Singaraja et al., (2013a) carried out the northeastern part of study area affected by seawater intrusion. Hydrogeological and geophysical investigations to delineate aquifer zone was attempted by Singh et al., (2006).

There has been without published work on the chemical and isotope data combined with hydrogeological information of the study area. Therefore, the study on behavior of aquifer in the study area has attained most important importance. Hence, it has been proposed to characterize the chemical and isotope data processes activated in the study area. The goal of this study was to identify the mechanisms of groundwater salinization in southeast coastal aquifers of India using hydrogeochemical and isotopic data.

Study area

Tuticorin is situated in the southeast coast of Tamil Nadu, India. The study area is located between 8° 19′ to 9° 22′ N latitude and 77° 40′

to 78° 23′ E longitude (Fig. 1) covering an area of about 4621 Km². The area experiences a hot tropical climate. The average annual temperature from 23°C to 29°C and the annual rainfall is about 570 mm to 740 mm. The district receives the rain under the influence of both northeast and southeast monsoons. The drainage pattern is of mostly dendritic.

Geologically major units exist in this area covered by HBG is the dominant formations and, however, alluvial deposits occur on both sides of the river, which are composed of clay, silt, sand and gravel. Alluvial Marine and Fluvial Marine are dominant in the Eastern part of the study area. The charnockite patches are noted in the study area, in the central part of the study area and along the western margin. The study area is also represented by intrusions of granite, quartzite and patches of sandstone. Major water bearing formations are Quaternary Alluvium, Tertiary formations, weathered fractured Pink Granites, Charnockites and Gneisses. Limited Fresh water availability is noted in sedimentary areas, the floating lenses of fresh water make the coastal tract vulnerable for water quality changes. Groundwater from alluvial /tertiary aquifer present in eastern part of the district is in hydraulic connection with the sea and hence it is vulnerable for saline water ingression (CGWB 2009).

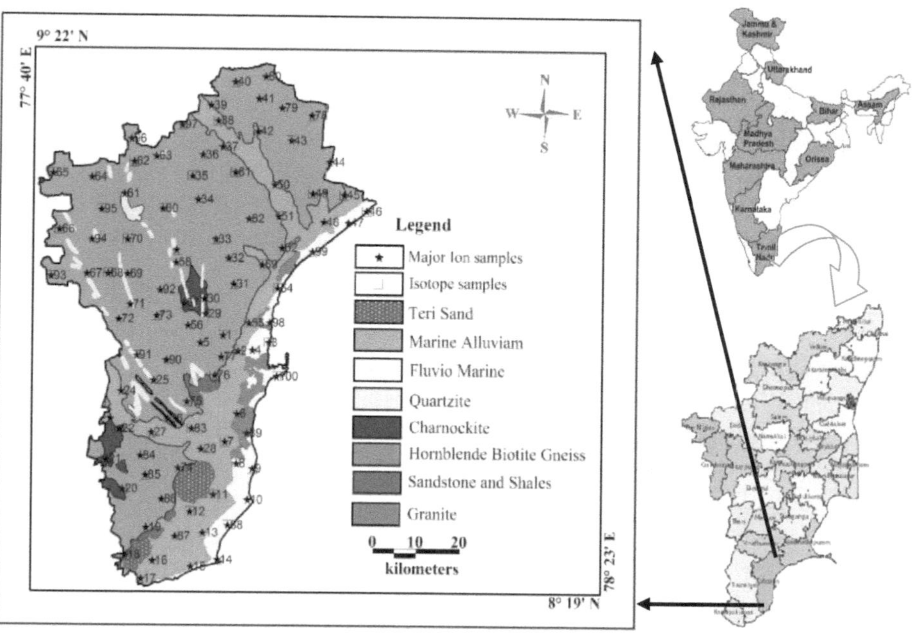

Fig. 1. Geology and Sample location map of the study area

Methodology

Field investigation was carried out during January to March 2016. A total of 100 groundwater samples were obtained from study area (Fig. 1). Electrical conductivity (EC), pH and Temperature were measured directly in the field. Major cation and anion concentrations (Na^+, K^+, Ca^{2+}, Mg^{2+},) were analyzed by ion chromatography (IC) (Dionex, Sunnyvale, CA, USA). Duplicate samples were analysed to check the analytical precision; it is ±5 % for all hydrogeochemical variables. The analytical precision for the measurements of ions was determined by calculating the ionic balance error, which is generally within ±10% (Mathhess, 1982; Domenico and Schwartz, 1990).

Stable Isotopic studies ($\delta^{18}O$ and δD) were carried out for 60 ground water samples and 15 rainwater samples collected from different locations and all the analysed by NIH Roorkee following standard procedures (IAEA). For oxygen isotopic analysis, about 2 ml of each groundwater sample was equilibrated with CO_2 by shaking for 6 h at 25°C (Epstein and Mayeda, 1953). For hydrogen isotopic analysis, metallic zinc was used to produce hydrogen gas in the zinc-reduction method (Coleman et al., 1982). Oxygen ($\delta^{18}O$) and Deuterium (δD) present in samples were analyzed by using Isotopic Ratio Mass Spectrophotometer (Finnigan Delta^{plus}Xp, Thermo Electron Corporation, Bermen, Germany, the standard deviation of our measurements is ±1.72% for Oxygen and ±0.8% for Hydrogen). All the measurements were carried out against laboratory substandard that were periodically calibrated against the international isotope water standards recommended by the IAEA (V-SMOW, GSIP and SLAP). The isotope results obtained are reported in terms of ä units (Permil deviation of the isotope ratio from the international standard V-Smow) δ being defined by

$$\delta = [(R_{sample} - R_{SMOW}) / (R_{SMOW})] \times 10^3$$

Where, R = D/H or $^{18}O/^{16}O$

Results and Discussion

Hydrogeochemistry

The physical parameters and chemical and isotopic compositions of $\delta^{18}O$, δD and d-excess are presented in Table 1, respectively. pH was found to be acidic to alkaline in nature in most of samples range between 6.80 to 9.40. The total dissolved solid (TDS) content of groundwater samples ranges from 253.94 to 19365.68 mgl^{-1}. Electrical Conductivity is found to vary from 461 to 19872μS/cm.

Table 1. Maximum, Minimum and Average of the Chemical and isotopic compositions of groundwater representing in the study area (All values in mgl^{-1} except EC in μS/cm and pH)

	Study area																	
	Temp	pH	TDS	EC	Ca^{2+}	Mg^{2+}	Na^{+}	K^{+}	F^{-}	Cl^{-}	HCO^{-}_{3}	NO^{-}_{3}	PO^{-}_{4}	SO^{2}_{4}	H_4SiO_4	δD (‰)	δ18O (‰)	D excess
Aver	31.82	7.67	1484	2357	81.37	50.04	365.27	25.42	0.46	636	338.88	4.34	0.21	37.78	12.21	-32.67	-4.47	3.07
Max	35.80	8.80	19365	30200	608	559.20	6812	218.50	2.43	10989	2940.20	57.00	3.21	324	15.20	-16.72	-1.80	13.04
Min	29.30	6.95	253	372	4.00	2.40	14.70	0.83	0.00	35.45	36.60	0.23	0.00	1	2.00	-52.12	-7.31	-5.15

Ca^{2+} concentration ranges from $29mgl^{-1}$ to $500mgl^{-1}$. The concentration of magnesium in groundwater samples in the study area varies from $9mgl^{-1}$ to $895mgl^{-1}$. Na^{+} concentration vary from $4mgl^{-1}$ to $4250mgl^{-1}$. In contrast to the concentrations of Ca^{2+}, Mg^{2+} and Na^{+} ions among the cations, the lower concentration of K^{+} is observed between $2mgl^{-1}$ to $213mgl^{-1}$ from the groundwate. Cl^{-} is ranges from $35.45mgl^{-1}$ to $9052.50mgl^{-1}$. Higher concentration of chloride in the coastal region may be due to sea water intrusion (Chidambaram et al., 2007; Singaraja et al., 2012). The value of HCO^{-}_{3} is observed from $50.4mgl^{-1}$ to $683mgl^{-1}$. The higher concentration of HCO^{-}_{3} in the water infers a dominance of mineral dissolution weathering (Stumm and Morgan, 1996). SO^{2-}_{4} values ranges from $2mgl^{-1}$ to $312mgl^{-1}$. The value of NO^{-}_{3} in the groundwater is observed between $0.7mgl^{-1}$ to $148.20mgl^{-1}$. The groundwater shows a very low content of NO^{-}_{3} near the coast and another region higher NO^{-}_{3} indication of anthropogenic pollution. The concentration of F^{-} and PO^{-}_{4} shows that lesser values are noted in the study area (Singaraja et al., 2013b). The geochemical trend of groundwater in the study area demonstrates that Sodium is the dominant cation ($Na^{+} > Mg^{2+} > Ca^{2+} > K^{+}$) Chloride is the dominant anion ($Cl^{-} > HCO^{-}_{3} > SO^{2-}_{4} > H_{4}SiO_{4} > NO^{-}_{3} > F^{-}$).

Geochemical Evolution

Piper plot shows (Fig. 2) majority of water samples fall in the field of Na-Cl type and these samples are found to be located near the coast, may be due to seawater intrusion (Prasanna et al., 2011). Some samples are also representing mixed water types (Na-Ca-HCO_3-Cl type, Ca-Mg-SO_4-C1 type, and Na-HCO_3-Cl) indicating anthropogenic contaminations (Srinivasamoorthy et al., 2011). It is interesting to note that large proportions of the groundwater showed Na-Cl type, which generally indicates a strong seawater influence (Chidambaram et al., 2007; Thilagavthi et al., 2013; Singaraja et al., 2016b). Chaddas plots of the major ion components of the groundwaters also demonstrate the four stages of geochemical evolution, as illustrated in the Piper diagram, which are controlled by cation exchange. The plot show (Fig 3) that majority of the samples of study area fall in Sea water (Na-Cl) region along the coastal region with followed by Base Ion-Exchange waters, Reverse Ion-Exchange Waters and Recharging Waters respectively. Subsequently, the resultant groundwater flushes the brackish aquifer to show geochemical characteristics of seawater freshening.

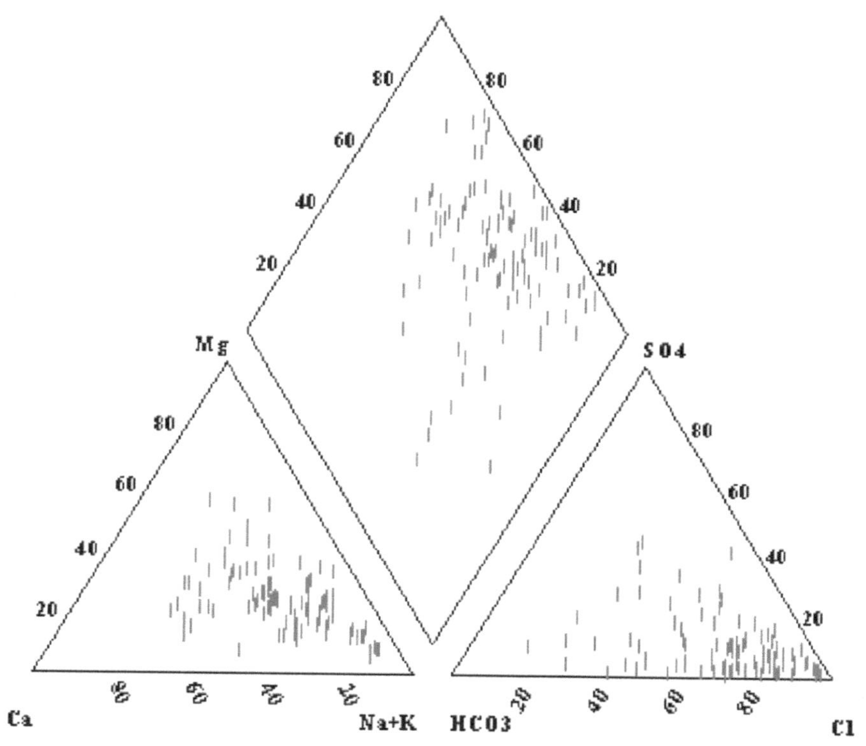

Fig. 2. Major facies representations in study area

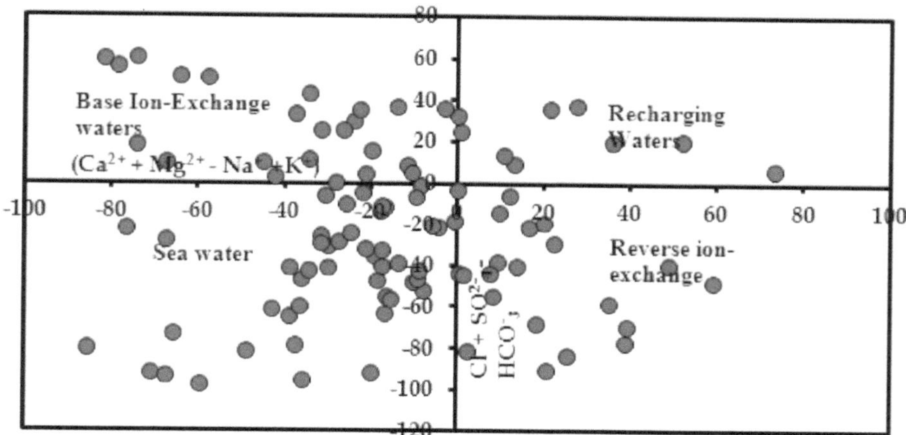

Fig. 3. Chadha's geochemical process evaluation for groundwater samples in the study area

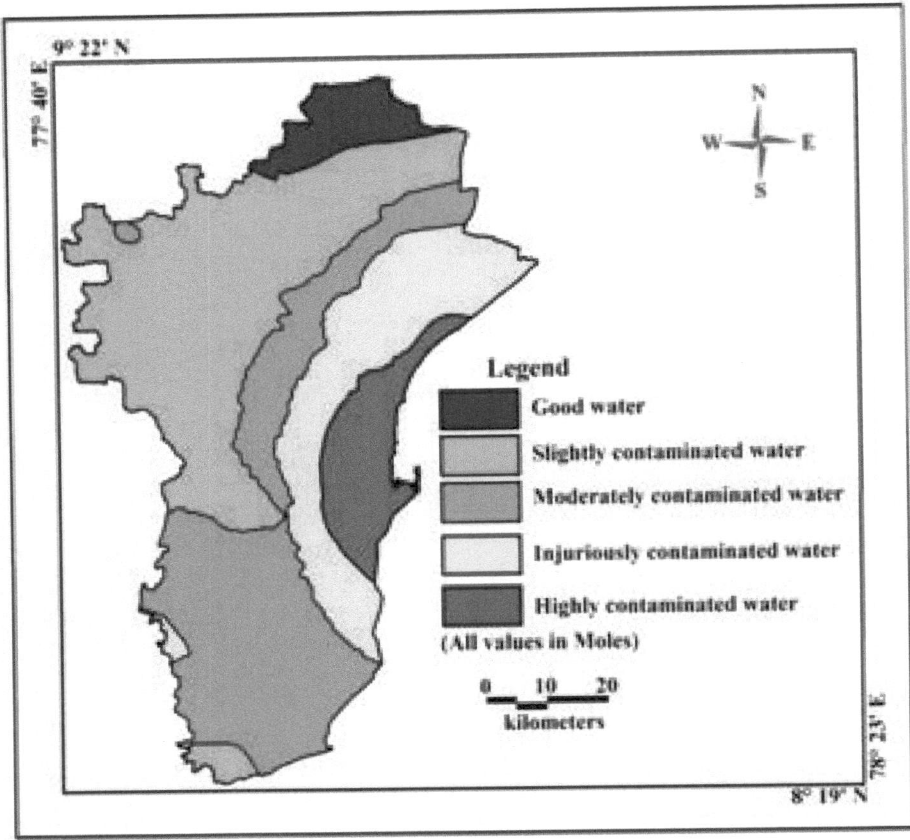

Fig. 4. Spatial distribution of Cl^-/HCO_3^- Moles ratio

Simpson classification includes five classes: <0.5 for good quality water, 0.5 to 1.3 for slightly contaminated water, 1.3 to 2.8 moderately for contaminated water, 2.8 to 6.6 injuriously contaminated and >15.5 for highly contaminated water. It is important in detecting seawater contamination in freshwater (Todd, 1959; Singaraja et al., 2016a). The spatial distribution of Cl^-/HCO_3^- shows higly contaminated water northeastern part of the study area along the coastal region (Fig. 4).

Environmental Isotopes

$\delta^{18}O$ verses δD isotopes

The plot of $\delta^{18}O$ and δD of the rainwater samples of the region (Fig. 5) was drawn to derive the local meteoric water line (LMWL). A regression of 0.98 was noted for various samples collected. The LMWL showed an equation of $\delta D = 7.25^* \delta^{18}O + 2.30$, against the global meteoric water line (GMWL) $\delta D = 8^* \delta^{18}O + 10$ of Criag (1961). This equation was substantially confirmed by (Yurtserver and Gat, 1971) and more recently by (Rozanskiet et al., 1993) who obtained the following equation considering the results obtained by IAEA, Vienna, $\delta D = (8.20 \pm 0.07) \delta^{18}O + (11.27 \pm 0.65)$. Similarly, the LMWL for Italy (Longinelli and Selmo, 2003) showed $\delta D = 7.61 \delta^{18}O + 9.21$. The present LMWL is almost near the GMWL derived by (Rozanski et al.,1993).

Isotopic composition of groundwater samples range from - 52.12 to -16.72% for δD and - 7.31 to -1.80% for $\delta^{18}O$ indicating that the isotopic composition of the groundwater ranges from lighter to heavier fractions. The comparison of the stable isotopes of the rain water with the groundwater samples show that majority of the samples (Group A) fall in line with the LMWL indicating the recharges from the local rainfall recharge (Chidambaram et al., 2010).There are most of the samples (Group B) indicative of evaporative enrichment/ sea water intrusion. It is also observed that the heavier isotopes are more in most of the samples along the coastal region which may be due to the precipitation near the shore or that of saline intrusion (Prasanna et al., 2009; Chidambaram et al., 2012; Erik et al., 2014). The salinity origin of ground water can generally be distinguished by using δD-$\delta^{18}O$ relationship; because the stable isotope fractionation is mainly related to temperature (Huddart et al., 1999; Luoma et al., 2015). This group B sample also shows that there is an end member of isotopic composition indicating sea water intrusion. It is evident that there is a clear mixing noted in between meteoritic samples and the sea water end member. The coastal region samples show the heavier isotopes of δD and $\delta^{18}O$ with maximum values Allison et al., (1984). Since the out of coastal region and river basin, easy infiltration of rainwater is depicted in the isotopic ratios. The groundwater has slightly less $\delta^{18}O$ and δD than the weighted mean for precipitation (Clark and Fritz, 1997). Minor deviation from the weighted average precipitation can also be attributed to the land use practices.

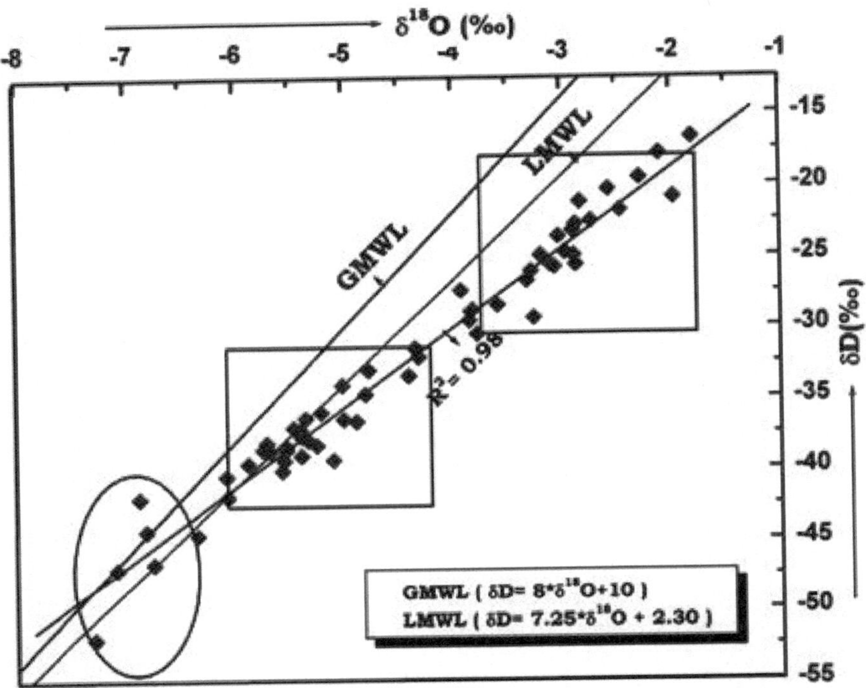

Fig. 5 Plot for $\delta^{18}O$ versus δD of both rain and groundwater samples in the study area

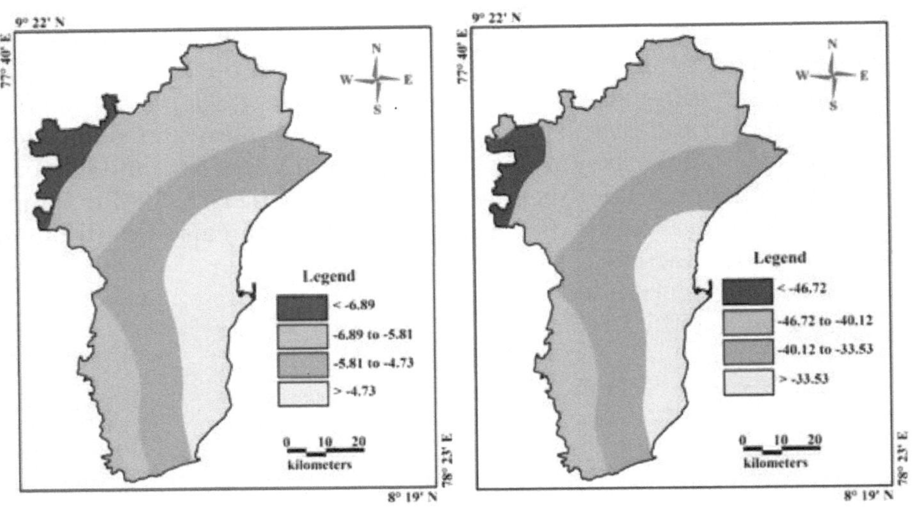

Fig. 6. Spatial variation of $\delta^{18}O$ and δD in the study area

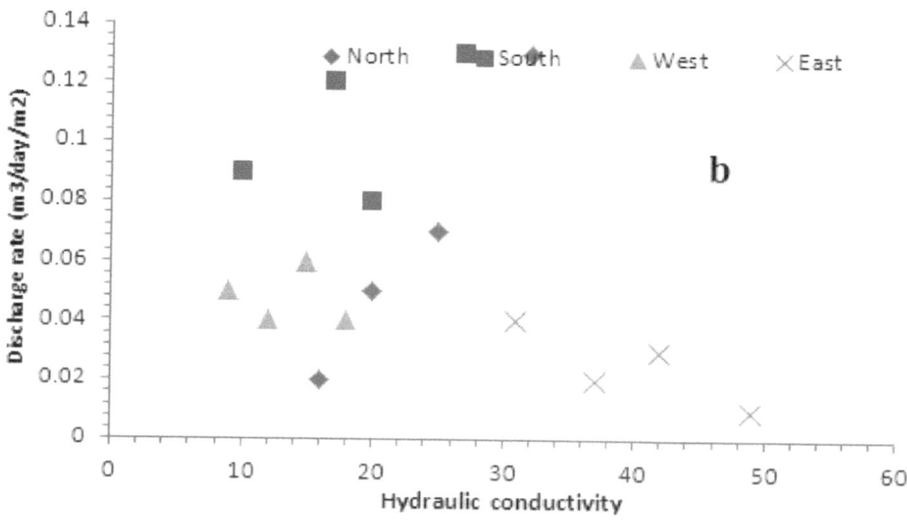

Fig. 7. Relationships between discharge rate and hydraulic conductivity.

The spatial variation of $\delta^{18}O$ and δD (Fig. 6) is represented in around the northeastern part of the study area and >5.65 is represented in southern part of the study area. The pattern of the distribution indicates that in the northeastern part of the study area is recharged by coastal region and Sothern part direct precipitation by river.

Hydrogeologic characteristics of salinization

Hydraulic conductivity and discharge rate were used to analyze the effect of hydrologic characteristics on groundwater salinization. In familiar, these parameters have theoretical correlation with one another when comparing various region of the study area like North, South and West and Eastern region. Eastern region with has high hydraulic conductivity. This observation is consistent with Darcy's Law in that an increase of hydraulic conductivity has an effect of decreasing discharge rate (Fig. 7). The discharge rate in the east region is comparatively lower than other regions, but the hydraulic conductivity is generally higher along the coastal region. The discharge rate ranges from 0.010 to 0.040 $m^3/day/m^2$ in the eastern region. Although higher hydraulic conductivities in the eastern region were observed, the groundwater discharge rates were comparatively lower than other regions (Fig. 7). The areas along the east coast in which saline groundwater occur are characterized by low discharge rate, and high hydraulic conductivity compared with other region of the study area.

Conclusion

From the present study, it can be concluded that hydrochemical characteristics based on major ions analyses show that changes of groundwater composition are mainly controlled by the salinization process linked to Ion-exchange reactions. Chaddas plot and Isotope data indicate that seawater intrusion/saltpan effluent along the coastal region. $\delta^{18}O$ versus δD also observed that the heavier isotopes are more in most of the samples along the coastal region which may be due to the precipitation near the saline intrusion and beyond coastal region and river zone, easy infiltration of rainwater. It is clearly indicates that in the northeastern part of the study area is recharged by coastal region and sothern part through precipitation by river. Spatial distribution of Cl^-/HCO^-_3, $\delta^{18}O$ and δD clearly show that seawater intrusion northeastern part of the study area. Hydrogeologic conditions, which improve the salinization in the study area, are as follows: (1) low discharge rate (2) high hydraulic conductivity.

Acknowledgements

This work was supported isotope analyses by Dr.R. Somesra Rao, NIT, Roorki. It was also partly supported by Dr.Al.Ramanthan JNU, New Delhi, India also helped to improve the quality of presentation.

References

Allison GB, Barnes, CJ, Hughes MW, Leaney FWJ (1984) Effect of climate and vegetation on oxygen-18 and deuterium profiles in soils. In: Isotope Hydrology. IAEA Symposium 270, Vienna 105-123

Caron ME, Grasby SE, Ryan MC (2008) Spring water trace element geochemistry: A tool for resource assessment and reconnaissance mineral exploration. Applied Geochemistry 23:3561-3578

CGWB (2009) South Eastern Coastal Region, District groundwater brochure, Thoothukudi district, Tamil Nadu

Chidambaram S, Bala Krishna Prasad M, Manivannan R, Karmegam U, Singaraja C, Anandhan P, Prasanna MV, Manikandan S (2012) Environmental hydrogeochemistry and genesis of fluoride in groundwaters of Dindiguldistrict, Tamilnadu (India). Environmental Earth Science 68(2): 333-342

Chidambaram S, Ramanathan AL, Prasanna MV, Karmegam U, Dheivanayagi V, Ramesh R (2010) Study on the hydrogeochemical characteristics in groundwater, post- and pre-tsunami scenario, from Portnova to Pumpuhar, southeast coast of India. Environmental Monitoring and Assessment 169:553-568

Chidambaram S, Ramanathan AL, Prasanna MV, Anandhan P, Srinivasamoorthy K, Vasudeven S (2007) Identification of Hrogeochemically Active Regimes in Groundwaters of Erode District, Tamilnadu A Statistical Approach. Asian Journal of water Environment and pollution 5(3):93-102

Clark ID, Fritz P (1997) Environmental isotopes in hydrogeology, New York, Lewis Publishers

Datta PS, Bhattacharya SK, Mookerjee P, Tyagi SK (1994) Study of groundwater occurrence and mixing in Pushkar (Ajmer) valley, Rajasthan, with ä18O and hydrochemical data. J Geol Soc India 43:449-456

Domenico PA, Schwartz W (1990) Physical and chemical hydrogeology. John Wiley, New York: pp.824

Erik O, Kari F, Justin S, Goldsmith GR, Todd D, Ronald A (2014) Oxygen isotope fractionation effects in soil water via interaction with cations (Mg, Ca, K, Na) adsorbed to phyllosilicate clay minerals. Journal of Hydrology 515(16): 1-9

Huddart PA, Longstaffe FJ, Crowe AS, (1999) äD and ä18O evidence for inputs to groundwater at a wetland coastal boundary in the southern Great Lakes region of Canada: J Hydrology 214:18-31

Longinelli A, Selmo E (2003) Isotopic composition of precipitation in Italy: a first overall map. J Hydrol 270: 75- 88

Luoma S, Okkonen J, Korkka-Niemi K, Hendriksson N, Backman B (2015) Confronting the vicinity of the surface water and sea shore in a shallow glaciogenic aquifer in southern Finland. Hydrol. Earth Syst. Sci 19:1353–1370

Mathhess G (1982) The properties of ground water (1st ed.). New York: Wiley

Mondal NC, Singh VP, Singh S, Singh VS (2011) Hydrochemical characteristic of coastal aquifer from Tuticorin, Tamil Nadu, India. Environ Monit Assess 175:531-550

Piper AM (1953) A graphic procedure I the geo-chemical interpretation of water analysis, USGS Groundwater Note no, 12

Prasanna MV, Chidambaram S, Shahul Hameed A, Srinivasamoorthy K (2011) Hydrogeochemical analysis and evaluation of groundwater quality in the Gadilam river basin, Tamil Nadu, India. J Earth Sci 120(1):85-98

Prasanna MV, Chidambaram S, Shahul Hameed A, Srinivasamoorthy K (2009) Study of evaluation of groundwater in Gadilam basin using hydrogeochemical and isotope data. Environ Monit Assess 168:63-90

Prasanna MV, Chidambaram S, ShahulHameed A, Srinivasamoorthy K (2010) Study of evaluation of groundwater in Gadilam basin using hydrogeochemical and isotope data. Environmental Monitoring and Assessment 168:63-90

Puthiyasekar C, Neelakandan MA, Poongothai S (2010) Heavy metal contamination in bore water due to industrial pollution and polluted and non polluted sea water intrusion in Thoothukudi and Tirunelveli of South Tamil Nadu, India. Bulletin of Environment Contamination and Toxicology 85:598-601

Rautio A, Korkka-Niemi K (2015) Chemical and isotopic tracers indicating groundwater/surface-water interaction within a boreal lake catchment in Finland. Hydrogeology Journal 23 (4): 687-705

Rautio A, Kivimäki AL, Korkka-Niemi K, Nygard M, Salonen VP, Lahti K, Vahtera H (2015) Vulnerability of groundwater resources to interaction with river water in a boreal catchment. Hydrol. Earth Syst. Sci. Discuss 12:2435–2476

Ravichandran S (2003) Hydrological influences on the water quality trends in Tamiraparani basin, South India. Environmental Monitoring and Assessment 87:293-309

Rozanski K, Araguas-Araguas L, Gonfiantini R (1993) Isotopic patterns in modern global precipitation. In Climate Change in Continental Isotopic Records, American Geophysical Union Monograph. 78:1-36

Selvam S, Venkatramanan S, Singaraja S, (2015) A GIS-based assessment of water quality pollution indices for heavy metal contamination in Tuticorin Corporation, Tamilnadu, India. Arab J Geosci DOI 10.1007/s12517-015-1968-3

Singaraja C, Chidambaram S, Anandhan P, Prasann MV, Thivya C, Thilagavathi R (2012) A study on the status of fluoride ion in groundwater of coastal hard rock aquifers of south India. Arabian Journal of Geosciences DOI 10.1007/s12517-012-0675-6

Singaraja C, Chidambaram S, Anandhan P, Prasanna MV, Thivya C, Thilagavathi R, Sarathidasan J (2013a) Hydrochemistry of groundwater in a coastal region and its repercussion on quality, a case study-Thoothukudi district, Tamilnadu, India. Arabian Journal of Geosciences doi:10.1007/s12517-012- 0794-0

Singaraja C, Chidambaram S, Prasanna MV, Thivya C, Thilagavathi R (2013b) Statistical analysis of the hydrogeochemical evolution of groundwater in hard rock coastal aquifers of Thoothukudi district in Tamil Nadu, India. Environ Earth Science doi:10.1007/s12665-013-2453-5

Singaraja C (2015) GIS-Based Suitability Measurement of Groundwater Resources for Irrigation in Thoothukudi District, Tamil Nadu, India. Water Qual Expo Health doi:10.1007/s12403-015-0159-5

Singaraja C, Chidambaram S, Noble Jacob, Ezhilarasan E, Velmurugan E, Manikandan M, Rajamani S (2016a)Taxonomy of groundwater quality using multivariate and spatial analyses in the Tuticorin District, Tamil Nadu, India, Environ Dev Sustain. DOI 10.1007/s10668-015-9654

Singaraja C, Chidambaram S, Noble Jacob, Selvam S, Prasanna M.V (2016b) Tidal effects on groundwater dynamics in shallow coastal aquifers-southeast coast of Tamilnadu, India Arabian Journal of Geosciences, 9(6):1-19.

Singh VS, Mondal NC, Singh S, Negi BC (2006) Hydrogeological and geophysical investigations to delineate aquifer zone around SIIL, Tuticorin, Tamilnadu, (p. 24). Technical report. no. NGRI-2006-GW- 564

Solai A, Suresh Gandhi M, Chandrasekaran K, Ram Mohan V (2012) Depositional environment in and around Tamiraparani estuary, and off Tuticorin, Tamil Nadu, India: Clues from grain size studies. Arabian Journal of Geosciences 6(7): 2419-2446

Srinivasamoorthy K,Vasanthavigar M, Vijayaraghavan K, Chidambaram S, Anandhan P, Manivannan R (2011) Use of Hydrochemistry and Stable Isotopes as Tools for Groundwater Evolution and Contamination Investigations Geosciences 1(1): 16-25

Stumm W, Morgan JJ (1996) Aquatic chemistry. John Wiley and sons Inc, New York. 1022

Thilagavathi R, Chidambaram S, Prasanna MV, Thivya C, Singaraja C (2012) A study on groundwater geochemistry and water quality in layered aquifers system of Pondicherry region, southeast India. Applied Water Science 2:253-269

Todd DK (1959) Seawater intrusion in costal aquifers. Trans. Am Geophys Union 34:749-754

Warner NR, Kresse TM, Hays PD, Down A, Karr JD, Jacksona RB, Vengosha A (2013) Geochemical and isotopic variations in shallow groundwater in areas of the Fayetteville Shale development, north-central Arkansas. Applied Geochemistry 35:207-220

Yurtsever Y, Gat JR (1971) In Stable Isotope Hydrology (eds Gat, J. R. and Gonfiantini, R.), IAEA Tech. Report. 103-142

Chapter **13**

Understanding the Distribution of Dissolved Organic Carbon in Layered Coastal Aquifers of Pondicherry Region

Thilagavathi[1] R., Chidambaram[1] S., Thivya[2] C. Pethaperumal[3] S., Tirumalesh[4] K. and Singaraja C.[5]

[1]Department of Earth Sciences, Annamalai University, Annamalai Nagar, India
[2]Department of Geology University of Madras. India
[3]State Groundwater Unit and Soil Conservation, Department of Agriculture Pondicherry, India
[4]Isotope Hydrology Section, Isotope Application Division, BARC, Mumbai, India
[5]Department of Geology, Presidency College, Chennai-5, India

ABSTRACT

Dissolved Organic Carbon (DOC) consisting of water soluble organic molecules and by products of organic decomposition and chemical degradation processes, constitutes a critical link in carbon and nutrient cycles. The quality, or chemical and physical character, of DOC affects the degree to which it is transformed by biological and chemical processes. The significance of DOC was studied in layered coastal aquifers of the Pondicherry region during four different seasons. Pondicherry region has varied geological set up ranging from Cretaceous to Recent formations. A total of 52 groundwater samples were collected from various aquifers like Alluvium, Tertiary(Upper Cuddalore, Lower Cuddalore) and Cretaceous for two different seasons of pre- monsoon(PRM), and post-monsoon (POM). The samples were analyzed for major ions and DOC. The range of DOC in the study area is 0-14 mg/l. Very high DOC concentrations were measured during PRM in the samples from Alluvium and Lower Cuddalore Formation. The samples of these formations were compared with pH, Ionic strength, log pCO_2 and carbonate minerals concentration. The alluvial aquifer shows that pH governed dissolution along the coastal tracts and in Upper Cuddalore formation it's based on interaction between aquifer matrix and groundwater.

Key Words: Groundwater, Dissolved Organic Carbon, Pondicherry, Fromation

Introduction

Increasing population density and changing agricultural practices in coastal areas have led to release of nutrients (and other contaminants) into the coastal environment from fertilizer use, industrial practices, and wastewater discharge. Most efforts to characterize natural organic matter (NOM) in drinking water sources have focused on dissolved organic compounds, which typically contain more than 90% of the NOM (Croue et al. 2000). DOM is a significant component of NOM it is a heterogeneous mixture of aromatic and aliphatic organic compounds containing oxygen, nitrogen, and sulphur functional groups (e.g., carboxyl, phenol, enol, alcohol, carbonyl, amine, and thiol) (Chen et al., 2003). DOM is generally quantified in terms of its dissolved organic carbon (DOC) content (Her et al., 2002; Korshin et al., 1997; Leenheer and Croue 2003; Perminova et al., 2003). Because DOC compounds have a wide range of molar masses and chemical structures, they can play a multifunctional role in the natural environment. DOC, can activate growth of microorganisms and promote growth of biofilm in water distribution systems or treatment units because it can serve as a carbon and energy source for microorganisms (Kim and Yu, 2005).

The concentration of DOC in ground water is generally lower than many surface waters, reflecting the chemical and biotic process of Dissolved Organic Matter (DOM) in the subsurface (Thurman 1985). Organic and inorganic wastes are being discharged into groundwater systems at an ever increasing rate (Goni and Gardner 2003; Santos et al. 2009) and these fluxes were considered significant components of coastal budgets. The presence of such wastes even at low concentrations can degrade water quality by causing the water to have objectionable taste, odour and color in some cases of DOC. Ground water degraded by waste discharge can be expected to remain in that condition for a long time, even though the input source of degradation has been stopped. Ground water movement will cause the zone of degradation to move, but not necessarily at the same rate as the ground water (Hughes and Robson, 1973). Studies have also proved that the hydrology of the region also controls the DOC biogeochemistry (Fraser et al. 2001). DOC and its associated functional groups play a role in the complexation and transport of heavy metals and radionuclides, and may enhance the solubility of nonpolar organic contaminants (Reuter and Perdue 1977; Chiou et al. 1986; McCarthy et al. 1993; Artinger et al. 1998; Zeh et al. 1999; Artinger et al. 2000b; Schussler et al. 2001; Hendry et al. 2004).

Since DOC posed the various hazards in water, it is necessary to understand the DOC chemistry in groundwater. Hence, a study to understand the mechanisms that govern DOC release to groundwater was carried out in the coastal Pondicherry region. The study region comprises a multilayered aquifer, and the DOC can enhance future changes in the chemistry and ecosystem functioning of aquatic environments in the region. The general groundwater qualities of this region based on the major ion distribution were reported by Thilagavathi et al. (2012) and Pethaperumal et al. (2010). A preliminary study of the DOC has been carried out in the study area by Thilagavathi et al. (2014). Detailed subsurface investigations have also been carried out in this region to further understand subsurface geology and geomorphology (Thilagavathi et al. 2013, 2014). Therefore, this present study was carried out to study concentration of DOC in the layered aquifers of the Pondicherry region with respect to different seasons.

Study Area

Pondicherry region is located on the east coast of South India which lies between 11°45' and 12° 0' North Latitude and 79° 37' and 79° 50' East longitudes (Fig 1). The region covers 293sq.km consisting of five Villages (Ariyankuppam, Bahour, Mannadipet, Nettapakkam and Villiyanur). The mean monthly temperature ranges from 22°C to 33°C. The relative humidity varies between 70% and 85%. The Pondicherry region in general is a flat plain with an average elevation of about 15 m above the mean sea level (MSL). The three major physiographic units generally observed are coastal plain as (younger and older), alluvial plain and uplands (National Assessment of shoreline change: Mani Murali 2013; Thilagavathi et al. 2013).

The alluvial plain, which occupies major portion of the Pondicherry region, is formed due to two major rivers namely Gingee and Ponnaiyar River. The alluvial plain, which occupies a major portion of the Pondicherry region, is formed due to two major rivers, namely Gingee and Ponnaiyar River. The region is covered by sedimentary formations, ranging in age from Cretaceous to Recent. The oldest sedimentary formations are the Cretaceous sediments exposed in the northwestern part of the study area and north of the Sankarabarani River, except a small extent of Archean formations in the northwestern part of the region. The Lower Tertiary is represented by the Paleocene Kadaperikkuppam and Manaveli Formations. The Upper Tertiary is represented by the Mio-Pliocene Cuddalore Formation. The Recent is represented by Quaternary laterites and alluvium. The laterites occur as thin cap over the Cuddalore Formation. Alluvium consists of shale and Kankar (nodular impure limestone).

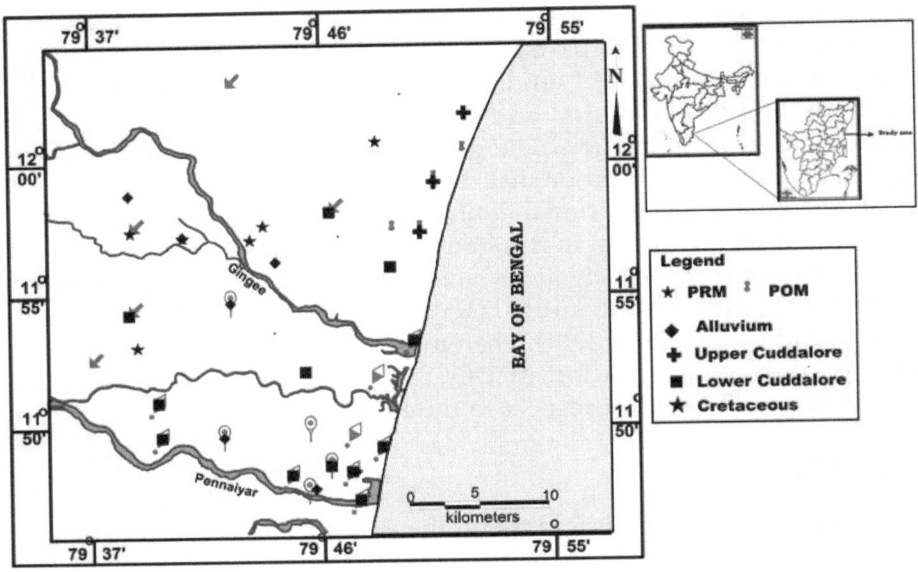

Fig. 1. Location map of the study area along with the Sampling locations

Methodology

A total of 52 groundwater samples were collected during pre-monsoon (PRM), and post-monsoon (POM). Samples cover the different formations in the study area: Alluvium Formation, Upper Cuddalore sandstone, Lower Cuddalore sandstone and Cretaceous Formation.

The samples were analyzed for major cations, anions, by using standard procedures (APHA 1998). DOC data were collected as non-purgeable organic carbon (NPOC) measured using "Total Organic Carbon Analyzer TOC-VCSH" (Shimadzu Corporation, model -ASI-V/TOC-VCSH). A small amount of hydrochloric acid was added to acidify the sample and the sample was purged with purge gas. This converted all Inorganic Carbon (IC) in the sample to CO_2 and drives the CO_2 out of the sample solution. The TOC concentration was determined by measuring the Total Carbon of the sample after the Inorganic Carbon was eliminated. NPOC measurement is equivalent to TOC by adopting defined acidification and sparging (Inorganic Carbon elimination) methods (Environmental Protection Agency (EPA), American Society for Testing and Materials (ASTM) etc.). As any purgeable organic compounds may be lost from the sample during sparging, the TOC measured by this method can be called NPOC (Chung et al. 2010). The geochemical reaction simulation model is done by using WATQ4F (Plummer et al. 1976).

Result and Discussion

Distribution of Dissolved Organic Carbon (DOC)

The data of DOC with respect to seasons and lithology regardless of seasons are given in Table 1. Higher DOC concentrations were measured during PRM in the samples from Lower Cuddalore aquifer of 14.4mg/l followed by Alluvial aquifer12.8mg/l. The samples of Upper Cuddalore and Cretaceous have DOC concentrations of 3.3 and 2.9mg/l (Fig 2) respectively. The higher concentration of DOC during POM is noted in Cretaceous aquifers of 3.5mg/l followed by the Alluvium of 2.3mg/l. Higher concentration of DOC shows a level of 2.1mg/l in geological formation of Lower Cuddalore and in Upper Cuddalore is about of 1.4mg/l is observed during POM. The sample had a distinctive and a common characteristic of waters as if the aquifer was deposited is a swampy and marshy condition. The floral assemblage at the time of deposition of Cuddalore formation reveals that it has taken place in the near shore (coastal environment) in lagoonal or swampy conditions (Mandaokar et al. 2012). The order of dominance of DOC in groundwater in different seasons based on average concentration is shown in Table 2. The concentration of DOC from this well is similar to the range reported for swamps, wetlands (Thurman 1985), marshes and bogs. Drever (1997) also noted that ground water associated with swamps might have elevated DOC concentrations. The DOC levels are relatively high in Lower Tertiary aquifers. These aquifers consist of coal and organic shale as a part of their formation (Thilagavathi et al. 2013). Ground water associated with these types of sedimentary deposits may naturally have elevated DOC concentrations (Drever 1997).

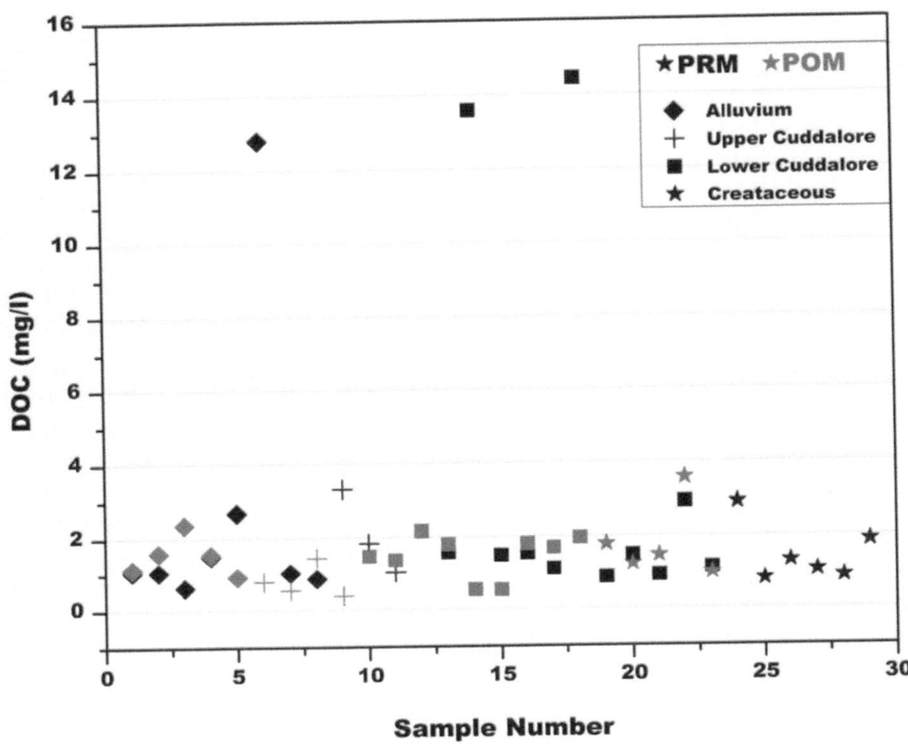

Fig. 2. Variations of Dissolved Organic Carbon in different aquifer

Table 1. Statistics of DOC concentration (mg/l) with respect to seasons and lithology

Formation		PRM	POM
Alluvium	Max	12.81	2.37
	Min	0.63	0.92
	Avg	3.51	1.55
Upper Cuddalore	Max	3.33	1.45
	Min	1.04	0.38
	Avg	2.12	0.83
Lower Cuddalore	Max	14.44	2.17
	Min	0.87	0.52
	Avg	4.19	1.46
Createceous	Max	2.92	3.58
	Min	0.80	0.98
	Avg	1.57	1.95

Table 2. Order of dominance of the DOC in ground water in different season based on the average concentration

PRM	Lower Cuddalore > Alluvium >Upper Cuddalore>Cretaceous
POM	Cretaceous>Alluvium>Lower Cuddalore>Upper Cuddalore

Hydrochemical Evaluation

The maximum, minimum and average concentration of cation and anion is given in table 3.The hydrochemical facies reflect the chemical processes operative in a specific lithological and geochemical environment. The general classification and trend of variation in the groundwater of the study area is exhibited in the Piper - diagram (Thilagavathi et al. 2012). Groundwater of few samples are dominated by alkali (Na^+ and K^+) over alkaline–earth cations (Ca^{2+} and Mg^{2+}) reflecting predominance of salinity. Weathering also results in the release of Na^+ and K^+ from the K feldspar and plagioclase alteration. On the other hand, the position of the groundwater in the anion triangle indicates dominance of the Cl and HCO_3, but a less abundance of sulfate in these waters. Hydrogeochemical facies interpretation is an useful tool for determining the flow pattern, origin of chemical histories of groundwater masses (Chidambaram et al. 2012; Karmegam et al. 2010).

Three major types of groundwater were identified by the piper facies, during pre-monsoon, Ca-Na-HCO_3, Na-HCO_3 type and Na-Cl type (Fig. 3). All the Upper Cuddalore Sandstone aquifer samples were fall in the Na-Cl type along with few representations of Cretaceous and Alluvial aquifers. The Lower Cuddalore Sandstone aquifer samples along with majority of the Alluvial and Cretaceous aquifers samples were fall in Na-HCO_3 facies. The dominant facies in all the aquifers is Na-Cl and Na-HCO_3 facies. During POM most of the samples fall in Na-Cl and Na-HCO_3 facies (Fig. 3b). Few samples of Alluvial and Cretaceous aquifers were fall in Ca-HCO_3 facies and that of Lower Cuddalore Sandstone in Ca-Cl facies. It's to be noted that there is a significant variation of cation in post-monsoon and variation of anion in pre-monsoon. The samples with higher DOC are shown in Figure 4. Alluvium during PRM and POM are represented by Na-Cl type, Lower Cuddalore and Cretaceous in both the seasons are represented by Na-Cl water type, Cretaceous and Lower Cuddalore samples are represented by the Na-HCO_3 water type.

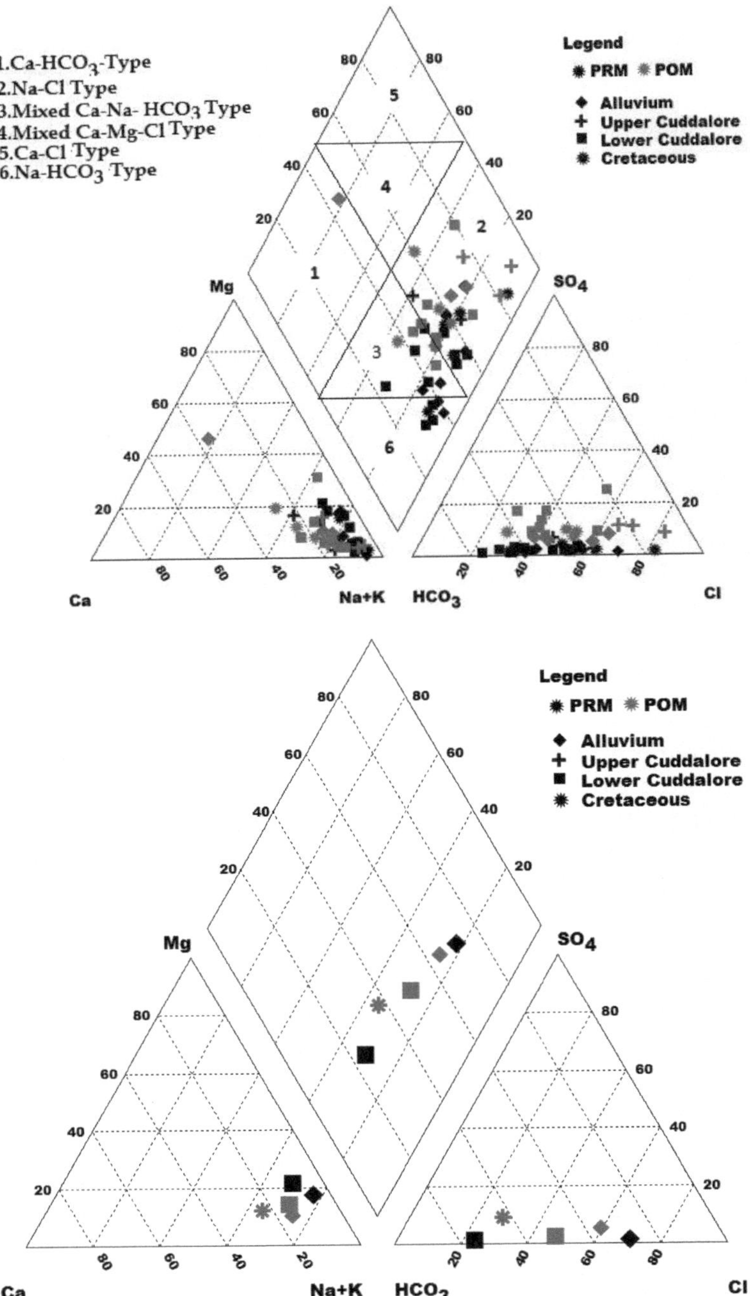

Fig. 3 (a) Piper facies diagram for groundwater samples for different season (b) Groundwater of higher DOC in different season and formation

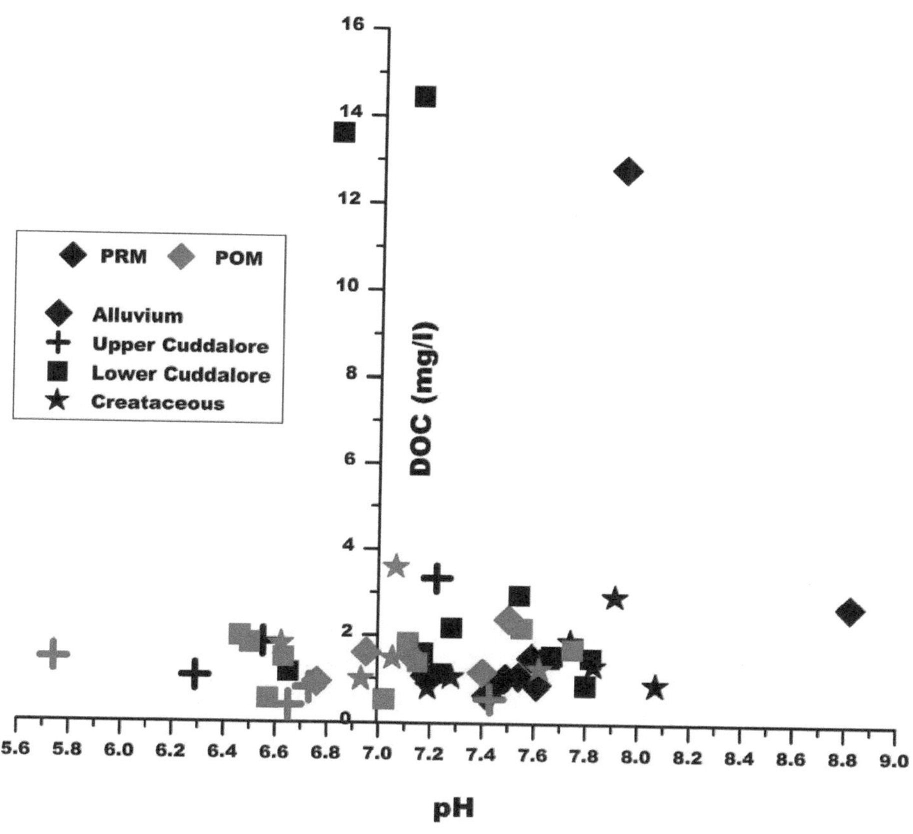

Fig. 4. Relationship between the pH and DOC for the groundwater samples

Table 3. Statistics of chemical composition of groundwater (all values are in mg/l except for EC in µs/cm and pH).

PRM		pH	EC	TDS	Ca	Mg	Na	K	Cl	HCO_3	H_4SiO_4	SO_4	PO_4	NO_3
Alluvium	MAX	8.8	2360.0	1780.0	56.0	105.6	892.3	6.3	497.0	390.4	240.0	18.0	12.4	23.3
	MIN	7.2	769.0	578.0	32.0	2.4	430.8	3.1	88.6	158.6	30.0	4.0	0.9	2.0
	AVG	7.8	1265.4	995.0	43.2	45.4	631.2	4.1	212.8	291.6	110.8	8.9	7.3	9.8
Upper Cuddalore	MAX	7.2	313.0	217.0	24.0	7.2	151.3	3.1	70.9	97.6	200.0	10.5	8.6	13.0
	MIN	6.3	209.0	106.5	16.0	0.0	45.7	3.0	53.2	48.8	20.0	2.5	0.2	8.5
	AVG	6.7	266.2	163.6	20.0	3.4	88.7	3.1	60.3	73.2	98.0	5.9	4.4	10.7
Lower Cuddalore	MAX	7.8	1440.0	1175.0	56.0	55.2	706.4	4.4	248.2	366.0	330.0	11.5	15.0	22.8
	MIN	6.7	349.0	228.0	16.0	4.8	161.3	3.1	35.5	97.6	20.0	3.0	0.1	0.3
	AVG	7.3	715.4	556.2	31.4	21.6	417.3	3.4	108.9	216.1	155.4	5.9	7.4	6.1
Createceous	MAX	8.1	2020.0	1760.0	80.0	74.4	883.0	11.5	531.8	500.2	160.0	15.0	8.3	72.6
	MIN	7.2	1000.0	848.0	36.0	12.0	514.8	3.1	141.8	0.0	0.0	7.0	0.1	1.8
	AVG	7.7	1351.3	1174.4	53.5	32.4	669.4	5.4	270.3	250.1	73.8	10.3	5.9	24.8
POM		**pH**	**EC**	**TDS**	**Ca**	**Mg**	**Na**	**K**	**Cl**	**HCO_3**	**H_4SiO_4**	**SO_4**	**PO_4**	**NO_3**
Alluvium	MAX	7.5	1888.6	1322.0	88.0	67.2	321.8	17.7	390.0	463.6	124.0	46.5	0.5	6.3
	MIN	6.8	1128.6	790.0	32.0	0.0	39.1	15.2	159.5	207.4	69.8	40.5	0.1	1.1
	AVG	7.1	1595.9	1117.1	54.3	26.1	200.0	16.6	253.2	359.0	94.5	43.7	0.3	3.1
Upper Cuddalore	MAX	7.4	868.6	608.0	20.0	14.4	229.9	17.3	319.0	122.0	170.0	46.0	0.6	18.9
	MIN	5.7	144.0	100.8	8.0	0.0	116.8	13.1	195.0	61.0	71.0	43.5	0.1	2.2
	AVG	6.6	440.4	308.3	15.3	6.4	176.9	15.5	242.2	97.6	113.8	44.9	0.3	9.5
Lower Cuddalore	MAX	7.8	1157.1	810.0	28.0	48.0	206.9	18.8	230.4	317.2	113.0	144.1	0.9	5.0
	MIN	6.5	324.3	227.0	16.0	0.0	69.0	15.1	53.2	146.4	3.0	22.5	0.0	0.3
	AVG	7.0	620.8	434.5	21.8	12.7	135.8	17.1	135.4	205.2	49.9	59.4	0.4	1.4
Createceous	MAX	7.6	1481.4	1037.0	56.0	24.0	229.9	18.6	248.1	414.8	99.8	49.0	0.7	5.6
	MIN	6.6	1072.9	751.0	16.0	0.0	117.0	11.8	106.3	256.2	48.0	43.0	0.1	1.1
	AVG	7.1	1310.2	917.1	38.3	11.7	173.0	15.6	169.6	338.1	73.8	45.9	0.4	2.7

Relationships Between DOC and pH

DOC mobility is greatly related to pH (Sommer et al., 2008) which affects the binding force of the humic substances. The study area shows increasing DOC concentration with increasing pH is noted during PRM in few samples from Alluvium and Lower Cuddalore formation with higher pH as 7.93 and the higher DOC 12.8 mg/l because of higher salt concentrations in the alluvial aquifer and the direct infiltration of domestic sewage and industrial effluent (Hossein and Ian Clark 2011). Upper Cuddalore and Cretaceous formation shows increasing the pH in PRM and POM, which may result in lowering of the DOC in most of (Fig.4.) the samples. This is due to the fact that in many location of this formation has higher pH which generally increases the binding capability of humic substance (Christensen and Christensen 2000) hence the increase of binding capability results in the decrease of organic carbon release into the groundwater. Marginal marine aquifer shows that the natural hydrogeochemical processes parental, connected mainly with organic matter decay in anaerobic environment and the influence of this process on the dissolution of geological environment components and ion exchange respectively (Sommer et al., 2008) Connectivity of shallow and deep aquifers is inhibited mainly due to the presence of lignite in the formation and more over there formation are shallow marine (Gassama et al., 2012) which may also induce DOC concentration.

Relationship of DOC and log pCO_2

Partial pressures of CO_2 (pCO_2) in water were calculated from measured pH and alkalinity data using the WATEQ4F program (Plummer et al., 1976). The higher log pCO_2values are noted in few samples in the Cuddalore Sandstone and Cretaceous aquifers during POM, It is mainly due to the presence of the peat beds and dissolution of silicate minerals (Chidambaram et al., 2011). Lower pH is also noted in these formations, which is evident from several model studies electrostatic interactions of the humic substances with organic ligands and cations (Tipping et al., 1992, Milne et al. 1995). They also found that groundwater pH decreases when log pCO_2 increases.

In both the seasons there is significant relation between LogpCO_2 and the DOC in all formations. Higher DOC range between -1.3 to -2.4 LogpCO_2 values in Alluvium, Lower Cuddalore formations in PRM The excess of Dissolved Inorganic Carbon with respect to alkaline earth metal ions are generally related to addition of CO_2 (Fig.5) which is generated by oxidation of organic matter during bacterially-mediated reduction Cortecci et al. (1999). The alluvium, and Lower Cuddalore aquifers are dominantly by quartz and K feldspar (Nair et al., 1971). Higher CO_2 levels may drive plant growth (Ainsworth et al., 2005), while warmer temperatures may speed organic decomposition and microbial activity (Andrews et al. 2000). The lesser amount DOC present in Alluvium, Upper Cuddalore, Lower Cuddalore, Cretaceous aquifers appears to be consistent with one or both of the following mechanisms: (1) removal of soluble organic compounds (e.g., fulvic acid) from the profile due to the relatively higher rates of flushing and/or (2) dilution of existing DOC due to mixing with larger fluxes of recharge waters (Fennell et al.,1998). This will result in higher pCO_2 and less DOC concentration in POM.

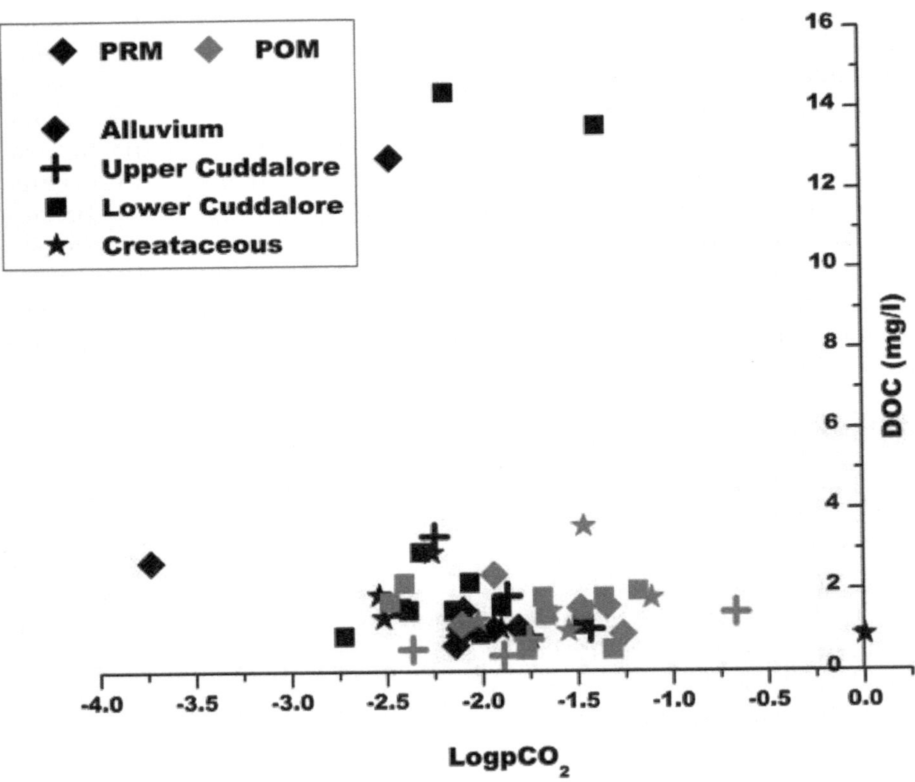

Fig. 5. Relationship between the LogpCO_2 and DOC for the groundwater samples

The higher concentrations of pCO_2 in irrespective of formation during POM may be due to the presence of Sulphate and Nitrate derived from the impact of fertilizers or cesspools or other anthropogenic activities. Indirectly the higher pCO_2 helps in the increase in dissolution of feldspars resulting in higher ionic concentrations (Venturelli et al., 2003). Hence the clay content in Alluvium, Lignite in Cuddalore and marine formation of Cretaceous formation may be the inherent factor for the DOC in groundwater.

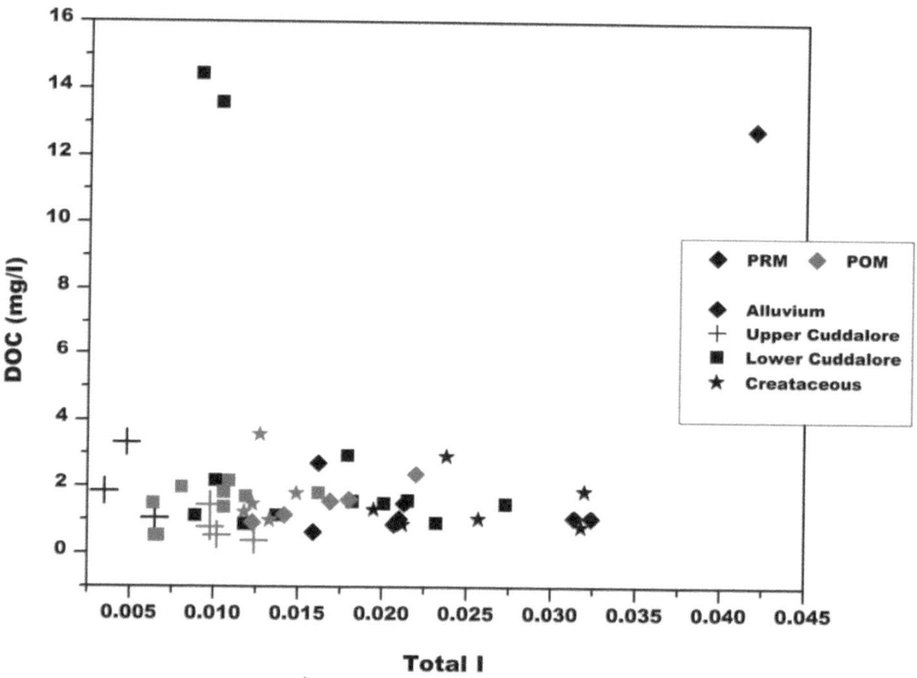

Fig. 6. Relationship between the ionic strength and DOC for the groundwater sample

Relationship of DOC and Ionic Strength

Ionic strength of a sample is directly related to the molality and charge of ions. In the study area, higher ionic strength was noted during PRM in two formations; Upper Cuddalore (0.032) and in Cretaceous formations (0.034), indicating lesser inflow of fresh water into the system. During POM higher ionic strength was noted in Alluvium (0.022). The higher ionic strength also indicates longer residence time or greater interaction with the host rock (Chidambaram et al., 2007). Low ionic strength was noted in other formations indicating higher inflow of fresh water/ recharge (Prasanna et al., 2010), which falls in line with the earlier observations (Thilagavathi et al., 2012).

During PRM few samples of Alluvial, Lower Cuddalore and Cretaceous aquifers (Fig. 6) exhibit ionic strength greater than 0.025 but in most of the samples range between 0.01 and 0.029. The dilute concentration of 0.005 ionic strength organic plume of a 60 year discharge of treated sewage into the aquifer (Leblanc et al., 1984) which may result in higher DOC. Higher ionic strength is noted in Alluvium and Cretaceous formation (Scholl et al., 1992) which shows the microbial transport occurs to a much greater extent within the organic contaminant plume (Pieper et al., 1997). During PRM few samples from Lower Cuddalore shows lesser ionic strength and the DOC concentrations increased, which may be due biological activity in the absence of toxic metals concentrations (Butler 2008) DOC is typically lowest in irrespective of seasons of PRM and POM in few samples of Alluvium with highest ionic strength which may be due to increase of sea salt concentrations (Evans et al., 2005).

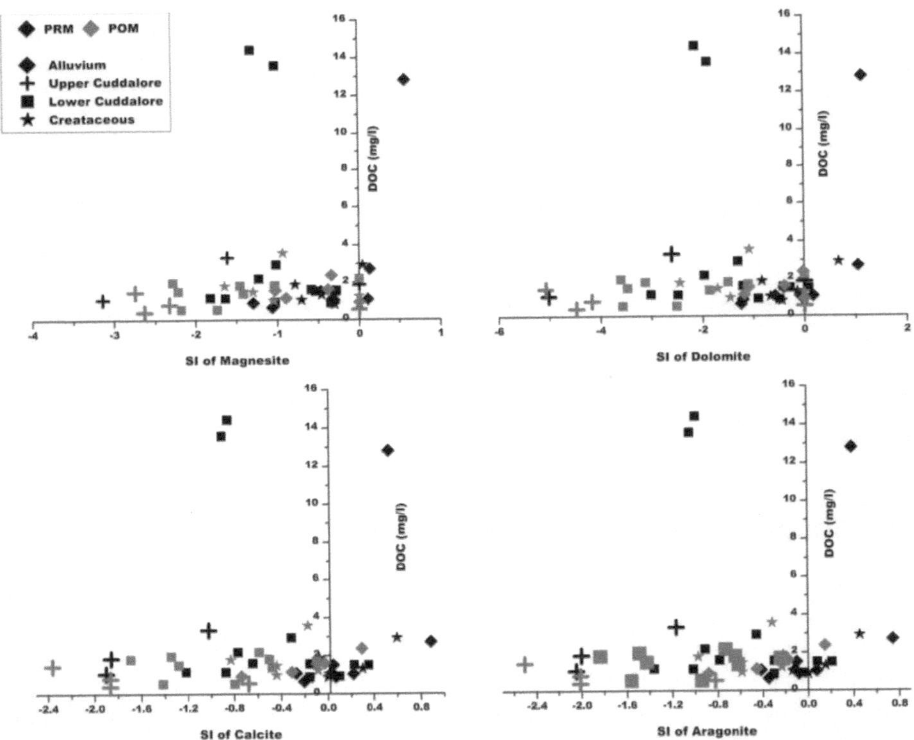

Fig. 7. Bivariate plot of Dissolved Organic Carbon and Carbonate minerals for groundwater samples of different seasons

Relationship of DOC and Carbonate Minerals

Bicarbonate derived by the pressure of weathering or from other source remains in the aqueous medium trying to equilibrate with different ions like Ca, Mg, or both to attain saturation with different minerals like Calcite (SI_C), Dolomite (SI_D), Aragonite (SI_{AR}) and Magnesite (SI_M). SI of carbonate minerals shows under saturation to over saturation state in all samples in irrespective seasons. The saturation state of carbonate minerals during PRM and POM are in the following order: $SI_C > SI_{AR} > SI_D > SI_M$. SI of these minerals was plotted against the concentration of DOC. Most of the samples in irrespective of formation during PRM and POM are saturated around 2mg/l of DOC (Fig. 7). SI_C show that few samples of PRM and POM samples are towards saturation from under saturation state in Alluvium, Upper Cuddalore, Lower Cuddalore and Cretaceous aquifers. This is mainly are due to the increase of the HCO_3 and Ca. During POM most of the samples from Upper Cuddalore and Lower Cuddalore samples indicates the under saturation to Saturation which is due to the removal of HCO_3 and Ca from the system.

PRM and POM samples in irrespective of formation exhibits from near saturation to over saturation state in Calcite (Fig. 7). Similar trend is noted in Dolomite (SI_D) and Magnesite (SI_M), it also attains saturation around 2mg/l DOC. They are nearly saturated to saturated state and it decreases with decrease of DOC. Alluvium and Lower Cuddalore samples clearly shows (Fig 6) that few samples during PRM exhibit an increase of DOC with increase of saturated state of the carbonate minerals, which is due to the fact that increase of groundwater recharge and dilution of the groundwater chemistry with respect to these minerals (Calcite and Dolomite) after the subsequent monsoon. Saturation state of Mg increases with increasing in magnesium concentration (Pethaperumal 2010). Similarly when Ca and Mg increases when the Saturation Index of Dolomite increases. It is noted that above 2mg/l of DOC, SI of these minerals which indirectly reflects that fact that above 2mg/l of DOC the concentration of Ca+Mg increases.

Conclusion

The Study of comparison of the average values of Dissolved Organic Carbon of different formation of different season and it is understood that during PRM Lower Cuddalore formation has higher average values followed by Alluvium, Upper Cuddalore, and Cretaceous formation and during POM Cretaceous has higher average values followed by the Alluvium, Lower Cuddalore and Upper Cuddalore. The hydrogeochemical facies reflect the higher DOC samples are represented by NaCl water type and mixed Na- HCO_3 water type. Higher log pCO_2 condition in Upper Cuddalore Sandstone and Cretaceous aquifers, is mainly due to the presence of the peat beds and dissolution of silicate minerals. Lower DOC value is observed in samples with high Log pCO_2 values in both the seasons and the reverse case is also noted in few samples. Higher DOC with high Log pCO_2 is noted in Cretaceous aquifers during PRM. The saturation state of carbonate minerals are in the following order: $SI_C > SI_{AR} > SI_D > SI_M$. Increase of the HCO_3 and Ca during PRM and POM most of the samples from Alluvium, Lower Cuddalore and Creatceous samples reflects the under saturation to saturation which is due to the removal of HCO_3 and Ca from the system. Few samples at Alluvium, Lower Cuddalore and Cretaceous, aquifers during PRM show reductions in DOC with increasing ionic strength. This process has also been proposed as a control on DOC mobilisation Higher DOC in Alluvium is due to the anthropogenic processes by the fertilizer, sewage and landfill leacheats. In Upper Cuddalore, DOC is results from the Marcasite and peat beds, in Lower Cuddalore, DOC dominance is mainly by the presence of lignite in the Cuddalore Formation and the shallow marine environment of deposition of Lower Cuddalore Formation indicates natural source of DOC. In Cretaceous formation, the dominance of the DOC is by the anthropogenic processes and the irrigation and diffusive transport of dissolved compounds can act to uncouple the distribution of organic carbon.

Acknowledgement

The authors wish to express thanks to Ministry of Water Resources (MOWR) (No. 29/INCGW-03/2010-R&D/30523062) for providing necessary nancial support to carry out this study, and author R.T wishes to express thanks to DST for providing the Inspire fellowship (DST/INSPIRE Fellowship/2010/[220] 18 March, 2011).

References

Ainsworth E A, Long S P (2005) What have we learned from fifteen years of Free Air Carbon Dioxide Enrichment (FACE)? A Meta-analytic review of the responses of photosynthesis, canopy properties and plant production to rising CO. New Phytologist 165: pp.351-372.

Andrews J A, Matamala R, Westover K M, Schlesinger W H (2000) Temperature effects on the diversity of soil heterotrophs and the ä[13]C International Journal of Earth Sciences and Engineering ISSN 0974-5904, Vol. 07, No. 04, August, 2014, pp. 1537-1550 of soil respired CO. Soil Biology & Biogeochemistry 32,pp. 699-706.

APHA (1998) Standard methods for the examination of water and wastewater, 19th edn. Washington, DC. American Public Health Association.

Artinger R, Buckau G, Geyer S, Fritz P, Wolf M, Kim J I *(2000a)* Characterization of groundwater humic substances: Influence of sedimentary organic carbon. Appl. Geochem., 15, 97–116.

Butler B A (2008) Effect of pH, ionic strength, Dissolved Organic Carbon, time, and particle size on metals release from mine drainage impacted streambed sediments. Epub 43(5):pp. 1392-402.

Chen W, Westerhoff P, Leenheer J A, Booksh K (2003) Fluorescence Excitation-Emission Matrix Regional Integration to Quantify Spectra for Dissolved Organic Matter. Environmental Science and Technology. 37(24): 5701-5710.

Chidambaram S, Prasanna M V, Singaraja C, Thilagavathi R, Pethaperumal S, Tirumalesh K (2012) Study on the Saturation Index of the Carbonates in the Groundwater Using WATEQ4F, in Layered Coastal Aquifers of Pondicherry, Journal Geological Society of India ,Vol.80, pp.813-824.

Chidambaram S, Vijayakumar V, Srinivasamoorthy K, Anandhan P, Prasanna M V, Vasudeven S (2007) A Study on variation in Ionic Composition of Aqueous System in Different Lithounits Around Perambalur Region, Tamil Nadu. Journal Geological Society of India, 70: 1061-1069.

Chidambaram S, Prasanna M V, Karmegam U, Singaraja C, Pethaperumal S, Manivannan R, Anandhan P, Tirumalesh K (2011) Significance of pCO_2 values in determining carbonate chemistry in groundwater of Pondicherry region, India. Front Earth Sciences,;5(2):pp. 197–206.

Chiou CT, Malcolm RL, Brinton TI, Kile DE *(1986)* Water solubility enhancement of some organic pollutants and pesticides by dissolved humic and fulvic acids. Environ. Sci. Technol. 20, 502–508.

Christensen J B and Christensen T H (2000) The effect of pH on the complexation of Cd, Ni and Zn by dissolved organic carbon from leachate-polluted groundwater. Water Research, 34 , pp.3743-3754.

Chung Chow Chan, Herman Lam, Xue-Ming Zhang. (2010). Practical Approaches to Method Validation and Essential Instrument Qualification. Published by John Wiley and sons Inc Hoboken , New jersey.

Cortecci G, Calafato A, Boschetti T (1999) New chemical and isotopic data on the groundwater system of Bagno di Romagna, Emilia-Romagna province, Italy. Miner. Petrogr. Acta (Bologna) ;42, pp.89– 101.

Croue J, Korshin GV, Benjamin M (2000). eds. Characterization of natural organic matter in drinking water. American Water Works Assoc.,2000.

Drever JI (1997) The Geochemistry of Natural Waters. Prentice-Hall, Upper Saddle River, 436p.

Evans C D, Monteith D T, Cooper D M (2005) Long-term increases in surface water Dissolved Organic Carbon: Observations, possible causes and environmental impacts. Environmental Pollution 137:pp. 55-71.

Fennell J, Bentley L R (1998) Distribution of sulfate and organic carbon in a prairie till setting: Natural versus industrial sources. Water Resources Research, 34(7):pp. 1781–1794.

Fraser C J D, Roulet N T, Moore T R (2001) Hydrology and dissolved organic carbon biogeochemistry in an ombrotrophic bog. Hydrological Processes, 15, 3151–3166. DOI: 10.1002/hyp.322

Gassama N, Dia A, Violette S (2012) Origin of salinity in a multilayered aquifer with high salinization vulnerability. Hydrological Processes, 26(2): pp. 168-188.

Goni M A and Gardner R L (2003) Seasonal dynamics in dissolved organic carbon concentrations in a coastal water-table aquifer at the forest–marsh interface. Aquatic Geochemistry, 9: pp.209–232.

Hendry M J, Wassenaar L I (2004) Transport and geochemical controls on the distribution of solutes and stable isotopes in a thick clay-rich till aquitard, Canada. Isotopes Environ. Health Stud, 40(1), 3–19.

Her N, Amy G, Foss D, Cho J (2002) Variations of Molecular Weight Estimation by HP-Size -Exclusion Chromatography with UVA Versus Online DOC Detection. Environmental Science and Technology. 36(15): 3393-3399.

Hossein Mohammadzadeh and Ian Clark (2011) Bioattenuation in Groundwater Impacted by Landll Leachate Traced with d^{13} C. Ground Water, International Journal of Earth Sciences and Engineering, ISSN 0974-5904, Vol. 07, No. 04, August, 2014, pp. 1537-1550 49(6): pp.880-890.

Hughes J L, Robson S G (1973) Effects of waste percolation on groundwater in alluvium near Barstow, California, in underground waste management and artificial recharged by jules Braunstein, Internat.Symposium,2d, New Orleans, La., Am. Assoc. Petroleum Geologist .1,pp.91129.

Karmegam U, Chidambaram S, Sasidhar P, Manivannan R, Manikandan S, Anandhan P (2010) Geochemical characterization of groundwaters of shallow coastal aquifer in and around Kalpakkam, South India. Res J Environ Earth Sci 2(4):170– 177.

Kim H, Yu M (2005) Characterization of Natural Organic Matter in Conventional Water Treatment Processes for Selection of Treatment Processes Focused on DBPs Control. Water Res. 39(19): 4779-4789.

Korshin GV, Benjamin M M, Sletten RS (1997) Adsorption of Natural Organic Matter (NOM) on Iron Oxide: Effects on NOM Composition and Formation of Organo Halide Compounds during Chlorination. Water Res. 31(7): 1643-1650.

Leenheer J A, Croue J (2003) Characterizing Aquatic Dissolved Organic Matter. Environmental Science and Technology. 37(1): 18-26.

LeBlanc D R (1984) Movement and Fate of Solutes in a Plume of Sewage-Contaminated Ground Water, Cape Cod, Massachusetts U.S. Geological Survey Open File Report 84–475, Reston, VA.

Mani Murali R, Ankita M, Amrita S, Vethamony P (2013) Coastal vulnerability assessment of Puducherry coast, India using analytical hierarchical process Nat. Hazards Earth Syst. Sci. Discuss., 1, 509–559.

Mandaokar B D, Debi Mukherjee (2012). Palynological Investigation of Early Miocene Sediments Exposed At Panruti, Cuddalore District, Tamil Nadu, India. International Journal of Geology, Earth and Environmental Sciences, 2 (3), 157-175.

McCarthy J F, Williams T M, Liang L, Jardine P M, Jolley LW, Taylor D, Palumbo AV, Cooper L W (1993) Mobility of natural organic matter in a sandy aquifer. Environ. Sci. Technol. 27, 667–676.

Milne C J, Kinniburgh D, De Wit J C M, Van Reimsdijk W H, Koopal K (1995) Analysis of proton binding by a peat humic acid using a simple electrostatic model. Geochim Cosmochim Acta, 59(6):pp.1101– 1112.

Nair K M, Rao V P (1971) Result of shallow drilling in the area north of Pondicherry. Unpublished ONGC field season report of 1969-1970.

Perminova I V, Frimmel F H, Kudryavtsev AV, Kulikova N A, Abbt-Braun G, Hesse S, Petrosyan V S (2003) Molecular Weight Characteristics of Humic Substances from Different Environments as Determined by Size Exclusion Chromatography and their Statistical Evaluation. Environmental Science and Technology. 37(11): 2477-2485.

Pethaperumal S (2010) Study on groundwater chemistry in the Pondicherry region. Ph.D. thesis, Annamalai University, India.

Pieper A P, Ryan J N, Harvey R W, Amy G L, Illangasekare T H, Metge D W (1997) Transport and recovery of bacteriophage PRD1 in a sand and gravel aquifer: effect of sewage-derived organic matter. Environ. Sci. Technol, 31: pp.1163–1170.

Plummer LN, Jones BF, Truesdell AH (1976) WATEQF – A Fortran IV Version Of WATEQ, A Computer Program For Calculating Chemical Equilibrium Of Natural Waters. US Geol Surv Water Resources Investigations Rept, 76 – 13, 61p.

Prasanna M V, Chidambaram S, Shahul Hameed A, Srinivasamoorthy K (2010) Study of evaluation of groundwater in Gadilam basin using hydrogeochemical and isotope data. Environ Monit Assess 168:63–90.

Reuter J H, Perdue E M (1977) Importance of heavy metal-organic matter interactions in natural waters. Geochim. Cosmoch. Acta, 41, 325–334.

Santos I R, Burnett W C, Dittmar T, Suryaputra I G N A, Chanton J (2009) Tidal pumping drives nutrient and dissolved organic matter dynamics in a Gulf of Mexico subterranean estuary. Geochim. Cosmochim. Acta 73: 1325–1339, doi:10.1016/j.gca.2008.11.029

Scholl M A, Harvey RW (1992) Laboratory investigations on the role of sediment surface and groundwater chemistry in transport of bacteria through a contaminated sandy aquifer.Environmental. Science and. Technology, 26:1410-1417.

Schussler W, Artinger R, Kim J I, Bryan N D, Griffin D (2001) Numerical modeling of humic colloid borne americium(III) migration in column experiments using the transport/speciation code K1D and the KICAM model. J. Contam. Hydrol., 47: 311–322.

Sommer M (2006) Influence of soil pattern on matterntransport in and from terrestrial biogeosystems: A new concept for landscape pedology. Geoderma, 133(1/2):107–123.

Thilagavathi R, Chidambaram S, Prasanna MV, Thivya C, Singaraja C (2012) A study on groundwater geochemistry and water quality in layered aquifers system of Pondicherry region, southeast India. Applied Water Science, 2, 253-269.

Thilagavathi R, Chidambaram S, Ramanatahan A L, Rao M S, Prasanna MV, Tirumalesh K, Pethaperumal S (2013) An Attempt to evaluate the influence of geomorphology on the hydrogeology of coastal aquifer. Indian Journal of Geomorphology, 18(2): 103-114.

Thilagavathi R, Chidambaram S, Thivya C, Prasanna MV, Pethaperuamal S, Tirumalesh K (2014) A Study on the Behaviour of Total Carbon and Dissolved Organic Carbon in Groundwaters of Pondicherry Region, India. International Journal of Earth Sciences and Engineering, 07, 1537-1550.

Thurman E M (1985) Organic Geochemistry of Natural Waters. M. Nijhoff; Distributors for the U.S. and Canada, Kluwer Academic, Dordrecht ,Boston, Hingham, MA, USA

Tipping E, Hurley M A (1992) A unifying model of cation binding by humic substances. Geochim Cosmochim Acta, 56(10): 3627-3641.

Venturelli G, Boschetti T, Vittorio D (2003) Na carbonate waters of extreme composition: Possible origin and evolution. Geochem J, 37: 351–366.

Zeh P, Kim J I, Marquardt CM, Artinger R (1999) The reduction of Np(V) in groundwater rich in humic substances. Radiochim. Acta, 87: 23–28.

Chapter **14**

A Study on the Lithological Influence on Uranium Concentration in Groundwater of Hard Rock Aquifers of Madurai District, Tamil Nadu

Thivya C., Chidambaram S.[1], Thilagavathi R.[1]
Prasanna M.V.[2] and Singaraja C.[3]

Department of Geology University of Madras, India
[1] Department of Earth Sciences, Annamalai University, India
[2] Faculty of Engineering and Science, Curtin University, CDT 250, 98009 Miri Sarawak, Malaysia
[3] Department of Geology, Presidency College, Chennai-5, India

ABSTRACT

Groundwater is most favored as a source of drinking water in all over world. It is often thought to be cleaner and easier to treat when compared with surface water and as a result, many wells have been sunk. Uranium and other radionuclides in water can lead to health problems if present in higher concentrations. In order to study the behavior and to understand the influence of lithology in groundwater of Madurai district, a study was conducted by collecting a total of 216 samples of different seasons (Premonsoon, South west monsoon, Northeast monsoon and Post monsoon) with respect to lithology and the number of samples was depending upon their spatial coverage in the study area. The groundwater samples were analysed for U, Ca^{2+}, Mg^{2+}, Na^{+}, K^{+}, HCO_3^{-}, Cl^{-}, SO_4^{2-}, PO_4^{3-}, NO_3^{-}, F^{-}, H_4SiO_4, pH, EC, TDS. It was identified that the higher concentration of Uranium (156 ppb) were noted in the northern part of the study area. They are chiefly restricted to the Granitic terrain and nearby Charnockitic regions. The weathering process induces the release of higher U ions from the granitic terrain and the evaporated recharge water from the tank / reservoir also induces the release of U ions from the rock matrix, it may also be due to fertilizer enriched agricultural return flows. The key source for all these processes and hydrogeochemical processes is identified to be lithology which inturn controls all the other processes in the study area.

Key words: Groundwater Uranium, Madurai, Granite, Charnockite.

Introduction

Uranium concentration in groundwater is important in understanding the radiological impact valuation to secure the standard of life. A number of studies have been carried out in different places in order to study uranium concentrations in drinking waters (Chau and Michalec, 2009; De Camargo and Mazzilli 1996; Killiari and Pashalidis 2010). The concentration of uranium in various rivers of India such as Yamuna (0.09–3.61 ppb) and Chambal (0.2– 1.74 ppb) (Rengarajan et al 2006), Bhagirathi (2.11–3.96 ppb) and Alakanda (1.86 ppb) (Sarin et al 1992) have been studied earlier. Springs and streams in Himachal Pradesh, India had 0.07 to 4.65 ppb of uranium (Singh et al 1999). Similarly uranium concentration in groundwater has been reported in several parts of India – Kumaun and Siwalik (1.08 ± 0.02 to 35.83 ± 0.09 ppb) (Ramola et al 1988), Punjab (11.71 ± 0.15 to 113.70 ± 0.46 ppb) (Singh et al 1995), Kolar district (0.3 to 1442.9 ppb) (Babu et al 2008) and Upper Siwaliks (1.08 ± 0.03 to 19.68 ± 0.12 ppb) (Singh et al 2009). The study also was carried out in Nalgonda district, Andhra Pradesh, southern India where occurrence of unconformity related uranium deposits have been reported (Sinha et al 1995). The presence of indoor radon/thoron concentration in this area was reported by Reddy et al (2003). Singh et al (2002) studied the uranium concentration (0.5 to 410 ppb) in groundwater in Palnadu sub-basin. The causes for the presence of high U concentration in groundwater of parts of Nalgonda district was reported by Brindha et al (2011). The higher uranium levels in granitic terrains of Madurai district of Tamilnadu has been studied by Thivya et al. (2014, 2015). Groundwater comprises concentrations of most constituents depend upon their contact with host rock and their concentration varies with the residence time in the hard rocks. Hydrogeochemical processes are responsible for the change of chemical composition of groundwater thus varies with the space and time.

The Kidneys and bones are the major locates for the accumulation of uranium in the body (Stevens et al., 1980; Morrow et al., 1982; Zamora et al, 1998) and skeleton is the major site for the long term storage of Uranium (Wrenn and Singh, 1982). Even lower concentration of Uranium in drinking water also causes the nephrotoxic effects (Kurttio et al. 2002). Hence the present study has been carried out in Madurai district of Tamilnadu to understand the influence of lithology on uranium levels in groundwater.

Table 1. Maximum, Minimum and Average concentration of Uranium with lithologies in groundwater

Seasons	Statistics	Fissile hornblende Biotite gneiss	Charnockite	Quartzite	Granite	Floodplain alluvium
PRM	Maximum	29.30	8.80	13.11	113.00	4.10
	Minimum	0.20	0.20	3.19	3.20	0.30
	Average	5.75	2.72	6.56	32.78	1.34
SWM	Maximum	13.85	20.40	4.73	156.84	16.59
	Minimum	0.56	0.60	2.83	1.77	1.42
	Average	3.72	3.25	3.84	52.62	6.10
NEM	Maximum	13.15	37.20	5.72	46.70	17.31
	Minimum	1.75	1.17	4.41	3.29	2.06
	Average	4.42	5.77	5.21	25.00	6.68
POM	Maximum	17.26	116.32	3.49	37.72	15.81
	Minimum	2.14	1.28	1.40	3.35	3.91
	Average	5.90	10.45	2.41	13.64	7.70

The WHO (2006) standard limit for Uranium in groundwater is 15 ppb. In the study area 8% of PRM, SWM and POM samples and 9% of NEM samples are above the permissible limit (Fig. 2).

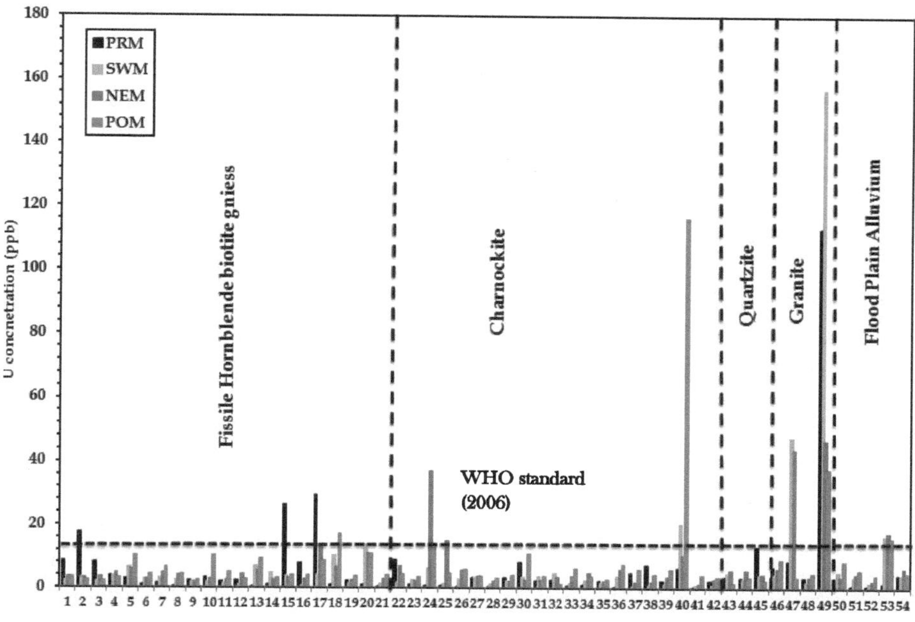

Fig. 2. Uranium concentration in different lithologies compared to WHO (2006) standard

The comparison between the U concentration and groundwater level indicate that the uranium concentration varies primarily due to recharge and discharge. The rainfall recharge increases the groundwater level which in turn results in the dissolution of uranium and increasing concentration in groundwater in SWM (Thivya et al. 2015). The recharging water reacts with the weathered rocks in the unsaturated zone and thus leaches out uranium increases uranium concentration in groundwater. However, as the recharge continues concentration of uranium in groundwater begins to reduce due to dissolution and anthropogenic activities during NEM.

Table 2. The orders of dominance of U in different lithologies in various seasons

Seasons	Lithology
PRM	Granite > Quartzite > Fissile hornblende biotite gneiss > Charnockite > Flood Plain alluvium
SWM	Granite > Flood Plain Alluvium > Quartzite > Fissile hornblende biotite gneiss > Charnockite
NEM	Granite > Flood Plain alluvium > Charnockite > Quartzite > Fissile hornblende biotite gneiss
POM	Granite > Charnockite > Fissile hornblende biotite gneiss > Flood plain alluvium > Quartzite

In the study area the U concentration in Granitic rock terrains shows a maximum of 113ppb, 156.84ppb, 46.70ppb and 37.72ppb in PRM, SWM, NEM and POM respectively (Table.2). The higher concentration of U is observed in Granitic rocks in all seasons except POM. In Flood plain alluvium U shows a maximum of 4.10 ppb in PRM, 16.59 ppb in SWM, 17.31ppb in NEM and 15.81 ppb in POM. U in groundwater has been reported upto 47 ppb from Holocene and Pleistocene alluvial aquifers in Bangladesh (BGS and DPHE 2001). In Charnockite U shows a maximum of 8.80 ppb, 20.40 ppb, 37.20 ppb and 116.32 ppb whereas in Quartzite it is 13.11 ppb, 4.73 ppb, 5.72 ppb and 3.49 ppb in PRM, SWM, NEM and POM respectively. The concentration of U in groundwater of Srilanka from Charnockite and quartzite formations ranges from 0.017–6.4 ppb (BGS 2005). In Fissile Hornblende biotite gneiss U has a maximum concentration of 29.30 ppb, 13.85 ppb, 13.15ppb and 17.26 ppb in PRM, SWM, NEM and POM respectively. The higher concentration is observed in SWM mainly due to infiltrating rainfall recharge flushes the U which results in increase of uranium in groundwater along with the rise in water level (Brindha and Elango 2012). The study area is also reported comprise pink granitic intrusions (Ingo Braum 2006) and the zone of higher U concentration is found around the granitic terrains enhances the leaching of U to the groundwater system (Munazzamali et al. 2011).

Monazite, apatite and sphene maintain consistent U concentrations during weathering, even following mechanical break-up of monazite grains and replacement of sphene by rutile and clays. This confirms that these primary minerals incorporate U by crystal lattice substitution (Fowler 1981; Speer et al. 1981). Subsequent alteration results in increased U concentrations by adsorption and ionic exchange. It has been reported that U is mainly hosted by the magnetite and allanite in the study area and present as independent grains with flaky graphite and also as inclusions in quartz (Pandey and Krishnamurthy 1995). These heavy minerals accumulate U and other 'incompatible' elements from magma in the later stages of crystallization (Porcelli and Swarzenski 2003). The Tungsten mineralization (Skarn type) in Calc-granulites associated with Neoproterozoic granites is reported in the Karungalakudi area (GSI 1995).

Effect of pH with U

In general, U solubility is highest at a pH of 2 (U aqueous concentration ~0.01 M), decreasing rapidly as pH increases to 4 (U aqueous concentration ~10^{-8} M), leveling out up to pH 8, then increasing slowly as pH climbs to 10 (U aqueous concentration ~10^{-6} M, Langmuir, 1997). As pH increases, the stability of the mineral species becomes more significant until precipitation occurs. As in the aqueous phase, most of the hexavalent U minerals contain U in the UO_2^{2+} form. The high uranium concentrations at almost neutral pH are related to the strong complexation of uranyl ion with aqueous carbonate. Other strong complexants of uranyl ion include phosphate and fluoride.

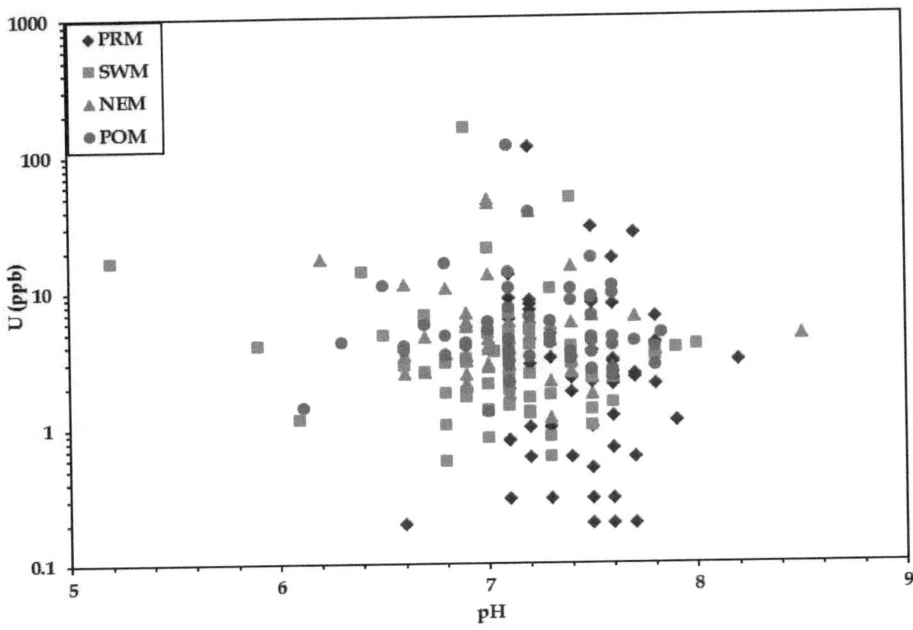

Fig. 3. Plot of U vs pH in groundwater samples of all seasons

Generally, decreasing pH results in the dissolution of soil minerals and since uranium is adsorbed or forms solid solution with the geological matrix, decreasing pH results in increasing uranium concentration in solution. In the study area it is noted that U decreases with the increase in pH as observed by Ioannidou et al. (2011) irrespective of seasons (Fig.3). Uranium concentration in each season has different ranges of pH which indicates that pH may be a governing factor for the distribution of U in groundwater. Higher concentration of U is observed in the near neutral pH. Jurgens et al. (2009) suggests that the adsorption of U is strongest above neutral pH values and consequently high pH values tend to affect the adsorption of U which is again a function of pCO_2 of groundwater. This higher pH reduces the U solubility in water. It is also noted that higher concentration of U occurs in a specific range of pH in the granitic terrain.

Effect of U with EC

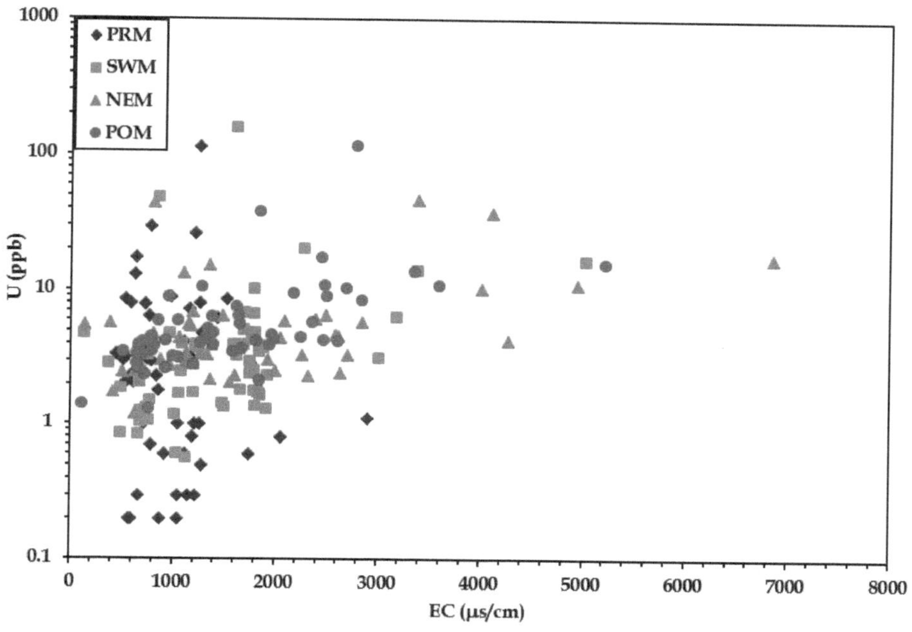

Fig. 4. Plot of U vs EC in groundwater samples of all seasons

The relationship between U and EC is plotted in the Fig.4. It is interesting to note that U concentration increases with increase of EC in SWM, NEM and POM, which may be attributed to the increase in residence time of water (Ioannidou et al. 2011). In PRM there are two trends are noted, they are

1. U increases with decrease in EC
2. U increases with increase in EC

The study area being a hard rock terrain, the residence time of water is generally higher in cracks and fissures if they are not interconnected sufficiently, which ultimately increases the U concentration. The increase of EC may also be due to pH induced weathering in the aquifer (Brasser et al. 2002), as U concentrations in groundwater decreased with flow direction due to U-retention by sorption processes.

Effect of pCO_2 with HCO^-_3 and U

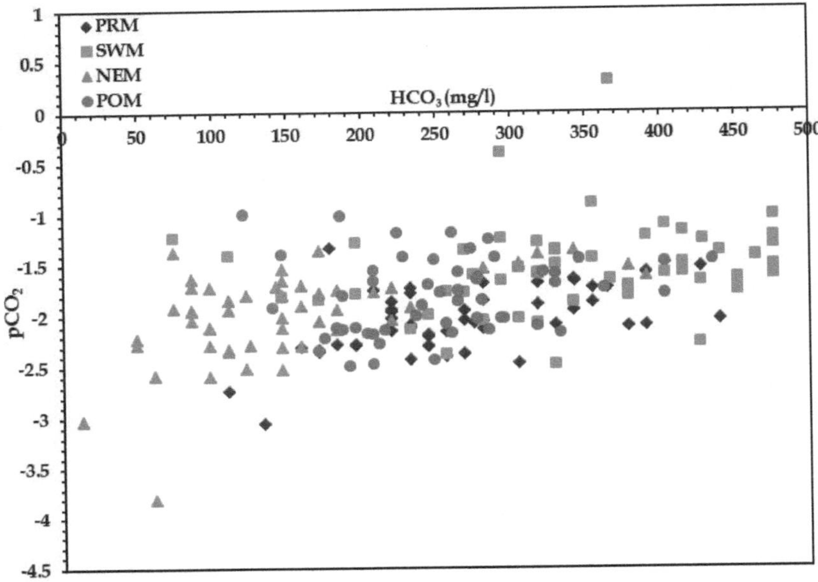

Fig. 5. Plot of pCO_2 Vs HCO^-_3 in groundwater samples of all seasons

pCO_2 is a substantial parameter and significant factor for the enhancement of solubility in carbonate rocks. pCO_2 in the subsurface is enhanced by the dissolution of minerals like mainly plagioclase, as a result it increases the pH, further formation of bicarbonate and results in accumulation of more cations to the remaining water. Subsequently, higher pCO_2 increases the weathering products. In the study area, the Log pCO_2 ranges from -0.38 to 0.2. Fig. 5 shows that there is an increase in pCO_2 with respect to the increase of bicarbonate. There is an increasing trend in all the samples irrespective of seasons. The higher value of pCO_2 also indicates longer residence time (Prasanna et al., 2008). Study area being a hard rock terrain residence time increases the leaching of U ions into groundwater. Higher pCO_2 increases the bicarbonate concentrations in the groundwater which in turn enhances the mobility of U and their species in groundwater of hard rocks, where the atmospheric interaction is low. Uranium is reported to be highly soluble in carbonate/ bicarbonate-rich oxidizing groundwater condition (EPA, 2000b).

Weathering of Fissile hornblende biotite gneiss and Granite (mainly chemical weathering) play an important role in leaching, distribution and dispersion of uranium from the source rock (Gueniot et al. 1988). The higher concentration of U with log pCO_2 is observed between -1 to -

1.5 (Fig.5). Water with high pCO_2 of around -1.5 indicates deep circulation of groundwater with the lesser amount of atmospheric contact or higher saturation of carbonate minerals, resulting from the leaching and interaction of country rocks of the minerals through which it flows (Chidambaram et al. 2011). The source of uranium mineralization is erosion of uraniferous granites in the region followed by transportation of uranium in mildly oxidizing saline alkaline groundwater solutions in constricted drainages to semi-closed basins with variable evaporative conditions.

Spatial distribution of U and lineament

Generally lineaments are underlain by zones of localized weathering and increased permeability and porosity. Meanwhile, some researchers studied relationships between groundwater productivity and the number of lineaments within specifically designated areas or lineament density rather than the lineament itself (Hardcastle, 1995). Therefore, mapping of lineaments are closely related to groundwater occurrence and yield is essential to groundwater surveys, development and management.

The spatial distribution of U is given in fig.6 shows that the higher concentration is observed in north eastern part of the study area where it is covered by granitic terrains. The higher content of U in PRM, SWM and NEM lies in granitic terrain whereas in POM it lies in Charnockite. It is interesting to note that U concentration is also higher in lineament regions apart from the lithology. Uranium and radon in ground water mainly depend on the fractures their degree and the angle of inclination to the bedrock (Skeppstrom and Olofsson 2007). The high concentration of U in charnockite during POM may be due to the leaching of elements present in fractures. Temple (2002) hypothesized that high U concentrations were due to U-rich fractures in the granitic rocks. Only relatively deep wells contained high U concentrations, possibly because U had already been leached from shallow fractures as it behaves as a closed system.

High U concentrations have often been reported in groundwater in granitic terrains. A large range of U concentrations was observed in groundwaters from the Lac du Bonnet Granite of the Canadian Shield. The U concentrations were found to be generally highest at shallow depths (<60 m; depth range of boreholes up to 1000 m) in oxic environment groundwaters showing evidence of active groundwater circulation. The dominant controls on U concentration were redox potential and the formation of soluble anionic U-carbonate species. U increases with the accumulation of alteration phases along fractures and around the primary

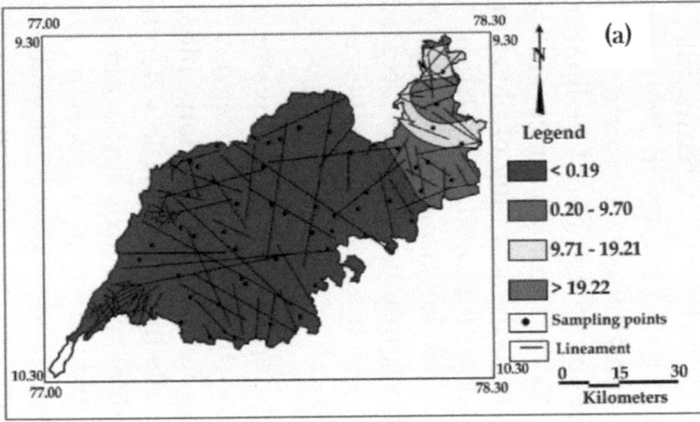

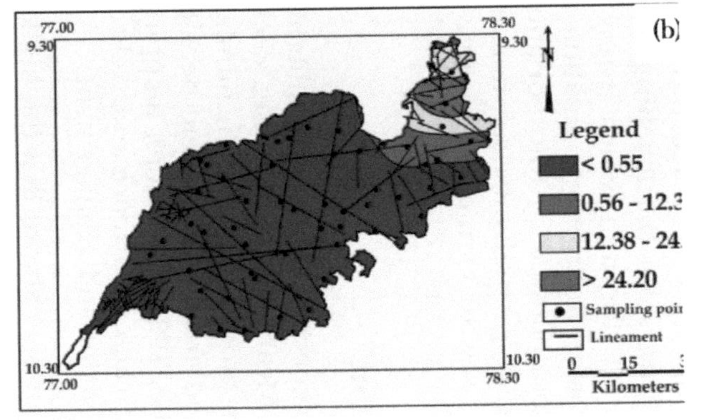

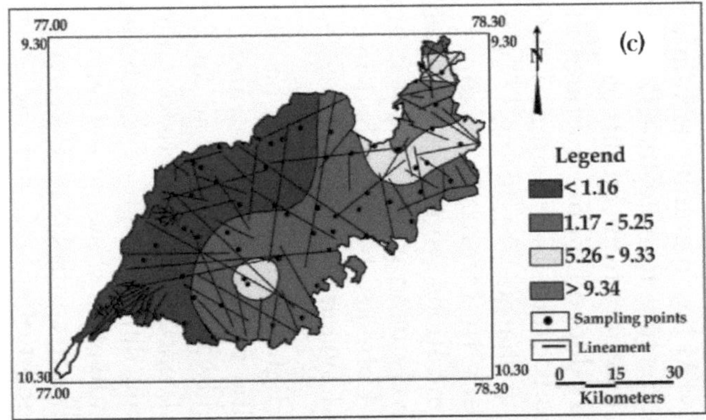

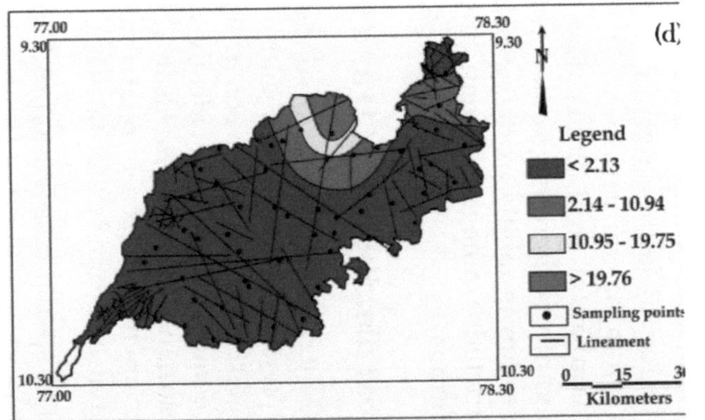

Fig. 6. Spatial distribution of U (ppb) and lineaments for all seasons a) PRM, b) SWM, c) NEM and d) POM Correlation matrix

phases. As weathering progresses, this U incorporated into the Interstitial spaces of biotite, feldspar and early alteration phases accumulate within the secondary interstitial aggregates and lose their discrete character (Langmuir 1978). High-U ground waters appear not to be a widespread feature of the granites of the region, perhaps because of the short residence times of the ground waters and their slightly acidic, low-alkalinity compositions. The sporadic nature of the U mineralisation may also be a factor.

The correlation of U with other ions indicates that the concentration of HCO_3 was positively correlated with the dissolved U concentrations suggesting that the higher concentrations and increased mobility were the result of complexation with HCO_3 (Fig.7). There was a positive correlation between uranium and alkalinity indicating the possible formation of anionic uranyl carbonato complexes, which maintain uranium in soluble form. It is well established that carbonate complexes of uranium do not strongly sorb to minerals and metal oxides (Katsoyiannis and Katsoyiannis 2006). Local association of U with HCO_3 and F suggests the possibility due to the mixing of ground waters. Higher correlations of U and F are generally reflected in the basement rock reported in ground waters draining felsic igneous rocks (Giblin 1994). Carbonate rocks produce alkaline ground waters and carbonate complexes tend to facilitate the transport of U (Boyle 1982). Negative correlation of pH resembles the redox chemical reactions in groundwater (Garrels 1960). The correlation between major ions and U indicates the weathering and leaching processes enhances the U in groundwater.

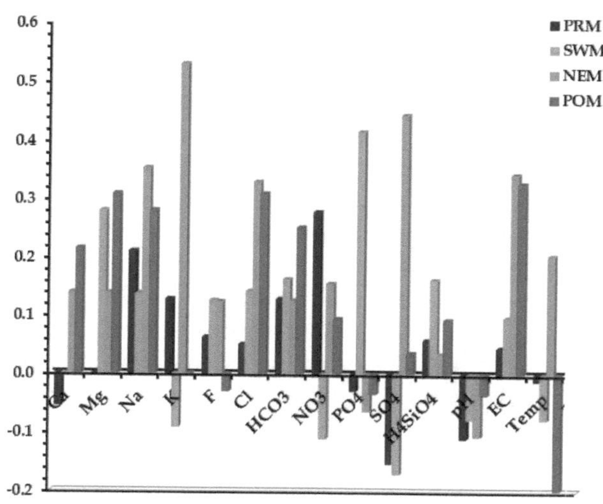

Fig. 7. Correlation coefficients of U with other parameters

The higher correlation between PO_4 and U in SWM suggest that it may be due to agricultural activities. In addition to the input derived from uranium rich rocks, use of phosphatic fertilizers for agricultural activity in this area may also add up to the uranium content in groundwater. Ioannides et al. (1997); Roselli et al. (2009) have reported the natural radioactivity in phosphatic fertilizers. The range of uranium content in the phosphate rock is about 0.005–0.02%. These phosphate rocks serve as a source of phosphatic fertilizer to enhance the soil productivity in addition to other fertilizers (Thivya et al. 2015). The total uranium resource in phosphate rock is estimated at 9 × 106 metric tonnes of uranium (Ragheb 2008). Hence, the phosphorous fertilizers manufactured form phosphate rocks may also contribute uranium to groundwater in the agricultural regions. Studies have shown that phosphate fertilisers possess uranium concentration ranging from 1 mg/kg to 68.5 mg/kg (Kawabata et al 2006). Hence this also will add up to the uranium concentration in groundwater that resulted due to uranium rich rocks. The correlation between EC, Cl and NO_3 with U reveals that in shallow groundwater is contaminated with anthropogenic stress resulting high U in oxidized groundwater.

Conclusion

The study conducted by collecting samples from different lithologies in different seasons to achieve the aforesaid objectives. The study show that the average concentration of U increases with monsoon and highest is observed during SWM. The higher content of U in PRM, SWM and NEM lies in granitic terrain whereas in POM it lies in Charnockite. Higher U concentration and its relation to pH indicate that pH is the governing factor for the distribution of U and higher U is noted in near-neutral pH. The U concentration increases with the increase in EC up to 1500 μs/cm, indicating longer residence time of water in the rock matrix facilitating preferential leaching of U into groundwater. It is interesting to note that U concentration is also higher in lineament regions apart from the lithology. The presence of uranium associated of allanite and magnetite in Charnockite are responsible for higher uranium in groundwater. Further in granitic, weathering predominantly with apatite enhances the uranium in groundwater. The higher concentration of U in charnockite during POM may be due to the leaching of elements present in fractures. Only relatively deep wells contained high U concentrations, possibly because U had already been leached from shallow fractures as it behaves as a closed system. saturated state. Higher pCO_2 increases the bicarbonate concentrations in groundwater, which enhances the mobility of U and their species in groundwater of this region. The correlation between the

H4SiO_4 and U indicates that the weathering of silicate minerals releases higher amount of bicarbonate into the groundwater. The weathering process and the lithology are major controls on the concentration of U in the study area.

Acknowledgments

The authors wish to express their thanks to University Grants Commission (UGC), India, for providing the necessary financial support to carry out this study with vide reference to UGC letter No. F: 39143/ 2010(SR) dated 27.12.2010.

References

APHA (1995) Standard methods for the examination of water and waste water, 19th edition, APHA, Washington DC, USASS.

Babu M N S, Somashekar R K, Kumar S A, Shivanna K, Krishnamurthy V, Eappen K P (2008) Concentration of uranium levels in groundwater Int. J. Environ. Sci. Tech., 5 (2), pp. 263-266.

BGS, DPHE (2001) Arsenic contamination of groundwater in Bangladesh, BGS Technical Report, Wc./00/19.

Boyle E A , Huested S S, Grant B C (1982) Chemical mass balance in the Amazon Plume II: copper, nickel and cadmium, Deep Sea Research, 29, pp.1355-1364.

Brasser T, Noseck U, Schonwiese D (2002) Uranium migration in argillaceous sediments as analogue for transport processes in the far field of repositories (Heselbach Site, Germany). Merkel, B. J., Planer-Friedrich, B. and Wolkerdorfer, C. (eds). Proceedings of the International Conference Uranium Mining and Hydrogeology III and the International Mine Water Association Symposium Freiberg, Germany, 15-21 September, pp. 139-146.

Brindha K, Elango L (2012) Impact of tanning industries on groundwater quality near a metropolitan city in India. Water Resour. Manag. 26(6), pp. 1747–1761.

Brindha, Elango L, Nair N R (2011) Spatial and temporal variation of uranium in a shallow weathered rock aquifer in southern India. J. Earth Syst. Sci., 120, No. 5, pp.911–920.

CGWB (2007) Ground water resources and development prospects in Madurai region.

Chau N D, Michalec B (2009) Natural radioactivity in bottled natural spring, mineral and therapeutic waters in Poland. J Radioanal Nucl Chem, 278, pp.121–129.

Chidambaram S, Karmegam U, Prasanna M V, Sasidhar P, Vasanthavigar M (2011) A study on hydrochemical elucidation of coastal groundwater in and around Kalpakkam region, Southern India Environmental Earth Sciences, Vol 64,1419-1431 ISSN 1866-6280.

De Camargo I M C, Mazzilli B (1996) Determination of uranium and thorium isotopes in mineral spring waters. J Radioan Nucl Chem, 212(4), pp. 251–258.

EPA (2000) Technical fact sheet: final rule for (non –radon) radionuclides in drinking water. EPA 815-F-00-013. Available from: www.epa.gov/ safewater/rads/technicalfacts.html

Fowler M B (1981) U content, distribution and migration in the Glendessary syenite, Inverness- shire. Mineral. Mag, 44, pp. 443-448.

Garrels R M, (1960) Mineral Equilibria: New York, Harper & Brothers, 254 p.

Giblin A (1994) Natural waters as sample media in drainage geochemistry. In Handbook of Exploration Geochemistry, 6 (ed. M. Hale and J. A. Plant), pp. 269-303. Elsevier Science.

GSI (Geological Survey of India) (1995) Geological and mineral map of Tamil Nadu and Pondicherry, 1:500,000.

Gueniot B, Munier Lamy C, Berthelin J (1988) Geochemical behaviour of uranium in soils, part II. Distribution of uranium in hydromorphic soils and soil sequences. Applications for surficial prospecting. Journal of Geochemical Exploration, Vol. 31, Issue 1, pp. 39–55.

Hardcastle K C (1995) Photo lineament factor: A new computer-aided method for remotely sensed fractured. Photogrammetric Engineering & Remote Sensing 61 (6), pp.739-747.

Ingo Braun (2006) Pan-African granitic magmatism in the Kerala Khondalite Belt, southern India Journal of Asian Earth Sciences, 28, pp. 38–45

Ioannidou A, Samaropoulos I, Efstathiou M, Pashalidis I (2011) Uranium in ground water samples of Northern Greece.J Radio anal Nucl Chem, 289, pp.551–555.

Jurgens B C, Fram M S, Belitz K , Burow K R, Landon M K (2009a) Effects of groundwater development on uranium: Central Valley, California, USA. Ground Water, 49, pp. 913–928.

Katsoyiannis I A, Katsoyiannis AA (2006) Arsenic and other metal contamination of groundwaters in the industrial area of Thessaloniki, Northern Greece. Environ Monit Assess, 123, pp.393–406.

Kawabata Y, Yamamoto M, Aparin V, Ko S, Shiraishi K, Nagai M, Katayama Y (2006) Uranium pollution of water in the western part of Uzbekistan. J. Radioanal. Nucl. Chem. 270(1), pp. 137–141.

Killiari T, Pashalidis I (2010) Simplified alpha-spectroscopic analysis of uranium in natural waters after its separation by cation- exchange. Radiat Measurem, 45(8):pp.966–968.

Kurttio P, Auvinen A, Salonen L, Saha H, Pekkanen J, Makelainen I, Vaisanen S B, Penttila I M, Komulainen H (2002) Renal effects of uranium in drinking water. Environmental Health Perspectives, 110, pp.337-342.

Langmuir D (1978) Uranium solution-mineral equilibria at low temperatures with applications to sedimentary ore deposits. Geochimica et Cosmochimica Acta 42, no. 6, pp. 547–569.

Langmuir D (1997) Aqueous Environmental Geochemistry. Prentice Hall, pp. 600.

Morrow P, Gelein R, Beiter H, Scott J, Picano J, Yuile C (1982a) Inhalation and intravenous studies of UF6/UO_2F_2 in dogs. Health Phys. 43, pp. 859-873.

Munazzam Ali, Aqeel Ahmed Shariff, Nasser Ali Qamar , Amanullah Laghari (2011) An appraisal of uranium source potential of granites, associated felsic rocks, kaolin and calcretes of Nagar Parkar area, Tharparkar Pakistan Journal of Himalayan earth sciences. Vol. 45(2).

Pandey, Krishnamurthy (1995) Uranium and thorium abundances in some graphite-bearing Precambrian rocks of India and implications. Research Communication, Current Science, Vol.68, No.8.

Porcelli D, Swarzenski P W (2003) The behavior of U- and Th-series nuclides in groundwater. 317-361 in Uranium-Series Geochemistry.Reviews in Mineralogy & Geochemistry, pp. 52.

Prasanna M V, Chidambaram S, Pethaperumal S, Srinivasamoorthy K, John Peter A, Anandhan P, Vasanthavigar M (2008) Integrated geophysical and chemical study in the lower sub basin of Gadilam River, Tamilnadu, India. Environmental Geosciences, v. 15, no. 4, pp. 145–152

Ragheb M (2008) Uranium resources in phosphate rocks, available from: https://netfilesuiucedu/mragheb/www/NPRE%20402%20ME%20405%20Nuclear%20Power%20Engineering/Uranium%20Resources%20in%20Phosphate%20Rocks.pdf.

Ramesh R, Anbu M (1996) Chemical methods for environmental analysis- water and sediments, pp.161.

Ramola R C, Singh S, Virk H S (1988) Uranium and radon estimation in some water samples from Himalayas; Nuclear Tracks and Radiation Measurement 15(1–4) pp.791–793.

Reddy K V K, Reddy B S, Reddy M S, Reddy Ch G, Reddy P Y, Reddy K R (2003) Baseline studies of radon/thoron concentration levels in and around the

Lambapur and Peddagattu areas in Nalgonda district, Andhra Pradesh, India; Radiat. Meas, 36, pp. 419–423.

Roselli C, Desideri D and Meli M A (2009) Radiological characterization of phosphate fertilisers: Comparison between alpha and gamma spectroscopy. Microchem. J. 91(2), pp.181–186.

Sarin M M, Krishnaswami S, Sharma K K, Trivedi J R (1992) Uranium isotopes and radium in the Bhagirathi–Alaknanda river system: Evidence for high uranium mobilization in the Himalaya. Curr. Sci. 62(12), pp.801–805.

Singh B, Singh G, Sandhu A S, Singh S (1999) Uranium estimation in water samples collected from some areas of Himachal Pradesh, India; Radiat. Meas. 31.pp. 683–685.

Singh J, Singh H, Singh S and Bajwa B S (2009) Estimation of uranium and radon concentration in some drinking watersamples of Upper Siwaliks, India. Environ. Monit. Assess, 154, pp.15–22.

Singh J, Singh L, Singh S (1995) High U-contents observed in somedrinking waters of Punjab, India J. Environ. Radioact, 26 (3), pp. 217-222.

Singh R V, Sinha R M, Bisht B S and Banerjee D C (2002) Hydrogeochemical exploration for unconformity related uranium mineralization: Example from Palnadusub-basin Cuddapah Basin, Andhra Pradesh, India. J. Geochem. Explor, 76, pp. 71–92.

Sinha R M, Shrivastava V K, Sarma G V G, Parthasarathy T N (1995) Geological favourability for unconformity-related Uranium deposits in northern parts of the Cuddapah Basin: Evidences from Lambapur uranium occurrences, Andhra Pradesh, India; Exploration and Research for Atomic Minerals 8, pp.111–126.

Skeppstrom K, Olofsson B (2007) Uranium and radon in groundwater. European water 17/18, pp.51-62.

Speer J A, Solberg T N, Becker S W (1981) Petrography of the uranium-bearing minerals of the Liberty Hill Pluton, South Carolina: Phase assemblages and migration of uranium in granitoid rocks. Econ. Geol, 76, pp. 2162-2175

Stevens W, Bruenger FW, Atherton D R, Smith J M, Taylor G N (1980) The distribution and retention of hexavalent ^{233}U in the beagle. Rad. Research, 83, pp.109-126.

Temple T (2002) personal communication via email, June 14.

Thivya C, Chidambaram S, Thilagavathi R, Tirumalesh Keesari, Rao MS, Prasanna MV, Nepolian M (2015) Occurrence of high uranium and radon in hard rock aquifers of south India- Evaluating the temporal and spatial trends. Groundwater for Sustainable Development, 68-77.

Thivya C, Chidambaram S, Tirumalesh K, Prasanna MV, Thilagavathi R, Nepolian M (2014) Occurrence of the radionuclides in groundwater of crystalline hard rock regions of central Tamil Nadu, India. Journal of Radio analytical and nuclear chemistry, 302:1349–1355.

WHO (World Health Organization) (2006) Guidelines for drinking-water quality. Third edition. Vol 1, Recommendations. Geneva, ISBN 92 4 154696

Wrenn M D, Singh N P (1982) Comparative distribution of uranium, thorium, and plutonium in human tissues of the general population. In: Natural Radiation Environment. Edited by Vohra, K.C. Wiley Eastern Ltd, New Delhi, pp. 144-154.

Zamora M L, Tracy B L, Zielinski J M (1998) Chronic ingestion of uranium in drinking water: a study of kidney bioeffects in humans. Toxicol Sci, 43: 68–77.

Chapter **15**

A Study on the Impact of Landuse Pattern on Groundwater in and Around Chidambaram Town

Vasudevan U., Chidambaram S., Thivya C.[1]
Thilagavathi R. and Devaraj, N.

Department of Earth Sciences, Annamalai University, Annamalai Nagar
Department of Geology University of Madras, India

ABSTRACT

Amongst all types of major land-use change, extending irrigated agriculture using a surface-water source has the greatest influence on groundwater - both significantly increasing recharge and changing water quality as excess irrigation water infiltrates into shallow aquifers. Hence, in order to understand the impact of the landuse on the groundwater system the study has been aimed to understand the process by adopting geophysical, geochemical and microbial analysis of groundwater samples. The study has been conducted in Chidambaram taluk, with a total area of 92.51Sq.km. A total of 39 groundwater samples were collected for the analysis of major cation and anion along with the microbial analysis, like Total Coliform, Total Viable Counts, Fecal coliforms, Vibrio cholera and Total Streptococci. The geochemical studies reveal that there is no correlation between these parameters and the microbial population. The migration of the water from the elevated regions (urban region) is witnessed and the concentration of the microbial population is higher in these regions except in the southern region which has impermeable clay and this arrest the movement of water. The source of the microbial population based on the land use pattern and the distribution of factor score analysis is believed to be the urban sewage and according to the migration of water it is noted to have higher levels in the north western region near the Vellar river, south eastern region near the landfill and then along the southern part of the study area.

Key words: Groundwater, Land Use, Geophysical, Geochemical, Microbes.

Introduction

Groundwater for a long time has been thought completely free of microbial contaminants and viruses, believing that vertical transport times are long enough and microbial survival too short to reach the aquifers. However, the risks of water contamination are obvious when having a look at the manifold Small and huge endemic outbreaks from pathogenic microbes and viruses in the last two centuries which could be linked to contaminated groundwater and drinking water consumption (Andersson and Bohan, 2001; OECD, 2003). Some relevant pathogenic microorganisms even became or always have been members of natural microbial communities. Others may survive for years and decades or even reproduce in the environment. There are multiple sources of contamination including wastewater entering different environmental compartments and manure applied to fields in agricultural areas. Pressures on terrestrial and aquatic environments may increase in the future due to global change. It may be suggested that pathogenic microorganisms and viruses will more often enter soil, river banks and aquifers along with irregular storms, flood and drought events (Foster and Chilton, 2003; Schwarzenbach et al., 2010). The contamination is due to: infiltration and percolation of untreated sewage from domestic or agricultural sources, livestock waste, latrines and septic tanks, use of cattle dung as fertilizer and poor maintenance of wells. The consumption of these waters exposes the population uses to serious health risks. The coexistence of other pathogens with the counted fecal indicators will affect the health of users. The fact that the consumption of water containing bacteria of fecal or other origin increases the number of the intestinal flora which engendered metabolic imbalances (Collignon and Butel, 2004). Indeed, a risk of gastroenteritis with a limited number of FS (3 to 10 bacteries/ 100ml) was highlighted (Zmirou, 1987).The bacteriological study of water is of particular significance in pollution studies as it is a direct measurement of deleterious effects of pollutants for human health. It is important to monitor not only the physiochemical characteristics of sea water but also the microbiological water quality for the overall assessment of water quality (APHA, 1998). Hence in order to have a comprehensive view of groundwater contamination in the coastal aquifer, a study relating the geological condition, geochemistry and the microbial population has been attempted.

Study Area

The study area falls in and around Chidambaram, in the North latitude of 11°27′42" and 11°21′51" and East Longitude of 79°40′13"

79°48′41" forming a part of Cuddalore district. The total aerial extent of the study area is 92.51Sq.km (Fig.1). As per the administrative setup the Chidambaram is divided into three blocks Bhuvanagiri, Parangipettai and Keerapalayam.

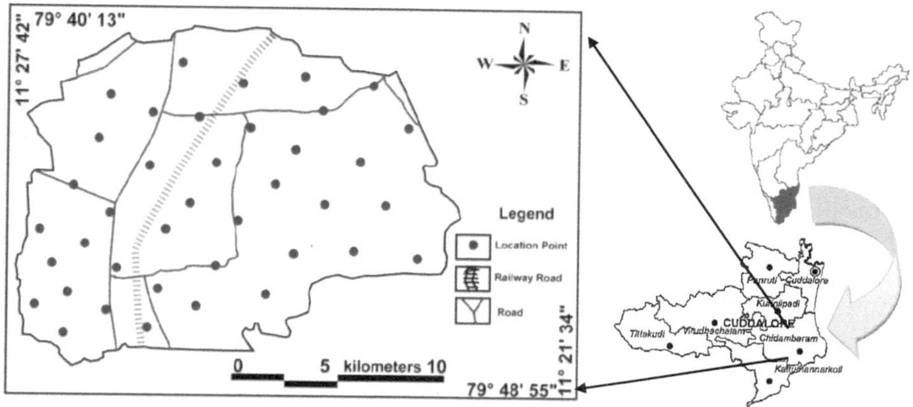

Fig. 1. Location map of the study area

The area has a hot tropical climate. The summer season, which is very oppressive, is from March and May. The temperature ranges from a maximum of 32.7 °C (90.9 °F) to a minimum of 24 °C (75 °F). The maximum and minimum rainfall received is 1400 mm and 1200 mm. Most of the agricultural lands are irrigated by canal water, nevertheless. There is average rainfall of 1300mm/yr. The soil types are alluvial and red those are conducive for crops like paddy, pulses and chili peppers. The soil types are classified as the black, red, and ferruginous and arenaceous. They are again subdivided into clays, loam and sand. Chidambaram area is drained by the river Kollidam which is a distribution branch of river Cauvery which is among the most sacred river in India. Cuddalore district is drained by Gadilam and Pennaiyar in the north Vellar and Coleroon in the south. All these rivers are ephemeral and carry floods during monsoon. The generally flow from west towards east and the pattern is mainly sub parallel. The area in a sedimentary formation composed of clays and sand of quaternary age. The area shows low undulating topography with the elevation of 5.75 m from the mean sea level. The study area comes under porous sedimentary formation type which holds large amount of water. The area contains mainly of aquifer and aquicludes. The first sub artesian aquifer found at a depth of 200 meters.

Methodology

The groundwater samples were collected in polyethylene bottles with inside stopper during the January 2014. Totally 39 groundwater samples were collected. One liter of water sample was collected in polyethylene bottle. Then it was sealed and brought to the laboratory for analysis and stored properly (4°C) before analysis. pH and electrical conductivity (EC) were determined in the field using Field Analysis Kit (Eutech Handheld Instruments). The collected groundwater samples were analyzed for major cations and anions. Calcium, magnesium, bicarbonate, and chloride were determined by titrimetric method. Sodium and potassium were analyzed through flame photometry (ELICO CL 378). The range of flame photometry is 1–100 ppm with an accuracy of ±1 digit. Silica, phosphate, and sulfate were determined by spectrophotometry (ELICO SL 171 minispec).

Plates for the analysis of microbes were prepared 5 days prior to sampling. The bacterial population in different samples was estimated by spread plating method on selective medium plates with 0.1 ml of suitable dilutions. All the media plates were incubated at 37°C at least for 24 to 48 h and final counts of colonies were noted. All trials were performed in triplicate. When very high counts of any specific group of bacteria were observed in any media at 24 h incubation, water/sediment samples held in a refrigerator were diluted 10 or 100 fold and re-examined as above by spread plating method. Isolation and enumeration of bacteria were made using selective growth media such as Nutrient agar for total viable counts (TVC), MacConkey agar for total coliforms (TC), M. Enterococcus Agar for total Streptococci (TS), M-FC agar for fecal coliforms (FCLO - FC), Thiosulphate citrate bile salts sucrose (TCBS) agar for Vibrio like organisms (VLO) and Vibrio cholera -like organisms (VCLO - VC).

Results and Discussion

Hydrogen ion concentration (pH) is closely related to free CO_2 complex which during present study ranged between 6.47 to 7.58 pH (Table.1) is controlled by the carbon dioxide bicarbonate-carbonate equilibrium system. Electrical conductivity of the groundwater is varying from 497 to 8100us/cm. The TDS of ground water varied from 295 to 2550 mg/l. Majority of the groundwater samples from the study area recorded, greater values than the permissible limit (1,000 mg/l) due to successive action of weathering and dilution processes. The counts of microbes are given in Table 2.

Table 1. Maximum, minimum and average of the Chemical constituents analyzed in groundwater (All Values in mg/l except EC in μscm^{-1}, and pH)

	pH	EC	TDS	Ca	Mg	Na	K	F	Cl	HCO_3	NO_3	PO_4	SO_4	H_4SiO_4
Maximum	7.59	8100.00	2550.00	168.00	117.00	631.00	91.00	0.20	1595.25	891.00	99.00	2.70	90.00	129.00
Minimum	6.48	497.00	295.00	16.00	7.20	25.20	3.00	0.04	53.18	122.00	0.50	0.01	0.60	21.50
Average	7.16	2591.90	1198.67	71.31	41.27	255.52	36.08	0.09	510.05	379.29	38.01	0.86	15.09	60.63

Table 2. TVC, TC, TS, FC and VC (CFU/ml) counts for groundwater samples

S.No	TVC	TC	TS	FC	VC	S.No	TVC	TC	TS	FC	VC
1	64000	2600	240	150	690	21	28000	920	100	80	170
2	27000	1800	120	80	380	22	31000	1600	200	170	330
3	8300	200	180	160	80	23	10800	330	40	30	70
4	14200	500	20	20	200	24	19700	690	90	80	190
5	49000	1800	70	60	500	25	61000	2400	280	230	470
6	53000	2000	240	180	500	26	67000	2600	230	240	560
7	66000	3100	180	200	660	27	91000	4200	510	480	540
8	6200	220	320	360	70	28	7800	290	30	30	80
9	2800	50	200	20	20	29	4800	50	0	0	20
10	14000	1300	0	0	260	30	22000	1900	240	230	220
11	36000	2900	140	150	420	31	36000	4200	500	340	390
12	11700	200	350	260	70	32	14100	280	40	30	70
13	62000	2800	30	20	620	33	88000	4300	510	400	1200
14	10100	780	320	250	160	34	180000	1110	100	120	140
15	2700	40	80	80	10	35	4900	70	0	10	20
16	18000	1300	0	0	220	36	27000	2600	230	200	250
17	26000	1600	210	140	250	37	31000	2000	200	200	210
18	12700	930	160	140	160	38	2200	40	130	100	180
19	3100	30	100	80	120	39	13000	1000	130	100	150
20	78000	3200	340	280	600						

Nitrate and Microbes

Higher concentrations occur in water polluted by sewage, fertilizers, and agricultural wastes containing organic nitrogen, free ammonia or ammonium salts (WHO, 2004). There is three set of patches observed between NO_3 and Microbes in groundwater of the study area (Fig.2). The TVC is highly associated with Nitrate. Lower nitrate concentration with elevated TVC activities and also increased of nitrate concentration with increase of TVC in groundwater. Presence of high numbers of faecal-indicator bacteria and elevated nitrate concentrations, which are particularly predominant in the residential areas. The unkempt disposal of wastes and organic matter from poultry in the river which finds their way into groundwater through infiltration and leaching had contributed to this high values in the samples. Total coliform bacteria show significant relationship with the nitrate (Fig. 2).

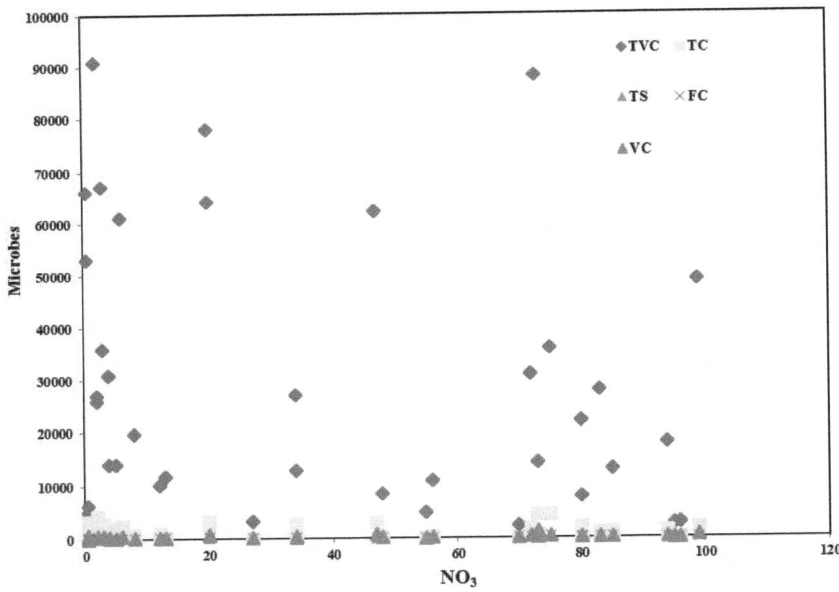

Fig. 2. Relationship of Microbes to NO_3 for groundwater samples

Few samples show that increasing the TC with nitrate (Fig. 3). Similarly, a negative relationship existed between nitrate concentration and coliform levels in some samples. They showed elevated levels of nitrate in the water which could be attributed coming from agricultural runoffs and household effluent (Kazmi and Khan, 2005). This creates as suitable environment for coliform growth as nitrate concentration becomes abundant. In turn, the nitrate concentration deceases as result of coliform uptake (Eukene et al., 2014). Total coliforms and nitrate results suggested that it was the groundwater source itself that contributed much of the microbiological and nitrate contamination rather than a storage or supply line contamination mechanism.

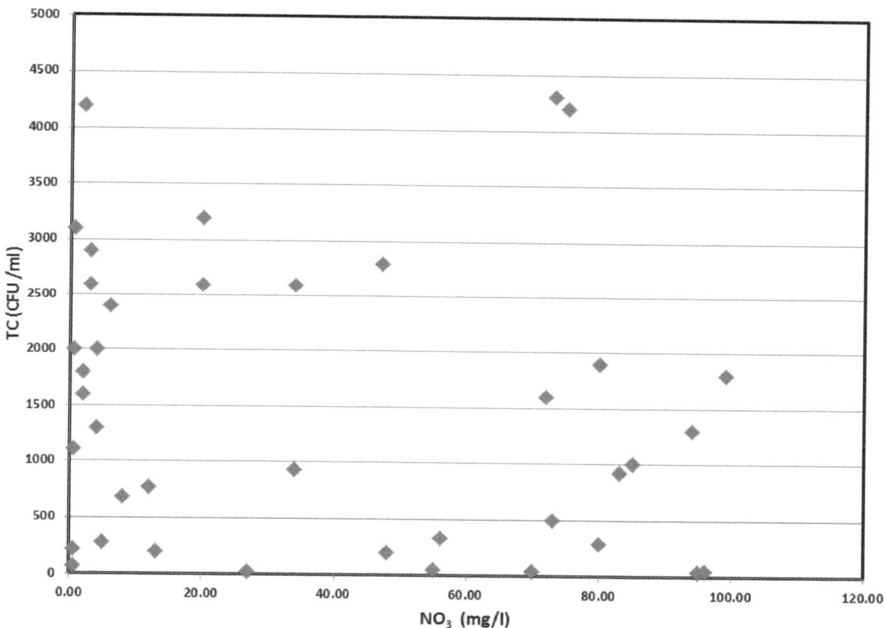

Fig. 3. Relationship of Total Coliforms to NO_3 for groundwater samples

Statistical Analysis

Correlation Matrix

A correlation matrix is established using Pearson's correlation coefficient between ion pairs. It is known as the best method of measuring the correlation, because it is based on the method of covariance (Swan and Sandilands, 1995). The correlation coefficient gives information about the degree of correlation and as well as the direction of correlation. The Pearson's correlation coefficient value lies between +1 to -1 and the degree of correlation is said to be perfect correlation if the correlation coefficient value is near ± 1. For values ranging between $\pm$ 0.75 and $\pm$ 1 it is said to be a high degree of correlation, similarly moderate degree of correlation for values between $\pm$ 0.25 and $\pm$ 0.75 and low degree of correlation for values between 0 and $\pm$ 0.25.

From the correlation Table 3 it is observed that there is no high degree of correlation is observed. But there is a moderate degree of correlation exists between Ca-TS, Mg-HCO_3, Na-K, Na-PO_4, K-NO_3, Cl-FC, HCO_3-H_4SiO_4, PO_4-H_4SiO_4, TVC-VC, TC-FC, TS-FC. Positively correlated bacterial groups indicate their common origin and they could be considered as faecal matter – microbial interaction factors. It disclosed the contribution of human fecal matter, animal droppings, and sewage sources to the microbial pollution in river and estuarine waters (Vignesh et al., 2013). The correlation between Ca and TS reflecting the control of different sources, like anthropogenic discharge associated with simple hardness (Nagvenkar and Ramaiah, 2009). The relationship between Cl-FC indicating the same geochemical origin of Allochthonous River processes. Cl may originate through anthropogenic activities like sewage disposal and effluents (Thivya et al., 2013) thus faecal coliform has a significant correlation with Cl which also originates from the sewage disposal and faecal matters. HCO_3 show significant correlation with Mg indicating the predominance of chemical weathering along with leaching of secondary salts during this season (Prasanna et al., 2008). PO_4-H_4SiO_4 relationship shows the influence of agricultural activities.

Factor Analysis

The purpose of factor analysis is to interpret the structure within the variance, covariance matrix of a multivariate data collection. The technique which it uses is extraction of the eigen values and eigen vectors from the matrix of correlations or covariance. The factor loadings were sorted according to the criteria of (Liu et al., 2003), i.e. strong, moderate and weak, corresponding to absolute loading values of >0.75, 0.75–0.50 and 0.50–0.30 respectively.

Factor 1 shows a strong positive loadings Mg and HCO_3 with a variance of 16.29% the association of these ions indicates leaching of secondary salts, because the evaporate waters contain relatively little Ca, Mg, Na and K cations, no complex sequences of clay reactions and ion exchange processes need be invoked (Tellam 1995). In this factor strong negative loadings obtained between the TVC and VC which suggests that these microbes may be evolved in different in origin (Brindha and Elango 2012) and strong negative loadings between SO_4 and pH. Factor 3 shows a positive loadings Ca, TS and FC (Table.4) with a variance of 11.17% indicating the anthropogenic sources with hardness (Vignesh et al., 2013). It can rapidly outgrow the native microflora leading to increased levels of indicator bacteria in natural water bodies. Pathogenic bacteria of human health concern have been studied mostly with respect to their survival in riverine environments (Sood et al., 2008; Moldovana et al., 2013). Factor 4 accounts a positive loadings K and NO_3 with a variance of 10% and strong negative loadings of TDS.

Table 3. Correlation analysis of various ions in groundwater samples collected to microbes

	pH	EC	TDS	Ca	Mg	Na	K	Cl	HCO_3	NO_3	PO_4	SO_4	H_4SiO_4	TVC	TC	TS	FC	VC
pH	1.00																	
EC	-.021	1.00																
TDS	.089	-.269	1.00															
Ca	.083	-.288	-.150	1.00														
Mg	-.103	-.016	-.423	.144	1.00													
Na	-.175	.145	-.075	-.214	-.109	1.00												
K	.029	.094	-.299	-.142	.089	**.273**	1.00											
Cl	-.081	.156	-.325	.112	-.001	-.207	-.171	1.00										
HCO_3	-.160	-.140	-.419	.248	**.626**	-.129	.090	.073	1.00									
NO_3	-.131	-.059	-.383	.226	.117	-.015	**.528**	.059	.161	1.00								
PO_4	-.207	-.237	.042	.018	-.067	**.283**	-.059	-.052	-.010	-.039	1.00							
SO_4	.332	-.044	-.121	.230	.106	-.155	-.045	.065	-.318	.186	-.291	1.00						
H_4SiO_4	-.483	-.182	-.062	.192	.232	.185	-.054	.118	**.489**	.039	**.414**	-.438	1.00					
TVC	.037	.071	.076	-.007	-.283	.161	-.015	-.103	-.161	-.119	.205	-.295	.116	1.00				
TC	.014	.149	-.019	.194	.080	-.200	-.023	.102	.186	.085	-.041	-.039	.216	-.113	1.00			
TS	.019	.101	-.019	**.302**	-.216	.048	-.011	.141	-.084	.137	.199	-.205	.203	.140	.114	1.00		
FC	.039	-.014	-.043	.244	-.159	-.032	.050	**.263**	.015	.090	.013	-.244	.145	.025	**.252**	**.572**	1.00	
VC	.095	.123	.153	-.110	-.199	.026	-.129	-.079	-.194	-.250	-.057	-.103	-.140	**.506**	-.348	.147	-.086	1.00

Table 4. Factor analysis for groundwater samples (Varimax rotated)

Rotated Component Matrix				
	Component			
	1	2	3	4
pH	-.200	-.589	.092	-.078
EC	-.228	-.130	-.018	.486
TDS	-.432	-.007	-.165	-.626
Ca	.270	-.052	.609	-.179
Mg	.770	.079	-.093	.093
Na	-.259	.421	-.309	.423
K	.074	.029	-.109	.781
Cl	.106	-.081	.483	.064
HCO_3	.721	.379	.128	.040
NO_3	.295	-.063	.249	.609
PO_4	-.106	.630	.011	-.114
SO_4	.180	-.749	-.032	.022
H_4SiO_4	.293	.802	.229	-.104
TVC	-.554	.335	-.028	.024
TC	.244	.004	.480	-.049
TS	-.384	.224	.700	.140
FC	-.214	.130	.763	.109
VC	-.564	.040	-.162	-.072

Factor Scores

The Factor scores are projections of data onto corresponding eigen vectors. They could be considering as the actual of station on the underlying factors. Factor scores were calculated for each sample. Factor scores on spatial diagrams and on the maps were plotted. Extreme positive factor scores (>+1) reflect areas most affected, and negative (<-1) reflect areas unaffected by the chemical process represented by the factor. Near-zero scores reflect areas affected to an average degree of the process or factor. The factor score were also estimated to find out the spatial variation of the factor representation and to identify the zone of its representation. They are commonly obtained by two approaches weighted least square method and the regression method. The regression method is used in the study to compute the factor scores. The positive zones indicate the dominance of that factor (Hydrogeochemical regime). In factor 1 higher score are located in the south western part of the region indicating the leaching of secondary salts (Fig. 4a). The higher scores of factor-2 are represented in the northern side (Fig. 4b) where the river flows and they may be indicating the influence of anthropogenic factors like use of fertilizer in the agricultural field (Chidambaram et al., 2008). In factor 3 the higher scores are observed in the central part and tend to move on the southern side (Fig.4c) which may be due to landfill leachates ad dumping of municipal waste in these regions The higher scores of factor-4 are represented in the central part (Fig. 4d) and also observed in patches in all the regions which may be due to the irrigational activities. In general K is a common constituent of soluble fertilizers or could come from livestock excrement (Conrad et al., 1999).

TVC, TC, TS, FC and VC Vs LULC

The comparison of TVC, TC, TS, FC, VC and LULC shows that the higher microbial activity is observed in the north western part of the region (Fig. 5, 6, 7, 8, 9) where the vellar river flows, southern part where the landfill sites are observed. Presence of coliforms organisms in water regarded as evidence of faecal contamination as their origin in the intestinal tract of human and other warm blooded animals (Atherholt et al. 2003 and Usharani et al., 2010). This clearly indicates that the bacterial contamination in the river water is chiefly caused by human excreta and domestic sewage. Also, Putheti and Leburu, (2009) reported that domestic and industrial wastewater, agriculture waste are sources of faecal bacterial contamination to rivers. The lower microbial activities are noted in the central part of the study area where it is covered by the crop lands.

Many factors influence the leachate composition including the types of wastes deposited in the landfill, composition of wastes, moisture content, the particle size, the degree of compaction, the hydrology of the site, the climate and age of the fill (Leckie et al., 1979) the soil provides a natural filtering action and adsorption site for the removal of bacteria and viruses (Corapcioglu and Haridas, 1985). Generally, the wells located in residential and cultivated areas have the highest content of bacteria than the wells located in the desert area. The variations of total bacterial count of the groundwater may be result from direct contamination, through the leaching from the septic tanks which are indirect contact with the water table in the study area (Chitanand, et al., 2008).

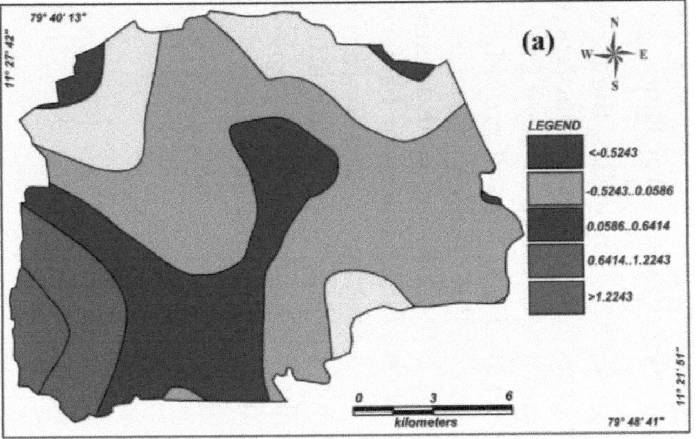

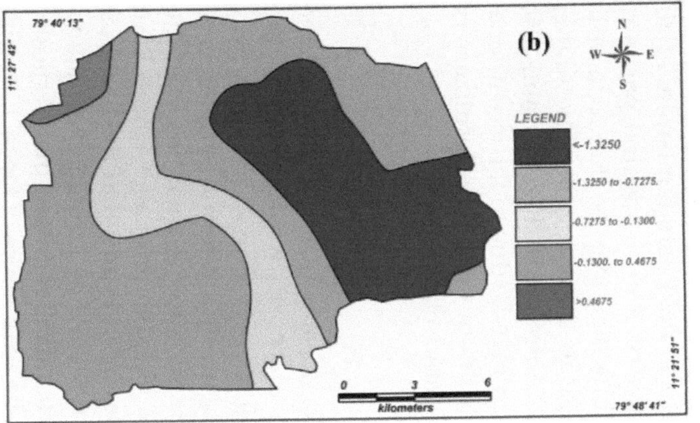

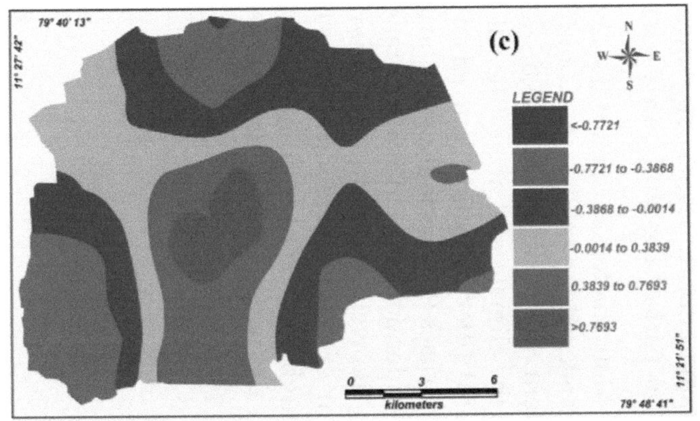

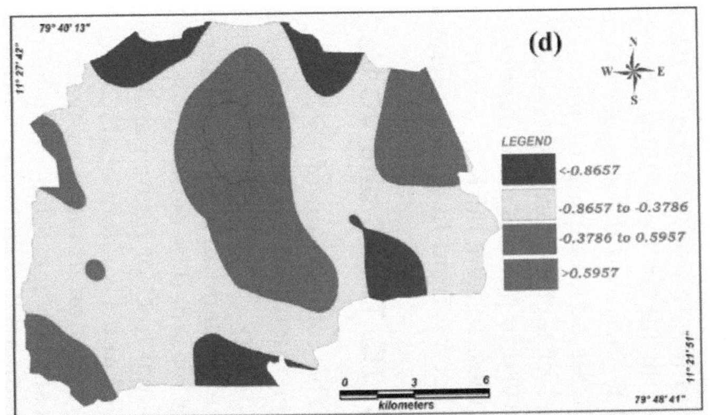

Fig. 4. Spatial distribution of factor scores (a) Factor-1; (b) Factor-2; (c) Factor-3 & (d) Factor-4

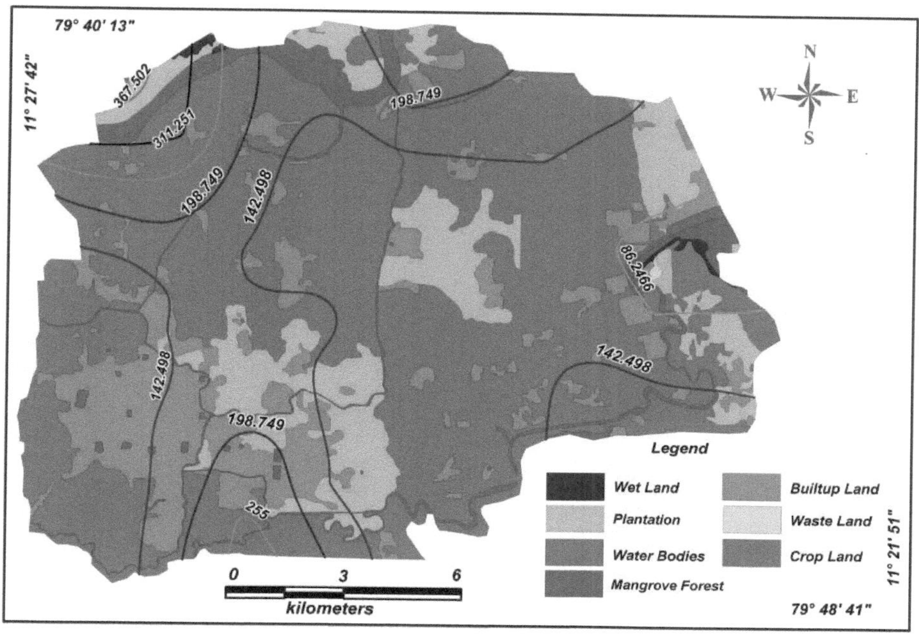

Fig. 5. FC overlay with LULC

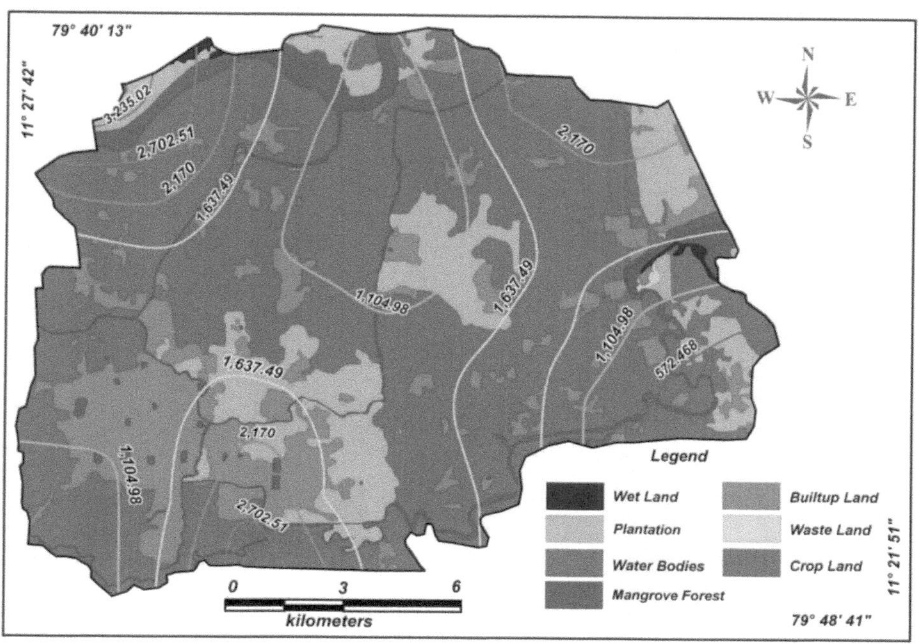

Fig. 6. TC overlay with LULC

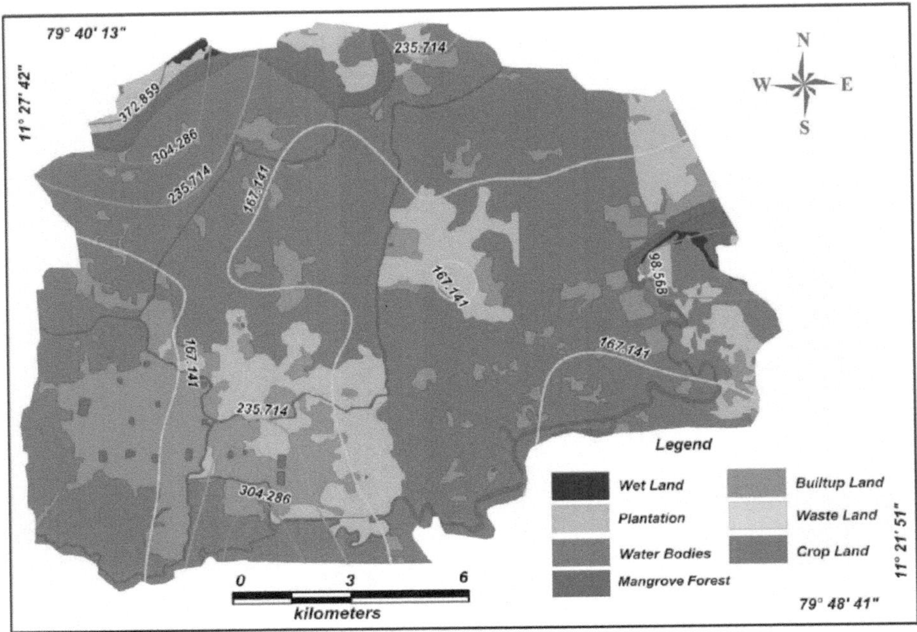

Fig. 7. TS overlay with LULC

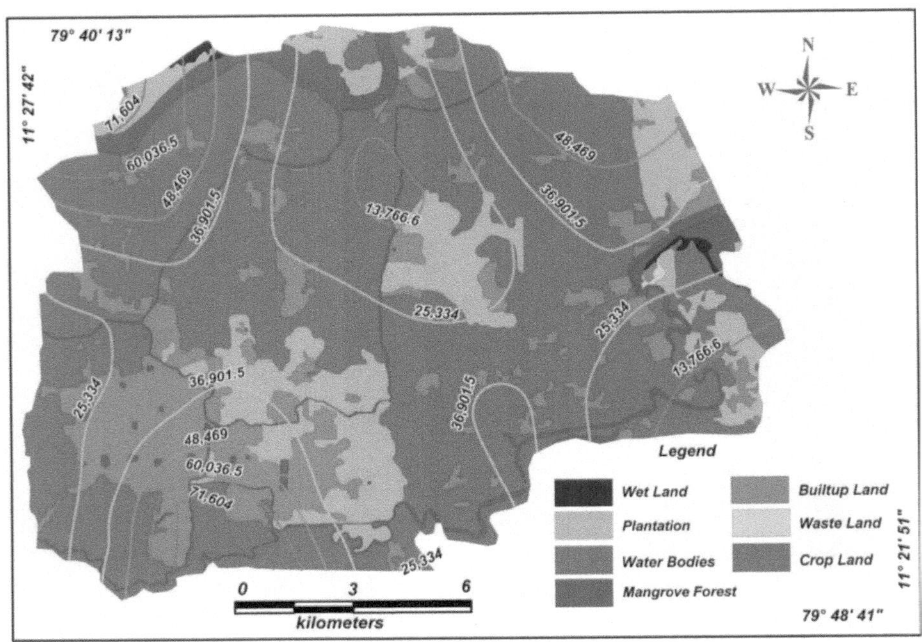

Fig. 8. TVS overlay with LULC

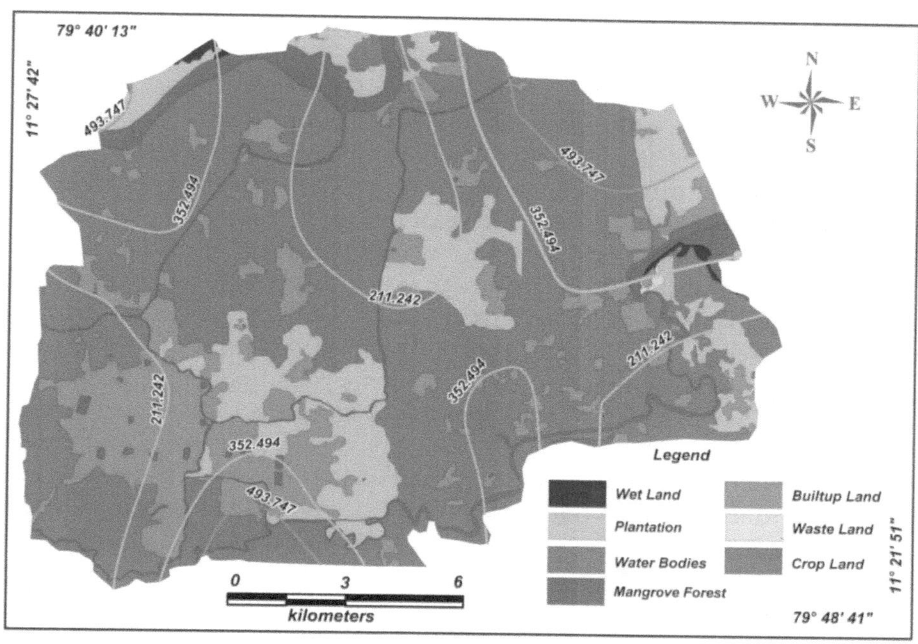

Fig. 9. VC overlay with LULC

Some of the reports (Ramaiah et al., 2005; Joachimsthal et al., 2004; Cabral and Marques, 2006) showed higher values of pathogenic microbes when compared to our study. The Enterococci group of bacteria has been most often suggested as alternative of coliform. The WHO (2004) also recommends the use of faecal streptococci (of which enterococci are a subgroup) as an additional indicator of faecal pollution. When combined with the measurement of E. coli, the interpretations could be afrmed with condence (Tyagi et al., 2006). The presence of TS and FS in most of the sites may be attributed to the the human/animal activities and sewage outfalls (Paul et al., 1995; Nevondo and Cloete, 1999; Lehloesa and Muyima, 2000).

Conclusion

There are multiple sources of contamination including wastewater entering different environmental compartments and manure applied to fields in agricultural areas. The leaching of clay minerals from the original landfill-soil mixture would leave behind the non-degradable fill material and siliceous matrix with average soil/rock grains of larger size fraction and of relatively higher resistivity than the parent landfill material. Generally, the wells located in residential and cultivated areas have the highest content of bacteria than the wells located in the desert area. The variations of total bacterial count of the groundwater may be result from direct contamination, through the leaching from the septic tanks which are indirect contact with the water table in the study area. Few samples show that increasing the TC with nitrate. Similarly, a negative relationship existed between nitrate concentration and coliform levels in some samples. They showed elevated levels of nitrate in the water which could be attributed coming from agricultural runoffs and household effluent. This creates as suitable environment for coliform growth as nitrate concentration becomes abundant. In turn, the nitrate concentration deceases as result of coliform uptake. The comparison of TVC, TC, TS, FC, VC and LULC shows that the higher microbial activity is observed in the north western part of the region where the Vellar river flows, southern part where the landfill sites are observed. Street and drain water flows into the rivers through a collecting system but direct input of sewage to the rivers is not known. Presence of coliforms organisms in water regarded as evidence of faecal contamination as their origin in the intestinal tract of human and other warm blooded animals. This clearly indicates that the bacterial contamination in the river water is chiefly caused by human excreta and domestic sewage.

References

Andersson Y, Bohan P (2001) Disease surveillance and waterborne outbreaks. In: Water quality: Guidelines, standards and health. L. Fewtrell and J. Bartram (ed.). London, UK, IWA Publishing.

APHA (1998) Standard methods for the examination of water and wastewater (19th ed.). Washington, DC: APHA.

Atherholt T, Feerst E, Hovendon B, Kwak J, Rosen J (2003) Evaluation of indicators of fecal contamination in groundwater. J. Am. Water Works Assoc., 95: 119-131.

Brindha K, Elango L (2012) Impact of tanning industries on groundwater quality near a metropolitan city in India. Water Resour Manag 26(6):1747–1761

Cabral B, Marques P (2006) Making exception handling work. Proceedings of the 2nd Conference on Hot Topics in System Dependability, 9. Seattle, WA.

Chidambaram S, Ramanathan AL, Prasanna MV, Anandhan P, Srinivasamoorthy K, Vasudevan S (2008) A statistical approach to identify the hydrogeochemically active regimes in groundwaters of Erode district, Tamil Nadu. Asian J Water Environ Pollut 5(3):93–102.

Chitanand MP, Gyanath G, Lade H S (2008) Bacterial assessment of ground water: a case study of Nanded city. Journal of Environmental Biology 29, no. 3: 315–318.

Collignon A, Butel MJ (2004) Etablissement et composition de la flore microbienne intestinale. In Flore microbienne intestinale , Rambaud JC., Buts JP. Et Corthier G., Flowie [Ed.], pp. 151-170.

Conrad JE, Colvin C, Sililo O, Gorgens A, Weaver J, Reinhardt C (1999) Assessment of the impact of agricultural practices on the quality of groundwater resources in South Africa. Water Research Commission Report 641/1/99.

Corapcioglu MY, Haridas A (1985) Microbial transport in soils and groundwater: Adv. Water Resour. 8, 188-200.

Eukene Oporto – Bensig, Mary Joyce L. Flores, Fleurdeliz F. Maglangi (2014)Water quality assessment of Bulacao River, Cebu, Philippines using fecal and total coliform as indicators, Journal of Biodiversity and Environment,al Sciences, Vol. 5, No.2, p.470-475.

Foster SSD, Chilton PJ (2003) Groundwater: the processes and global significance of aquifer degradation. Philosophical Transactions of the Royal Society of London, B, 358, 1957-1972.

Joachimsthal EL, Ivanov V, Tay J, Tay ST (2004) Bacteriological examination of ballast water in Singapore Harbour by flow cytometry with FISH. Marine Pollution Bulletin 49: 334-343, doi:10.1016/j.marpolbul.2004.02.036

Kazmi S, Khan A (2005) Level of Nitrate and Nitrite Contents in Drinking Water of Selected Samples Received at AFPGMI, Rawalpindi, Pakistan. Pakistan Journal of Physiology,1,1-2.

Leckie JO, Pacey JG, Halvadakis C (1979) Journal of Environmental Enjineering, Division American Society of Civil Engineers 105 (EE2), 337.

Lehloesa LJ, Muyima Nyo (2000) Evaluation of the impact of the household treatment procedures on the quality of groundwater supplies in the rural community of the Victoria district, Eastern Cape. Water SA 26 (2) 285-290.

Liu C, Lin K, Kuo Y (2003) Application of factor analysis in the assessment of groundwater quality in a black disease area in Taiwan. The science of the total environment, 313, 7789.

Moldovana OT, Melega IN, Leveib E, Terentec M (2013) A simple method for assessing biotic indicators and predicting biodiversity in the hyporheic zone of a river polluted with metals. Ecol Indic 24:412–420

Nevondo TS, Cloete TE (1999) Bacterial and chemical quality of water supply in the Dertig village settlement. Water SA 25 (2) 215-220.

Nagvenkar G S, Ramaiah N (2009) Abundance of sewage-pollution indicator and human pathogenic bacteria in a tropical estuarine complex, Environ Monit Assess, 155:245–256 DOI 10.1007/s10661-008-0432-1

OECD, W. (2003): Assessing microbial safety of drinking water - improving approaches and methods. Published on behalf of the World Health Organization and the Organisation for Economic Co-operation and Development by IWA Publishing: 1-295.

Paul JH, Rose JB, Jiang S, Kellog C, Shinn E (1995) Occurrence of faecal indicator bacteria in surface water and the subsurface equiter in Key Largo, Florida. Appl. Environ. Microbiol. 61 (6) 2235-2241.

Prasanna MV, Chidambaram S, Pethaperumal S, Srinivasamoorthy K, John Peter A, Anandhan P, Vasanthavigar M (2008) Integrated geophysical and chemical study in the lower subbasin of Gadilam River, Tamil Nadu, India. Environ Geosci 15(4): 145–152.

Putheti R, Leburu R (2009) Role of probiotics and their influence in different physico-chemical and microbiological studies of water – a case study. Intern. J. Fisher Aquacult., 1(1): 1-4

Ramaiah N, Paul J T, Fernandes V, Raveendran T, Raveendran O, Sundar D, Revichandran C, Shenoy D M, Manguesh G, Kurien S, Gerson V J, Shoji D T, Madhu N V, Sreekumar S, Lokabharathi P A, Shetye S R (2005) The September 2004 stench off the southern Malabar coast – A consequence of holococcolithophore bloom; Curr. Sci. 88(4) 551–554.

Schwarzenbach, R.P., Egli, T., Hofstetter, T.B., von Gunten, U. and Wehrli, B. (2010). Global water pollution and human health. Annual Review of Environment and Resources 35, 109–136.

Sood A, Singh KD, Pandey P, Sharma S (2008) Assessment of bacterial indicators and physicochemical parameters to investigate pollution status of the Gangetic river system of Uttarakhand (India). Ecol Indic 99:709– 717.

Swan ARH, Sandilands M (1995) Introduction to Geological Data Analysis. Blackwell Science: Oxford, 446 p.

Tellam J H (1995) Hydrochemistry of the saline groundwaters of the lower Mersey Basin Permo-Triassic sandstone aquifer, UK, J. Hydrol.,165 (1–4), 45–84.

Thivya C, Chidambaram S, Singaraja C, Thilagavathi R, Prasanna MV, Jainab I (2013) A study on the significance of lithology in groundwater quality of Madurai district, Tamil Nadu (India). Environ Development Sus. doi:10.1007/s10668-013-9439.

Tyagi P, Wu PC, Chancellor M, Yoshimura N, Huang L (2006) Recent advances in intravesical drug/gene delivery. Mol. Pharm. 3 (4), 369–379.

Usharani K, Umarani K, Ayyasamy PM, Shanthi K, Lakshmanaperumalsamy P (2010) Physico-chemical and bacteriological characteristics of Noyyal River and groundwater quality of Perur, India. J. Appl. Sci. Environ. Manage, 14(2): 29-35

Vignesh S, Dahms HU, Emmanuel KV, Gokul MS, Muthukumar K, Kim BR, James RA (2013) Physicochemical parameters aid microbial community? A case study from marine recreational beaches, Southern India. Environ Monit and Assess 186:1875-1887.

WHO (World Health Organization) (2004) Guidelines for drinking water quality recommendations, vol 1. WHO, Geneva, p 515

Zmirou D, Ferley J.P, Collin J.F, Charrel M and Berlin J. (1987) A follow–up study of gastro-intestinal diseases related to bacteriologically substandard drinking water. American journal of public health, 77:582-584.

Chapter **16**

Identification of Spatial Extent of Fluoride Contamination in Groundwater of Dindigul District, Tamil Nadu - A Statistical Approach

Manivannan R., Chidambaram S., Anandhan P., Sinagaraja C.[1] and Johnsonbabu G.[1]

Department of Earth Sciences, Annamalai University, Annamalai Nagar
[1]Department of Geology, Presidency College, Chennai, India

ABSTRACT

Geochemical processes occurring within the groundwater and reactions with aquifer minerals have a profound effect on water quality. Hydrogeochemical composition of groundwater can also be indicative of its origin and history of passage through underground materials which water has been in contact. Assessing groundwater quality and developing strategies to protect aquifers from contamination are necessary aspects for proper planning and designing water resources. Anthropogenic activities like explosion of population, industrial growth, inputs of fertilizer, pesticides, and irrigation has been a crucial factor for determining the quality of groundwater. In this scenario, 43 groundwater samples were collected from the Dindigul district of Tamilnadu, a Hard rock terrain. The samples wereanalyzed for major cations and anions along with heavy metals to derive the different quality parameters and to obtain the pollution indices of different samples. This study focuses to find out the spatial relationship between the analyzed fluoride concentration in groundwater and pollution indices derived from the heavy metal concentration in the groundwater of the study area and their variation with respect to season.

Key words: Groundwater quality, Fluoride, Anthropogenic, Hard rock terrain, Heavy metal.

Introduction

The increasing population and rapid development in the various water based fields, particularly in agriculture and industry in the last couple of decades this has led to ever increasing demand for groundwater to meet the requirement of water. Such demands are met with indiscriminate exploitation of groundwater, in the absence of any groundwater legislation. The only source of replenishment of this exploited resource is rainfall, which is limited to few monsoon months in a year, particularly in semi-arid regions of our country.The drinking water crisis that ensues in many villages in summer imposes serious health hazards to the rural masses and is responsible for the huge loss of livestock populations for want of drinking water and fodder. The quality of water is of vital concern to the mankind, as it has direct link with human health. As groundwater moves along its path from recharge to discharge areas, a variety of hydrogeochemical processes alter its chemical composition.In general, the quality of groundwater depends on the composition of recharge water, the interaction between the water and the soil, the soil–gas, the rock with which it comes into contact in the unsaturated zone, and the residence time and reactions that take place within the aquifer (Freeze and Cherry, 1979; Hem, 1985). Thus, the principal processes that influence the quality of water in an aquifer are physical, geochemical and biochemical. Changes in the concentrations of certain constituents due to natural or anthropogenic causes may alter the suitability of groundwater.The district of Dindigul in the state of Tamilnadu in India is known for industrial activities, e.g. leather tanneries, automobile, food processing, etc. (Varadarajan and Krishnamoorthy, 1993)(Table 1). Further, intensive agricultural practices account for changes in land-use/land-management practices and application of fertilizers for high crop yield result in changes in topography of area and release of toxic compounds to the natural environment. Several studies have demonstrated the environmental pollution in the Dindigul region (e.g.Bhaskaran 1977, Hardoy et al., 1993; Mondal et al., 2004, 2005). The tannery waste is characterized by its strong color (reddish–dull brown), high BOD, high pH and high dissolved solids. Being voluminous and highly puerile, when tannery untreated effluents discharged and itdamages the normal life of the receiving stream.The state government public health records demonstrate that moderately good quality water ($F^- \leq 1.5$ mg L^{-1}) is available for domestic activities whereas high levels of F^- (≥ 1.5 mg L^{-1}) are also reported from some districts. Among few fluoride affected districts, Dindigul district is the one with high levels of F^- affecting human health (PWD, 2008). Keeping all this factors in mind the study aims to bring out a complete picture about the seasonal variation of fluoride in groundwater of Dindugal district of Tamilnadu.

Study Area

The study area is a hard rock terrain with frequent failure of monsoons and increased growth of population. The rocks are deeply weathered and rainfall recharge may be scarce. Increased use groundwater and limited recharge have contributed to the lowering of the water table to a greater extent. This has resulted in the decrease of yields in many dug and tube wells substantially or even fallen to zero, particularly during the summer.

The study area of the Dindigul District (lat 9°91′–10°84′ N and long 77°24′–78°44′ E) is located in Tamilnadu (India). It is a hard rock terrain with a total area of 6,267 km² (Fig. 1).

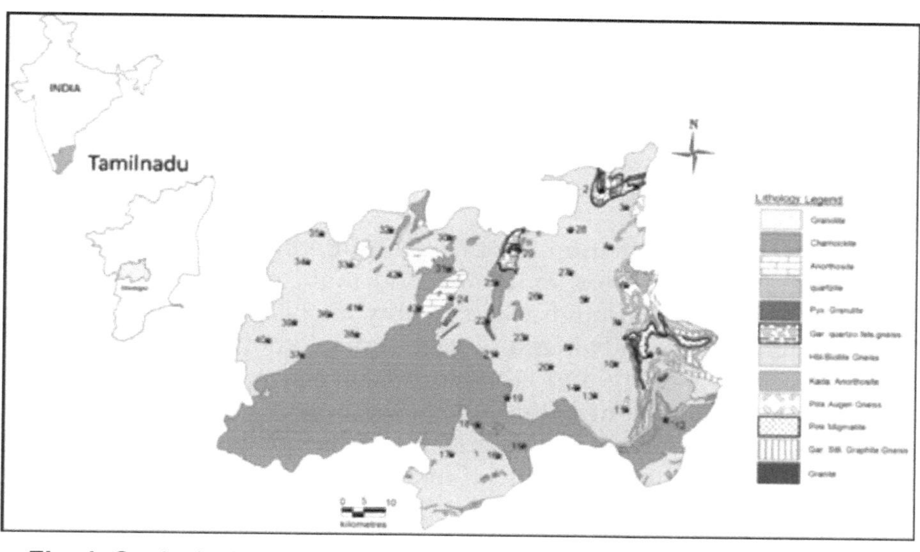

Fig. 1. Geological map of the Dindigul District with the sampling locations.

The climate in this area is generally tropical with an annual rainfall ranges between 700 and 1600 mm. The study is classified based on four main seasonsas pre-monsoon (PRM; April-June), southwest monsoon (SWM; July -September), northeast monsoon (NEM; October-December) and post-monsoon (POM; January-March). The study area receives higher rainfall during NEM and highest temperatures were noted during PRM. The relative humidity varies between 65 and 85%. Water in this area is stored in weathered Charnockite and Granite Gneisses. The depth of water level ranges from 0.12-15 mbgl (CGWB, 2008). The groundwater serves as the lifeline for drinking, domestic and irrigation purposes in the study area. The monsoonal rainfall is the main source for recharge; the vagaries in monsoon make this situation more critical. Moreover the presence ofdisseminated tanneries and their untreated effluents also make the water unsuitable for different purposes (Mondal et al. 2002; Mondal et al. 2005; Manivannan et al. 2010).

Sample Collection and Analysis

A total of 43 bore wells were selected to collect 172 water samples during four seasons of 2008 and 2009. The samples collected covers Pre-monsoon (May), South West Monsoon (August), North East Monsoon (November) and Post Monsoon (February). Immediately after collection, samples were stored at 4°C in an ice chest and transported to laboratory for the chemical analysis.The pH, electrical conductivity (EC) and fluoride (F^-) were immediately analyzed using the Orion Ion hand-held electrodes after collection. HCO_3^- by Titrimetry; Cl^-, SO_4^{2-}, PO_4^{3-}, NO_3^-, H_4SiO_4, Ca^{2+}, Mg^{2+}, Na^+ and K^+ by the ion chromatograph (IC, Metrohm 861). Duplicate samples were analyzed to check the analytical precision; it is ±5% for all hydrogeochemical variables.

Geospatial Analysis

Geographic Information System (GIS) offered such capabilities as they integrate multi sector, multilevel and multi period database. The accurate information through water quality analysis, scientific study on F^- distribution in groundwater and geochemical knowledge with spatial information on geology and climate to understand the source/cause, type and level of F^- contamination (Manivannan et al. 2010). Spatial distribution of dissolved metals and Fluoride ions were plotted using MAPINFO 8.5 coupledwith vertical mapper.

RESULTS AND DISCUSSION

a) Water chemistry

The geochemical trend of groundwater in the study area (Table.1) demonstrates that bicarbonate is the dominant anion, (HCO_3> Cl> SO_4> NO_3) and sodium is the dominant cation (Na> Ca> Mg> K), with a significant dilution during NEM and POM seasons. Fluoride concentration is higher in all seasons but more prominent during PRM and SWM and the concentrations during different seasons show that almost 70% of the samples fall within the permissible limit and rest of the samples fall in the range of 1-1.5 mg L^{-1}, more representations in this range are noted

Table 1. Water chemistry (range (avg. ±SD)) of the Dindigul groundwater of India. All parameters are in mg L[-1], except pH, temperature (°C) and EC (iS/cm).

Variable	PRM	SWM	NEM	POM
Temperature	[illegible]	30.8-33.9(32.17±0.64)	28-33.9(32±0.93)	31.3-33.9(32.2±0.61)
pH	7.8-8.9 (8.36±0.26)	6.2-8.5(7.22±0.42)	7.1-8.97(8.4±0.53)	6.86-8.67(7.41±0.32)
EC	620-3272 (1717±677)	632-5800(2240±1293)	465-5140(2064±1151)	546-2675(1519±503)
TDS	392-2600 (1214±533)	443-4500(1601±928)	320-3420(1364±741)	320-1986(1133±421)
HCO_3	122-500 (290±113)	98-671(427±136)	229-524(377±86)	170-732(401±110)
NO_3	1.01-11 (4.39±2.86)	1.28-7(4.63±1.64)	0.98-9.91(4.66±2.4)	0.68-9.63(4.64±2.59)
PO_4	0.03-2.7 (0.81±0.23)	0.03-26.2(9.33±5.57)	0.01-3(0.72±0.66)	0.07-1.35(0.24±0.19)
SO_4	8-194 (64±43)	4-168(33±31)	12.5-106(44±19)	10-156(48±39)
Cl	56-1258 (362±240)	36-974(316±223)	54-903(340±220)	53-585(250±141)
F	0.11-3.3 (0.99±0.78)	0.02-3.4(0.75±0.55)	0.04-2.3(0.73±0.49)	0.2-2(0.63±0.32)
H_4SiO_4	0.4-5.2 (2.78±1.08)	2-71(25±19)	1.5-9.9(6.64±2.24)	0.79-42(26±7.76)
Na	54-555(236±138)	26-647(243±155)	30-563(241±128)	20.3-398(145±91)
K	13-221 (57±43)	2.3-268(27±39)	0.65-161(29±20)	2.8-94(25±22)
Ca	12-148 (48±32)	12-124(51±22)	20-152(63±31)	20-126(64±25)
Mg	24-158 (67±27)	5-122(50±29)	5-98(40±24)	14.4-103(66±21)

during monsoon seasons (Table 2). Still very high concentrations ($\leq$ 1.5 mg L^{-1}) are also noted in PRM and NEM (Table 2). The dilution has played a significant role during the POM which resulted in the lesser F^-in groundwater. There is a wide range of the concentration of F^- is noted during monsoon seasons. According to WHO the permissible limit for F^- in drinking water is 1.5 mg L^{-1} (WHO 2008) whereas USPHS (1987) has set a range of allowable F^- concentration in drinking water for regions depending on their climatic conditions because the amount of water consumed and consequently the amount of F^- ingested is influenced primarily by air temperature. Where, the maximum allowable concentration of F^- decreases with increase in air temperature. The tropical regions with T >26°C has a maximum allowable limit of 1.4 mg L^{-1}. Intake of F^-concentrations above 1.0 mg L^{-1} may cause dental fluorosis and intake of F^- concentrations above 3.0 mg L^{-1} may result in skeletal fluorosis (Ripa 1993; USPHS 1987; WHO 1984).

b) Spatial and temporal variability in fluoride distribution

Understanding the spatial distribution of F^- in groundwater is very important to delineate the causative factors for enrichment of groundwater F^- and to adopt sustainable mitigation measures to protect the inhabitants (Ahmed et al. 2002). Higher concentration of this ion in observed in north, west and southeastern part of the district during PRM; northern and southern part in SWM; west, northeast and south eastern part in NEM and the same trend as that of the NEM was noted in the POM (Fig.2). In general, the spatial distribution of fluoride is noted during NEM compared to other seasons. The study area is entirely composed of hard rock; the distributions of F^- found in the hard rocks are generally related to the depth of the aquifer. The water table in the study area occurs in deeper conditions, this situation become more critical during the PRM. The higher F^- concentrations are typically observed in deeper wells as well as in granitic and gneissic aquifers, a trend which has been commonly recognized by many studies on groundwater in granitic setting (Chae et al. 2004). The increase of fluoride concentration towards deeper wells is possibly due to the increase of temperature and residence time with increasing depth, which enhances the dissolution of fluorine bearing minerals in rocks (Saxena and Ahmed, 2003). The geology of this area is chiefly composed of peninsular gneiss and Charnockites which consists of Biotite and Hornblende and Apatites (Fig.1). The Apaptites/Fluorapatites in Charnockites are also major contributors of F^- (Chidambaram 2000).

Table 2. Range of F⁻fluoride concentrations (mg L^{-1}) in groundwater of different seasons (n=43).

Season	0-1	1-1.5	>1.5
PRM	33 (76.70 %)	4 (9.3 %)	6 (14 %)
SWM	31 (72.09 %)	10 (23.25 %)	2 (4.65 %)
NEM	22 (51.16 %)	12 (27.91 %)	9 (20.93 %)
POM	39 (90.70 %)	3 (6.98%)	1 (2.32%)

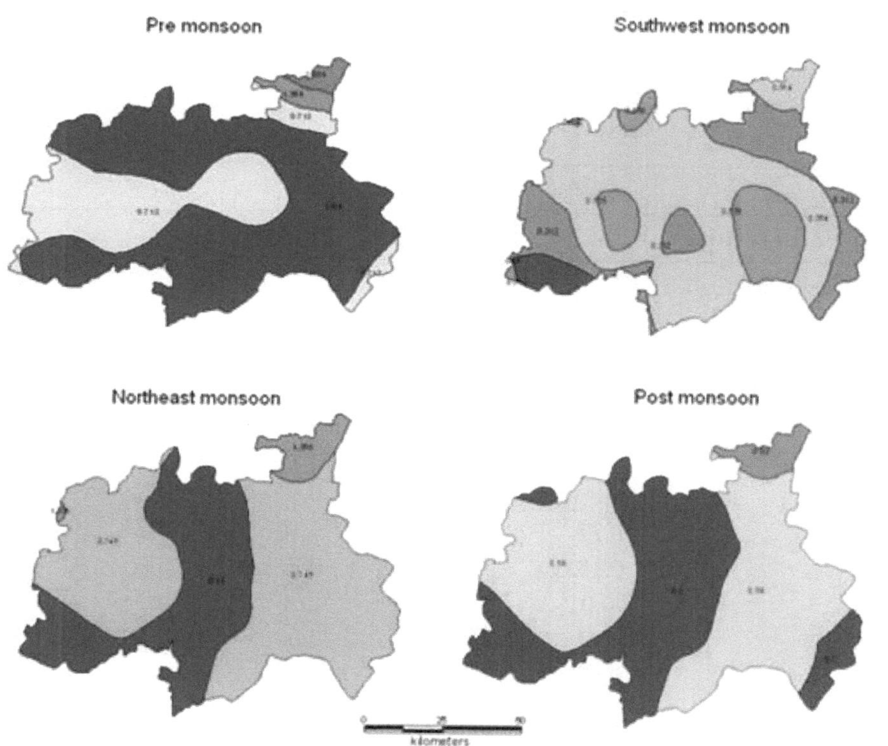

Fig. 2. Spatial and temporal distribution of F^- concentrations in the Dindigul district groundwater.

c) Factor analysis

The PCA analysis with varimax rotation is adapted to define the causative factors (called varifactors or VFs) which regulate the fluoride chemistry and distribution in the groundwater. The factor analysis was carried out for all the four seasons; five factors were extracted for PRM and NEM seasons and four for the other two seasons (Table 3). The representation of the total data variability (TDV) shows that higher representation of TDV was noted during PRM with 73% followed by POM (69.3%), NEM (69.2%) and SWM (65.3%). Liu et al. (2003) classified the factor loadings as 'weak', 'moderate' and 'strong' corresponding to absolute loading values of 0.3–0.5; 0.5–0.75 and >0.75, respectively. Among the different factors extracted, F^- was represented as the fourth factor during PRM with 11 percentage of variance (PV). SWM results reveal that the second factor with 16.86 PV was represented by F^-. The F^- factor was represented as the first factor with 22.55 PV during NEM and again as first factor with 22.03 PV during POM. It is interesting to note that F grows as a significant factor during and after monsoon.

Table 3. Summary of fluoride represented factors during different periods of the study

Seasons	PRM	SWM	NEM	POM
Total No. of factors	**5**	**4**	**5**	**4**
Fluoride Factor	4	2	1	1
pH	-0.08	0.23	-0.12	0.09
Ca	0.07	-0.24	0.32	0.46
Mg	-0.14	-0.13	0.27	0.64
Na	0.53	0.34	0.84	0.64
K	0.25	0.04	0.71	0.16
Cl	0.09	0.01	0.81	0.86
HCO_3	0.77	0.54	0.10	0.72
SO_4	0.15	0.60	0.57	0.06
NO_3	-0.01	0.78	-0.17	0.12
PO_4	0.14	0.64	0.08	-0.03
F	0.64	0.51	0.60	0.70
H_4SiO_4	-0.01	-0.10	0.11	0.15
% of Variance	11.67	16.86	22.55	22.03
TDV	73.01	65.3	69.22	69.33

Pre-monsoon

The spatial representation of factor scores of each sample for this season shows that the positive representation of this factor is represented in the northern and south western part of the study area. The northern part of the study area is chiefly composed of Hornblende Biotite Gneiss and few Granitic intrusions. The south western part is composed of Charnockite.

South-westmonsoon

This representation is clearly witnessed by positive values of the spatial distribution of the factor scores in the northern part. The presence of NO_3 and PO_4 indicates the anthropogenic influence mainly from agricultural fertilizers (Jha and Jha 1982). Since this region is agricultural intensive area and high quantities of fertilizers like super phosphate, sulfa NPK, Urea, etc., which generally contains F^-, are being applied to increase crop yield (SrinivasRao 1997) and unplanned fertilizers application leads to discharge of agricultural effluents with high F^- which subsequently pollutes the groundwater.

North–east monsoon

In monsoon, effluents might have been dispersed very quickly along with the precipitated water and hence affected F^- levels in the eastern, southern and western part of the study area.

Postmonsoon

The geographic position covered by the positive values of this factor is similar to that of the NEM but spatially it extends more than that of the NEM.

d) Pollution evaluation indices

Pollution evaluation index is a method of quantifying and numerically benchmarking thechemical quality of water in a particular area of interest (Edet and Offiong, 2002), and seasonaland spatial distribution of pollution indices are presented in Figure.4. These were computedbased on the Indian standards for most of metals and WHO, FAO and USEPA standards wereused for few metals. The range and mean values of HPI, HEI and Cd for pre-monsoon were 55-164 (93), 11-289 (29.56) and 0.059-271 (31.61), respectively, and 45-195 (115), 9.3-261 (41.23)and 1.51-243 (38.51), respectively for post-monsoon. In both seasons, high spatial variability in pollution indices is mainly because of regional differences in industrial activity, geologicalsetting and land-use pattern.

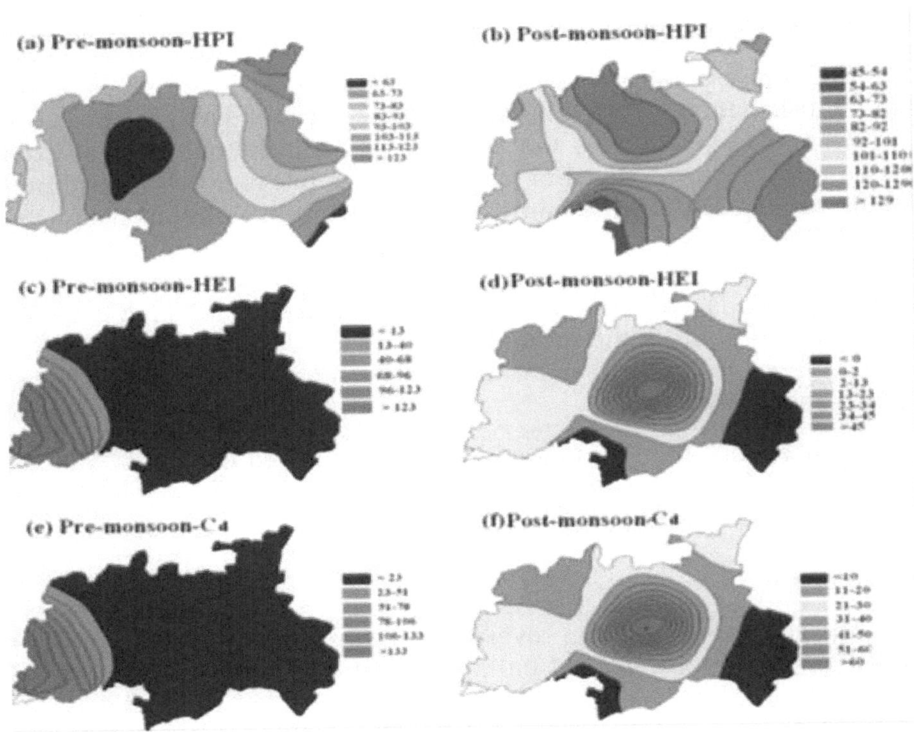

Fig. 4. Spatial distribution of pollution indices in the Dindigul groundwater.

Table 4. Classification of groundwater quality based on pollution indices (PRM-Premonsoon; POM- Postmonsoon).

Index	Category	Degree of pollution	No. of locations		Percentage (%)	
			PRM	POM	PRM	POM
HPI	< 50	Low	0	1	0	2.32
	50 - 100	Medium	26	14	60.46	32.56
	>100	High	17	28	39.53	65.12
HEI	< 40	Low	40	35	93.02	81.39
	40 - 80	Medium	0	4	0	9.30
	> 80	High	3	4	6.98	9.30
C_d	< 1	Low	9	2	20.93	4.65
	1 - 3	Medium	12	12	27.91	27.91
	> 3	High	22	29	51.16	67.44

HPI is a weighting method to describe the water quality and, therefore, has been used to complement with the criteria for various pollution indices. Following the approach developed byEdet and Offiong (2002) to categorize the groundwater metal pollution, HPI is classified intothree categories: low (HPI <50), medium (HPI = 50-100) and high (HPI > 100). In pre-monsoon, HPI values were distributed in the medium range (60%) and in post-monsoon HPI values were inthe high category (65%; Table 4). Similarly, HEI values were computed to harmonize the criteriafor pollution indices and also applied as a tool to evaluate the pollution extensiveness in thegroundwater systems. Like HPI, HEI was also categorized into three categories: low (HEI <40), medium (HEI = 40-80) and high (HEI > 80), and observed that 93% and 81% of stations fall inthe category low in pre-monsoon and post-monsoon seasons respectively (Table 4). The degree of contamination (Cd) was employed as a reference of describing theextent of metal pollution in the Dindigul groundwater. In order to describe the water quality, Cdcan be classified into three categories as described in Backman et al. (1997) and Edget andOffiong (2002): low (Cd <1); medium (Cd = 1-3) and high (Cd >3). Cd values in both seasonswere higher than 3 (Table 4), this clearly explains that the groundwater is highly polluted with metal contamination due to the industrial pollution and other human induced disturbances. Seasonal differences in pollution indices explain that metal pollution increases in post-monsoon;this may be recharge of groundwater with high concentrations of surface water after monsoon(Fig.4). However, further research is required to understand the surface water and groundwaterinteractions and dynamics.

Conclusion

The information technology is user friendly and it helps us to manage the resources of our country. It helps the decision makes and policy designers to admit the spatial variation of different parameters with respect to time and helps them to design the solution for the current context. This study reveals that the factors responsible for the variation of fluoride ions in groundwater is concentrated in the north eastern part of the study area fluoride ion vary spatially with season and the source of heavy metal varies according to the lithology of the study area and there is no much variation in their spatial distribution. There is a significant variation of the Pollution indices with season.

Acknowledgement

Authors would like to thank the DST- Fast track scheme (SR/FTP/ES-24/2006) for help extended by financial support.

References

Ahmed S, Bertrand F, Saxena VK, Subrahmanyam K, Touchard F (2002) Ageostatistical method of determining priority of measurement wells in a fluoride monitoring network in an aquifer. Journal of Applied Geochemistry, 4(2B), 576–585.

Backman B, Bodis D, Lahermo P, Rapant S, Tarvainen T(1997) Application of a Basin (southeastern Nigeria).Geo Journal 57, 295–304.

Bhaskaran TR (1977) Treatment and disposal of tannery effluents. Central Leather Research.

CGWB (2008) Ground water resources and development prospects in Dindugal region.

Chae GT, Kim K, Yun ST, Kim KH, Kim SO, Choi BY, Kim HS, Rhee CW (2004) Hydrogeochemistry of alluvial groundwaters in an agricultural area: an implication for groundwater contamination susceptibility. Chemosphere, 55, pp. 369–378.

Chae G T, Yun S T, Mayer B (2007) Fluorine geochemistry in bedrock groundwater of South Korea. Science of the Total Environment, 385, 272-283

Chidambaram S (2000) Hydrogeochemical studies of groundwater in Periyar district, Tamilnadu, India. PhD thesis, Department of Geology, Annamalai University.

Edet A E, Offiong O E (2002) Evaluation of water quality pollution indices for heavy metal, Geneva.

Freeze A R, Cherry J A (1979) Groundwater, Prentice – Hall, Inc Englewood cliffs, New Jersey, pp. 604.

Frengstad B, Skrede AK, Banks D, Krog JR, Siewers U (2002) The chemistry of Norwegian groundwaters: III. The distribution of trace elements in 476 crystalline bedrock groundwaters, as analysed by ICP-MS techniques.Sci. Total Environ, 246, pp. 21–40.

Hardoy J E, Mitlin D, Satterthwaite D (1993) Environmental Problems in Third World Institute (CLRI), Madras.

Hem J D (1985) Study and interpretation of the chemical characteristics of natural water. USGS water supply paper, pp.117–120.

Jha L B, Jha M (1982) Fluoride pollution in India. International Journal of J Environmental Studies,19, 225–230.

Manivannan R, Chidambaram S, Srinivasamoorthy K (2010) A statistical approach to study the spatial and temporal variation of Electrical conductivity of groundwater in Dindigul district, Tamilnadu-using GIS.

In: Chidambaram S (ed) Recent trends in water research, remote sensing and general perspectives. IK International, New Delhi, pp 186-196.

Mondal N C, Saxena V K, Singh V S (2005).Impact of pollution due to tanneries on groundwater regime. Current Science,88, 1988-1994.

Mondal N C, Saxena V K, Singh V S (2004) Assessment of groundwater pollution due to tannery industries in and around Dindigul, Tamilnadu. Environ. Geol. 48(2), 1149-157.

Mondal N C, Singh V S, Sarma M R K, Thangarajan M (2002) Impact of tannery effluent in groundwater and its control: A case study in and around Dindigul, Tamil Nadu, India. In: Proceedings of ICHWM-2002. BS Publisher, pp. 596–604.

PWD (Public Works Department) (2008) Incidence of fluoride in Tamilnadu State. Ground & Surface Water Resources Data Centre, Chennai.

Ripa LW (1993) A half-century of community water fluoridation in the United States: review and commentary. Journal of Public Health Dentistry, **53,** 17–44.

Saxena K, Ahmed S (2003) Inferring the chemical parameters for the dissolution of fluoride in groundwater. Environmental Geology, **43,** 731–736.

SrinivasaRao, N. (1997).The occurrence and behavior of fluoride in the groundwater of the lower Vamsadhara River Basin, India. Hydrological Sciences Journal, 42(6), 877–892.

USPHS (United States Public Health Services) (1987) Drinking Water Standards.Public Health Services Pub, Washington, D.C.

Varadarajan D B, Krishnamoorthy S (1993) Environmental implications of leather tanneries.

WHO (World Health Organization) (2008) Guidelines for water quality.3rd edition, Vol 1.Drinking Water Quality Control in Small Community Supplies, Geneva.

Colour Plates

Chapter 1: A Study on the Interrelationship of U and Rn in Groundwater of Hard Rock Aquifers

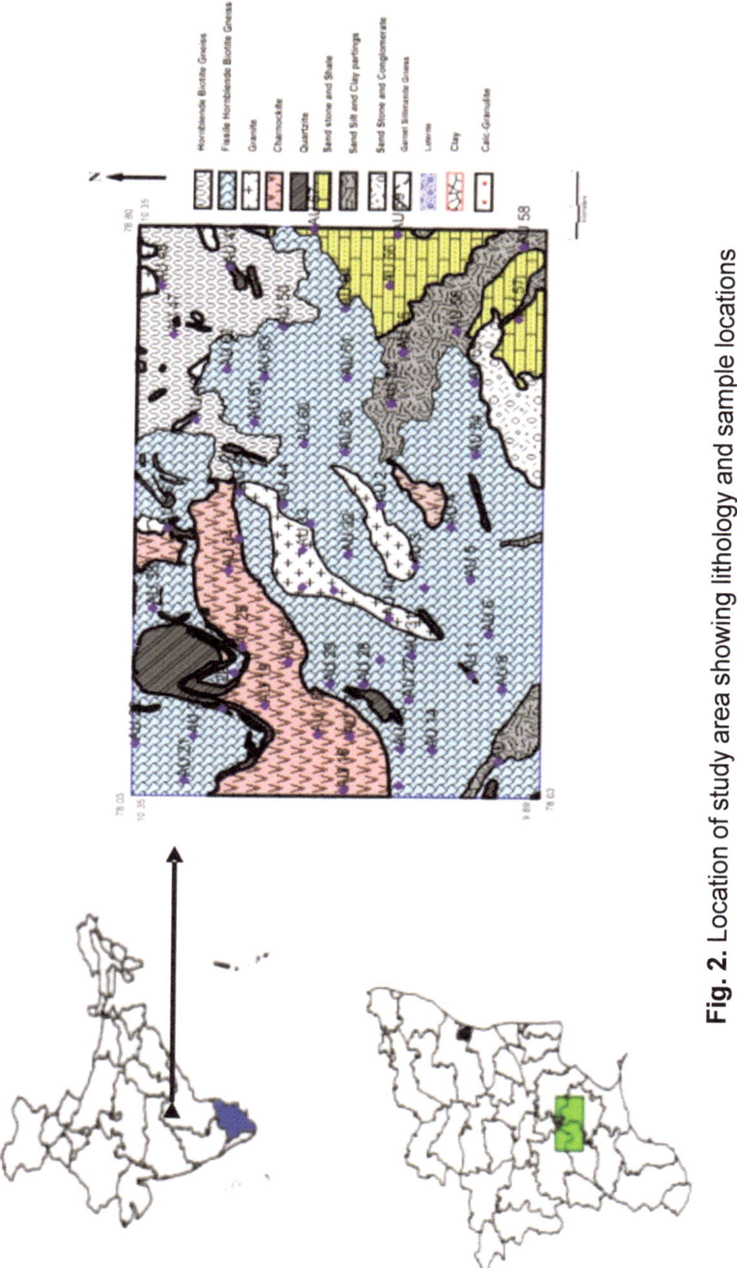

Fig. 2. Location of study area showing lithology and sample locations

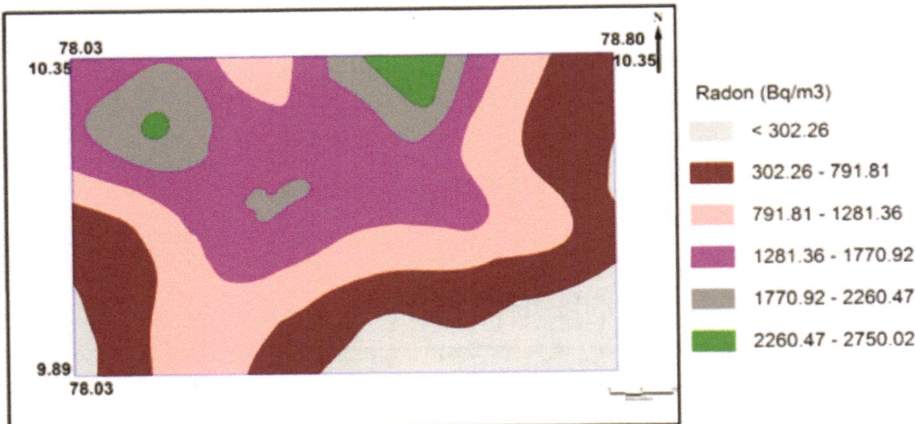

Fig. 3. Spatial distribution of Radon

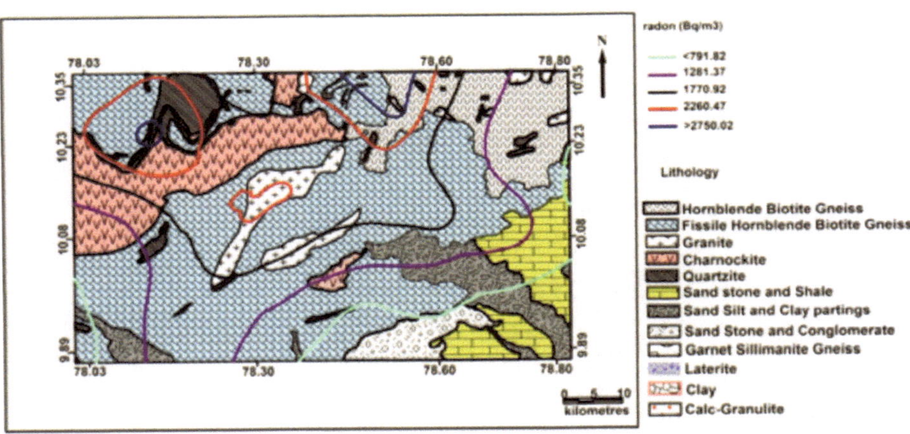

Fig. 4. Spatial distribution of Radon over lithology

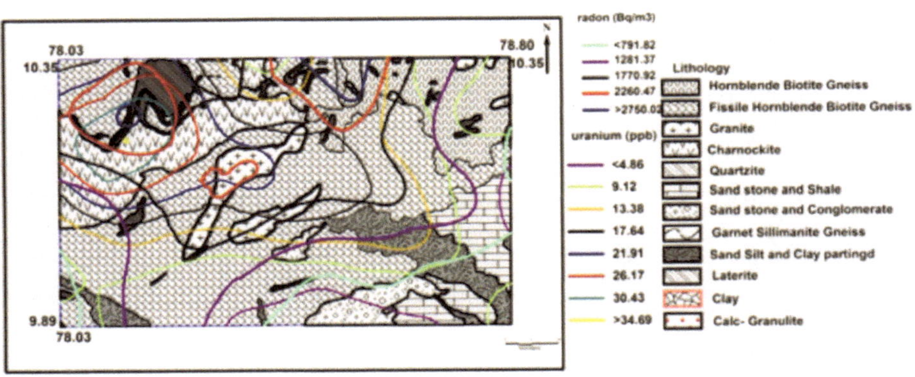

Fig. 5. Spatial distribution of U and Rn over Lithology

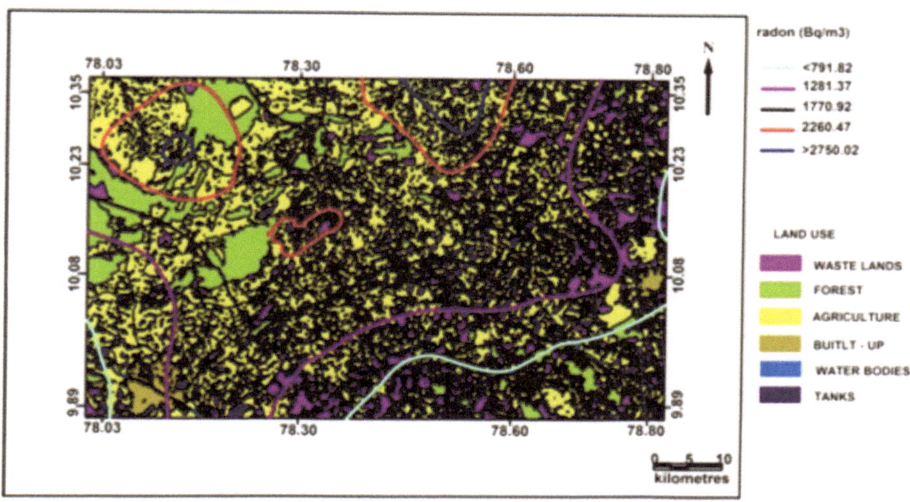

Fig. 6. Spatial distribution of Rn over Land use

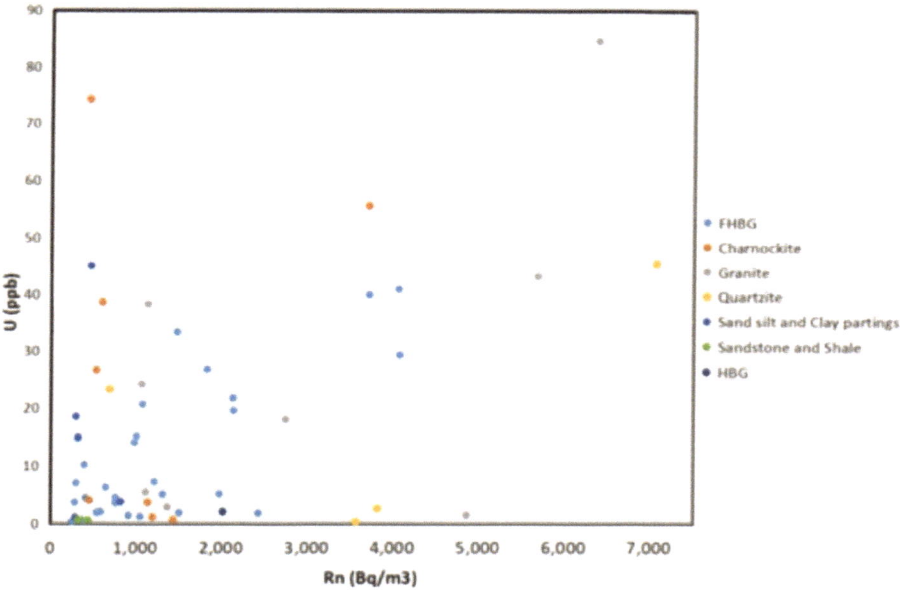

Fig. 7. Scatter plot showing inter relationship of U and Rn

Chapter 3: An Attempt to Determine the Mountain Front Recharge (MFR) by Using State of Saturation Index (SI) of Carbonate Minerals Along the Foothills of Courtallam, Tamil Nadu

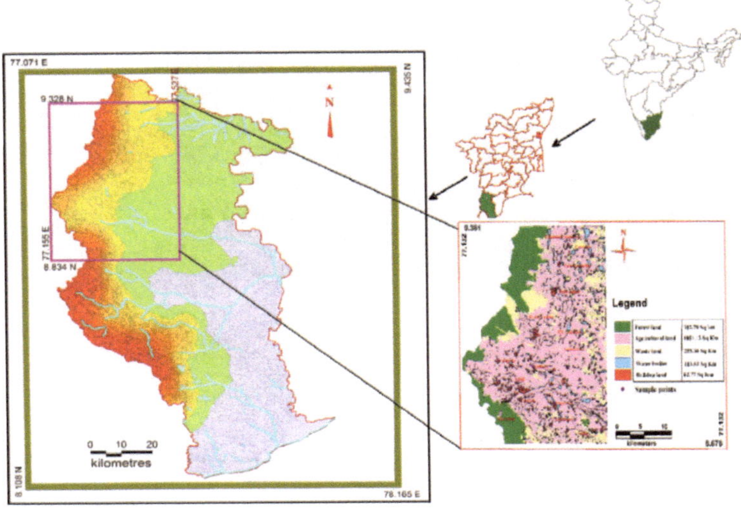

Fig. 1. Location map of the study area with land use pattern.

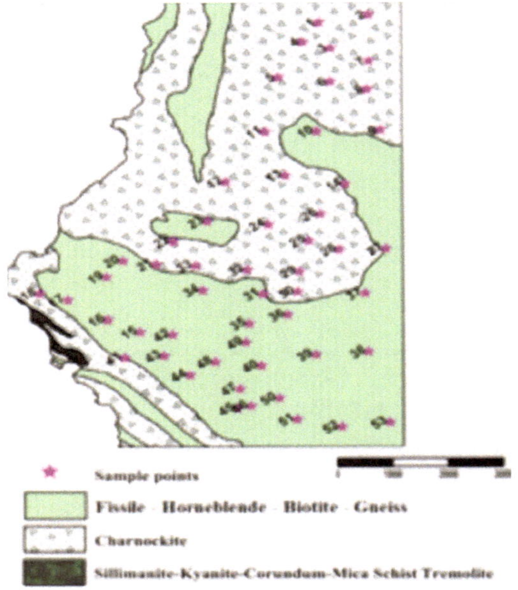

Fig. 2. Lithology map of the study area along with sampling points

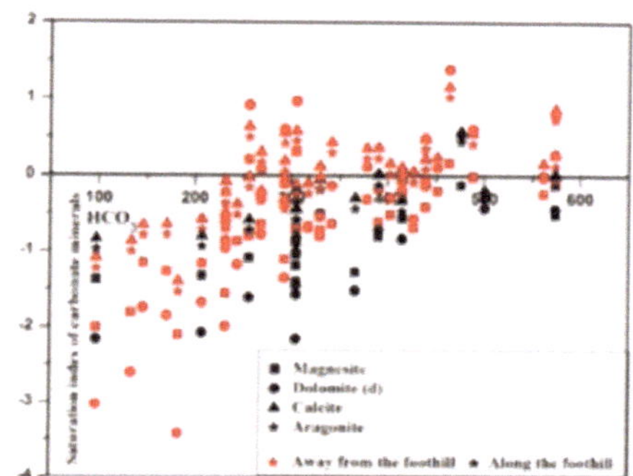

Fig. 3. The relation between the SI of carbonate minerals in all samples and their respective HCO_3.

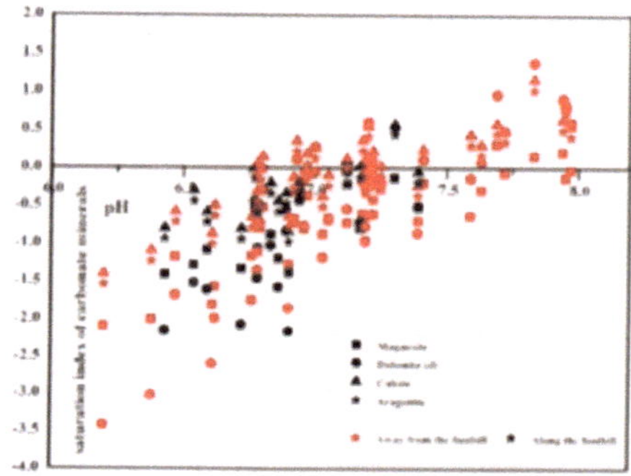

Fig. 4. The relation between the SI of carbonate minerals in all samples and their respective pH.

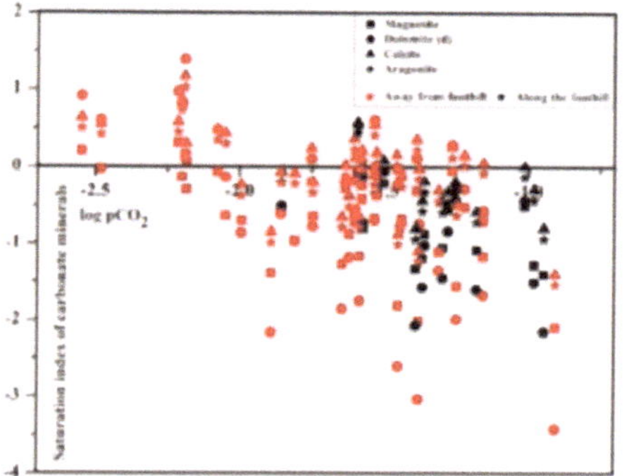

Fig. 5. The relation between the SI of carbonate minerals in all samples and their respective logpCO_2.

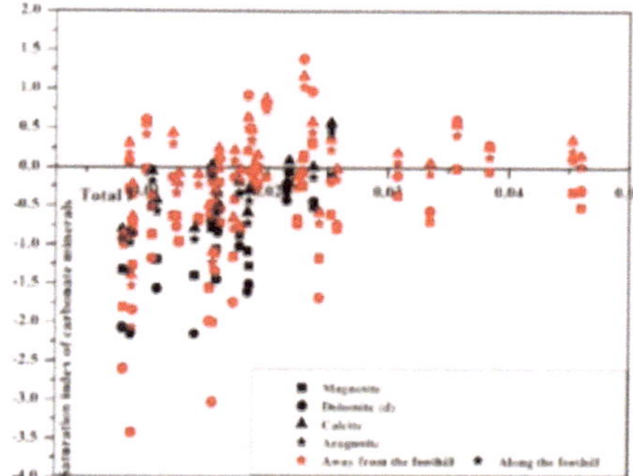

Fig. 6. The relation between the SI of carbonate minerals in all samples and their respective ionic strength.

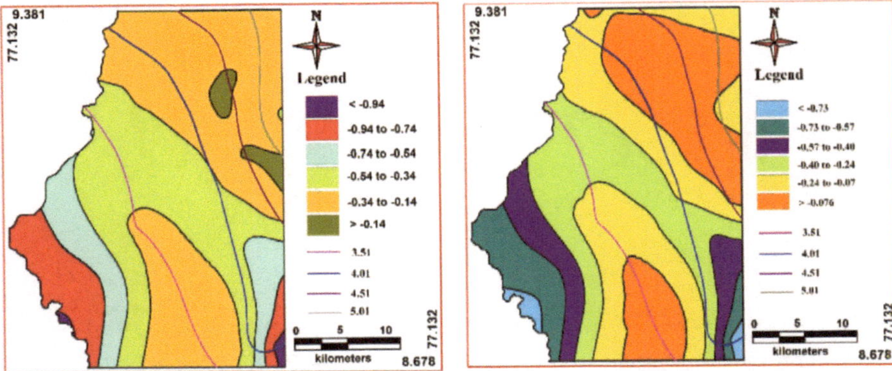

(a) Spatial distribution of SI of Aragonite. (b) Spatial distribution of SI of Calcite.

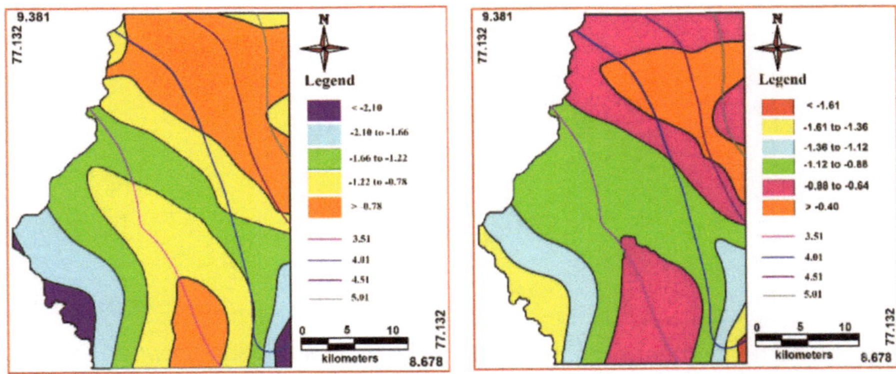

(c) Spatial distribution of SI of Dolomite. (d) Spatial distribution of SI of Magnesite.

Fig. 7. Spatial distributions SI of carbonate minerals with water level fluctuation in mbgl.

Chapter 4: Determination of Coastal Vulnerability Using Isotope and Geochemistry

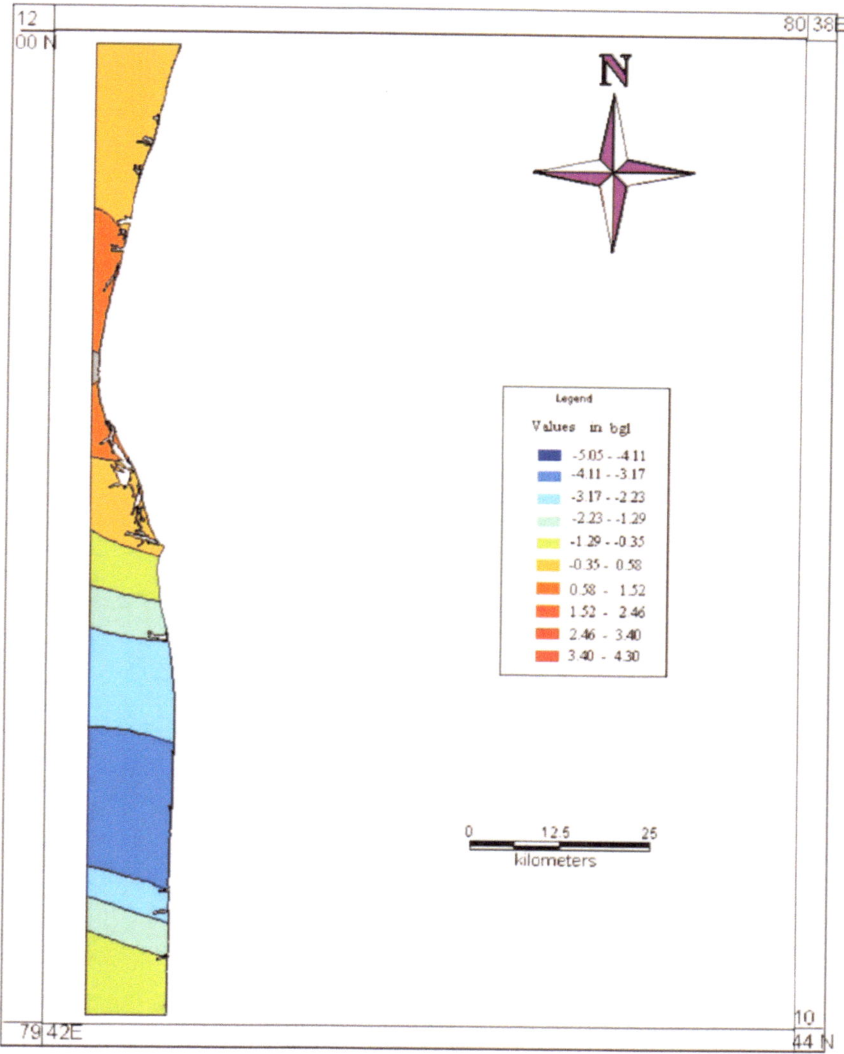

Fig. 1. Spatial distribution of reduced water level in Study area

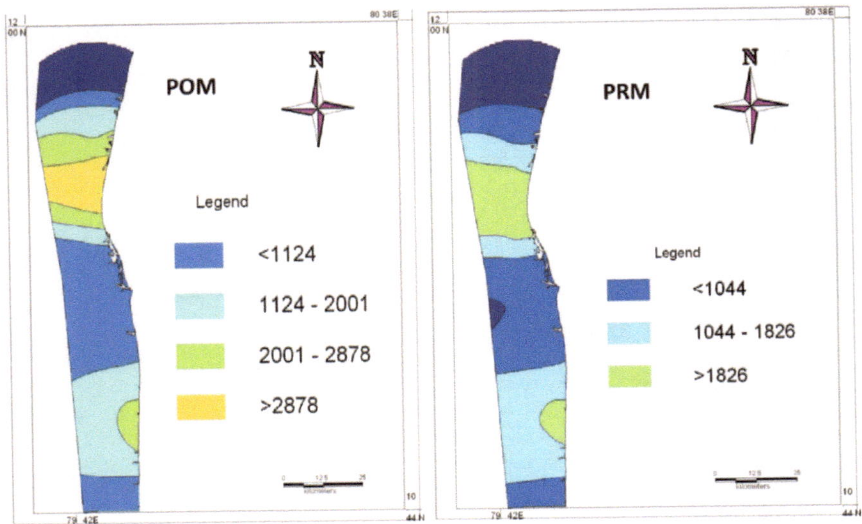

Fig. 3. EC for deep groundwater samples

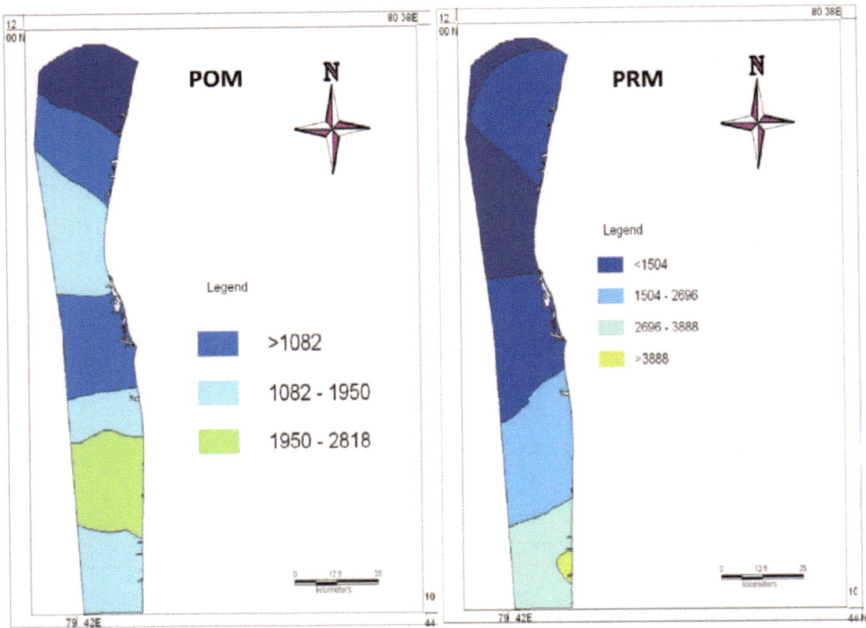

Fig. 4. EC for shallow groundwater samples

Chapter 6: A Study on the Removal of Fluoride in Groundwater by Using Natural Materials

Fig. 1. Model of FRT erected

Fig. 2. Natural materials used in the tank

Chapter 7: Assessment of Seawater Water Intrusion Mapping Using Electrical Resistivity Tomography in Coastal Tracts of Pondicherry, India

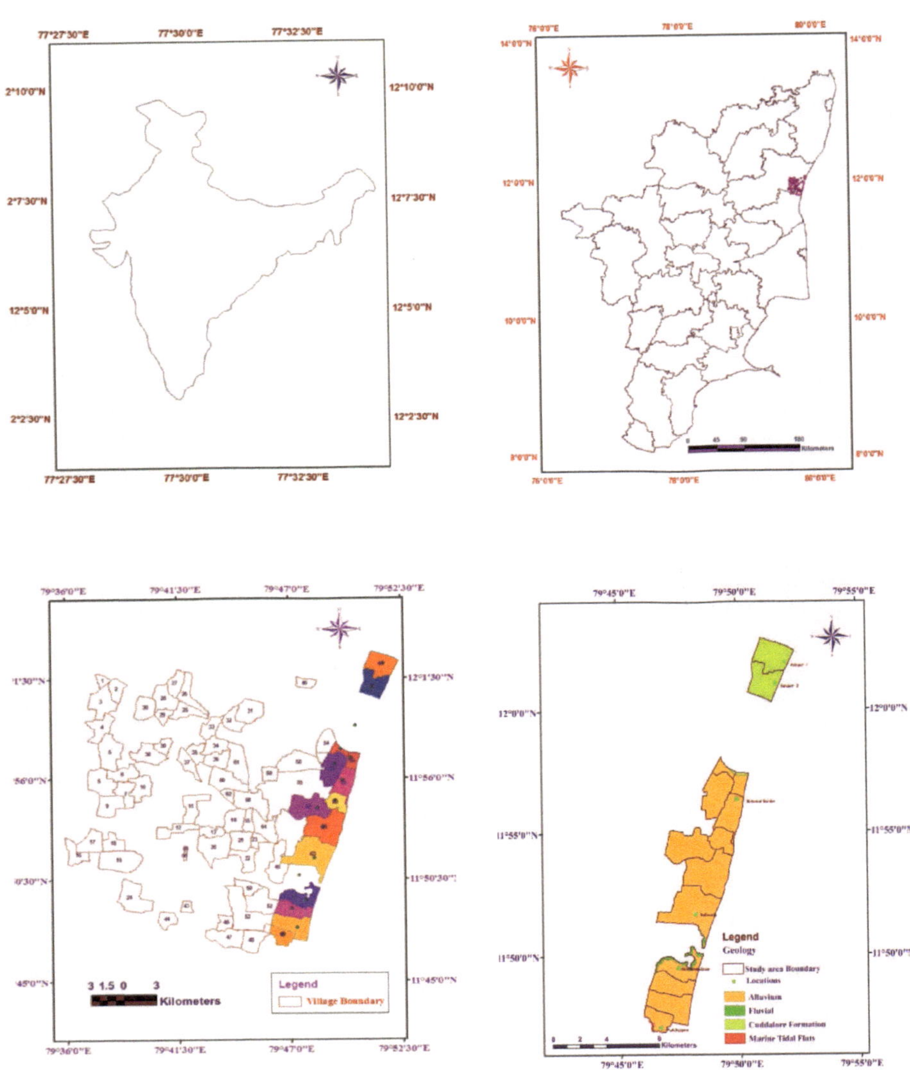

Fig. 1. Location, Geology and ERT soundings

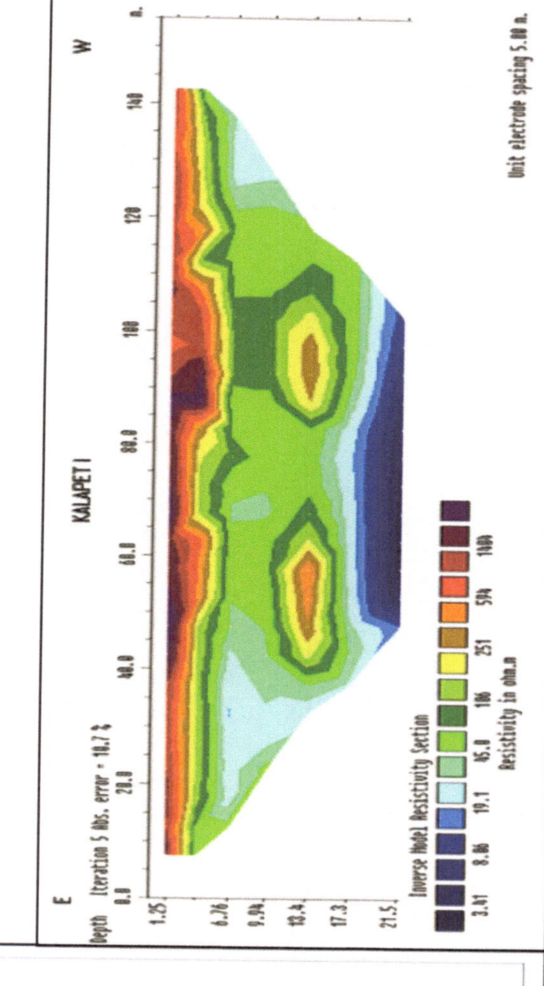

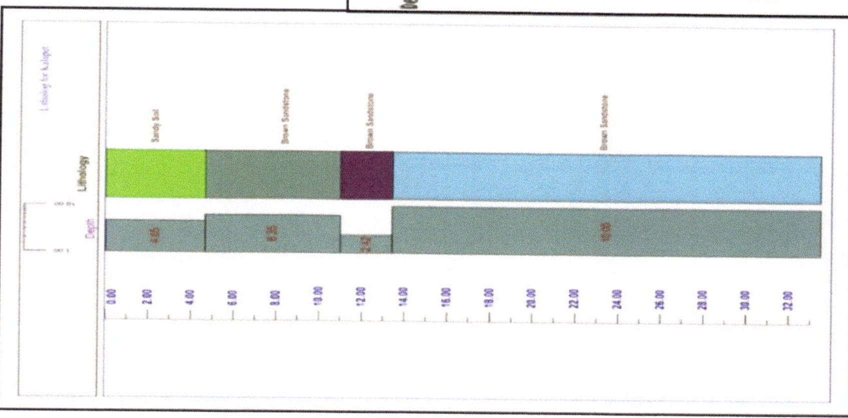

Fig. 3. Litho log and ERT profile of Kalapet - 1

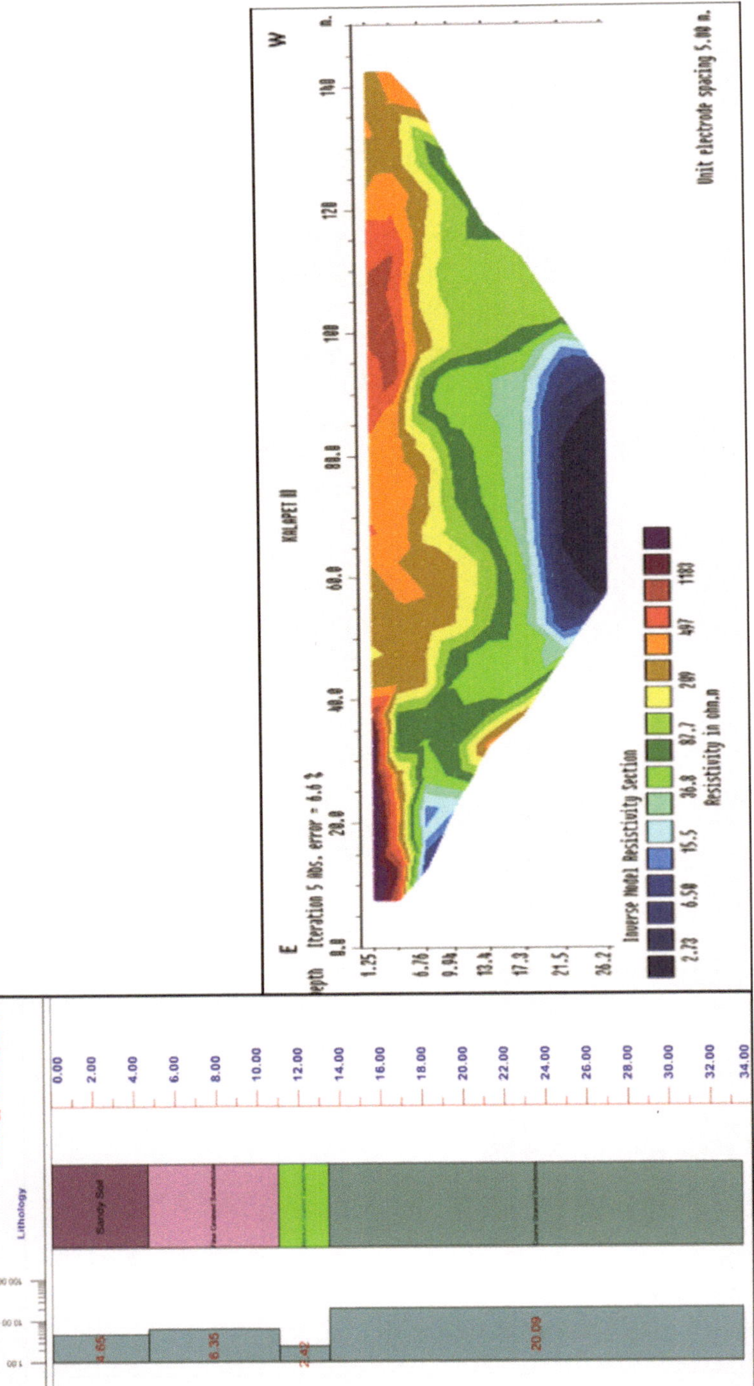

Fig. 4. Litho log and ERT profile of Kalapet - II

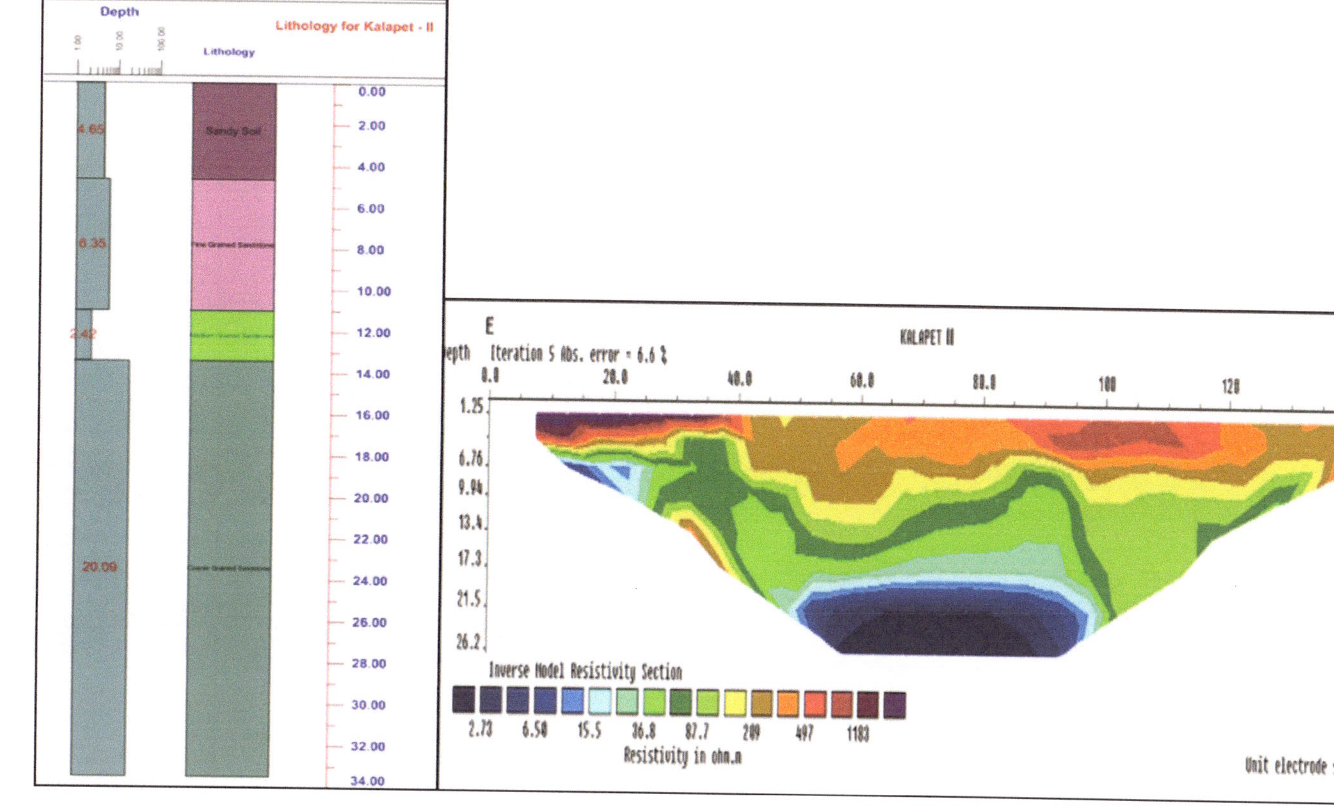

Fig. 5. Litho log and ERT profile of Botanical Garden

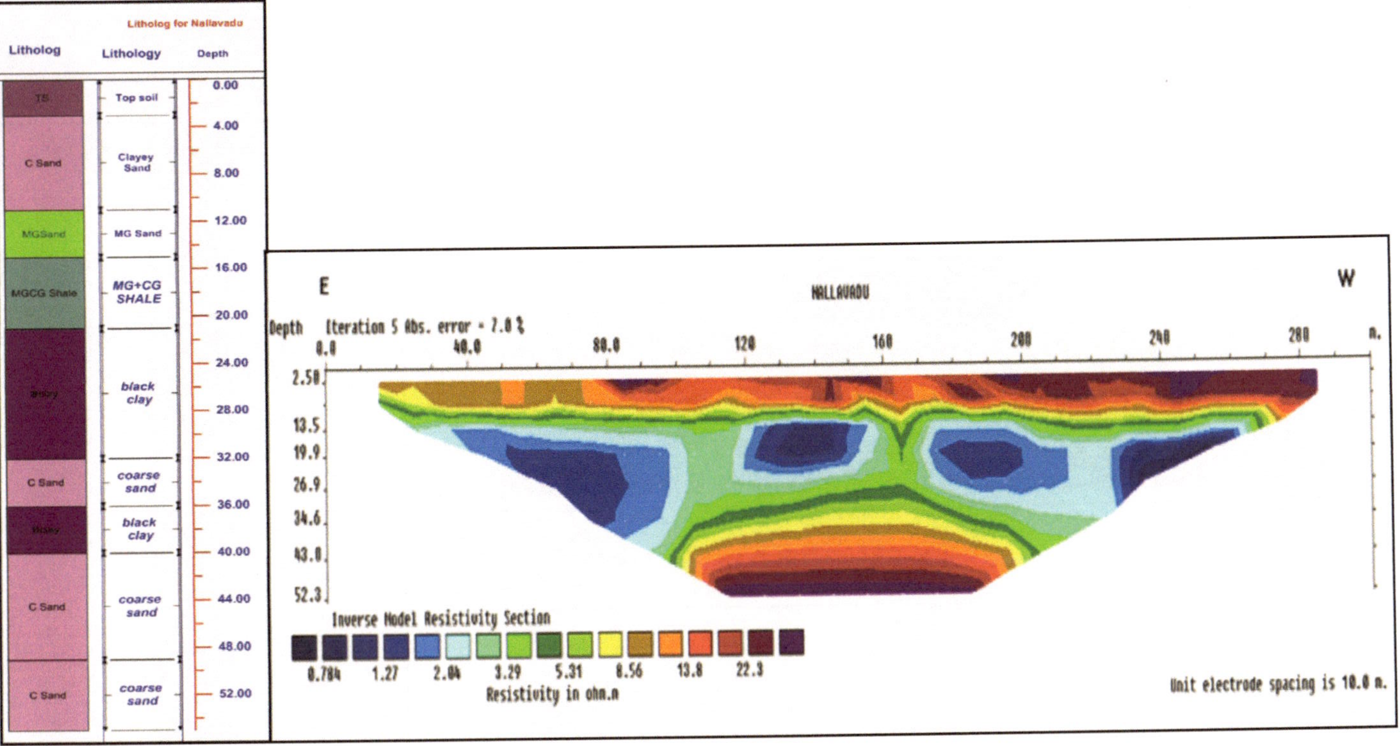

Fig. 6. Litho log and ERT profile of Nallavadu

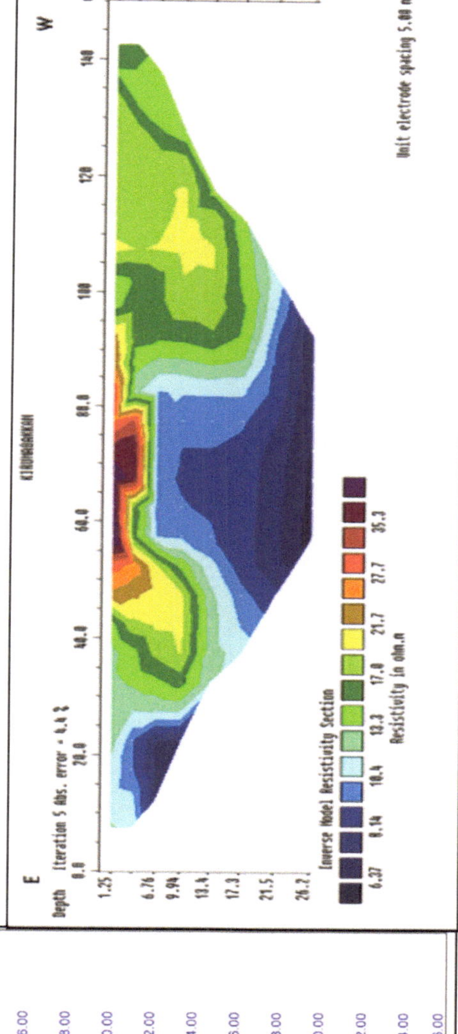

Fig. 7. Litho log and ERT profile of Kirumambakkam

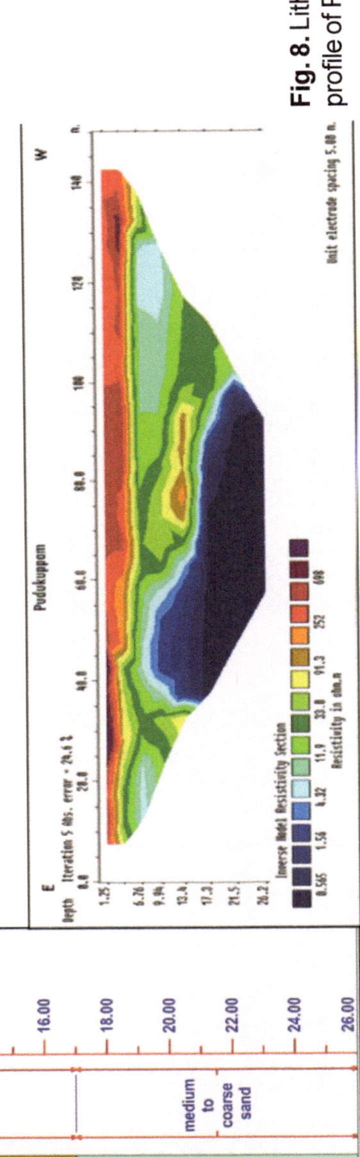

Fig. 8. Litho log and ERT profile of Pudukuppam

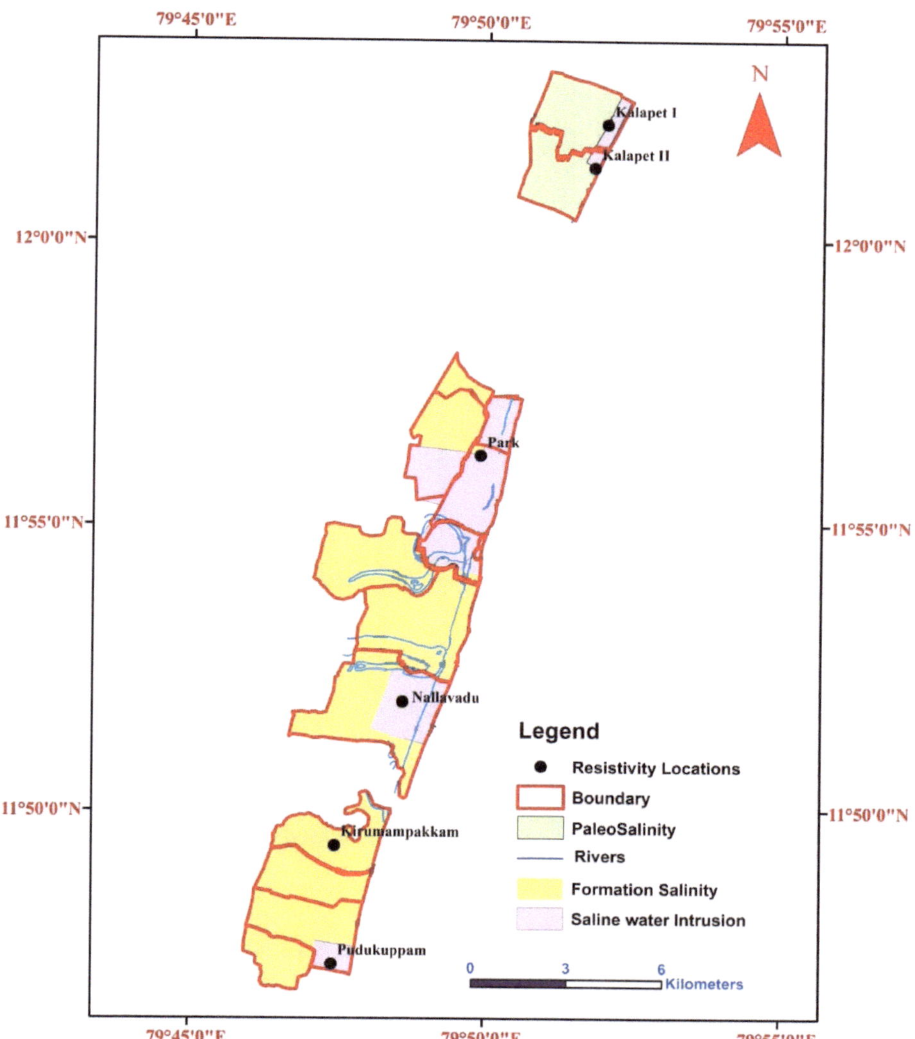

Fig. 9. Isolation of different saline sources in the study area

Chapter 10: Hydrogeochemical Characteristics of Groundwater at Various Lithological Conditions with Space and Time

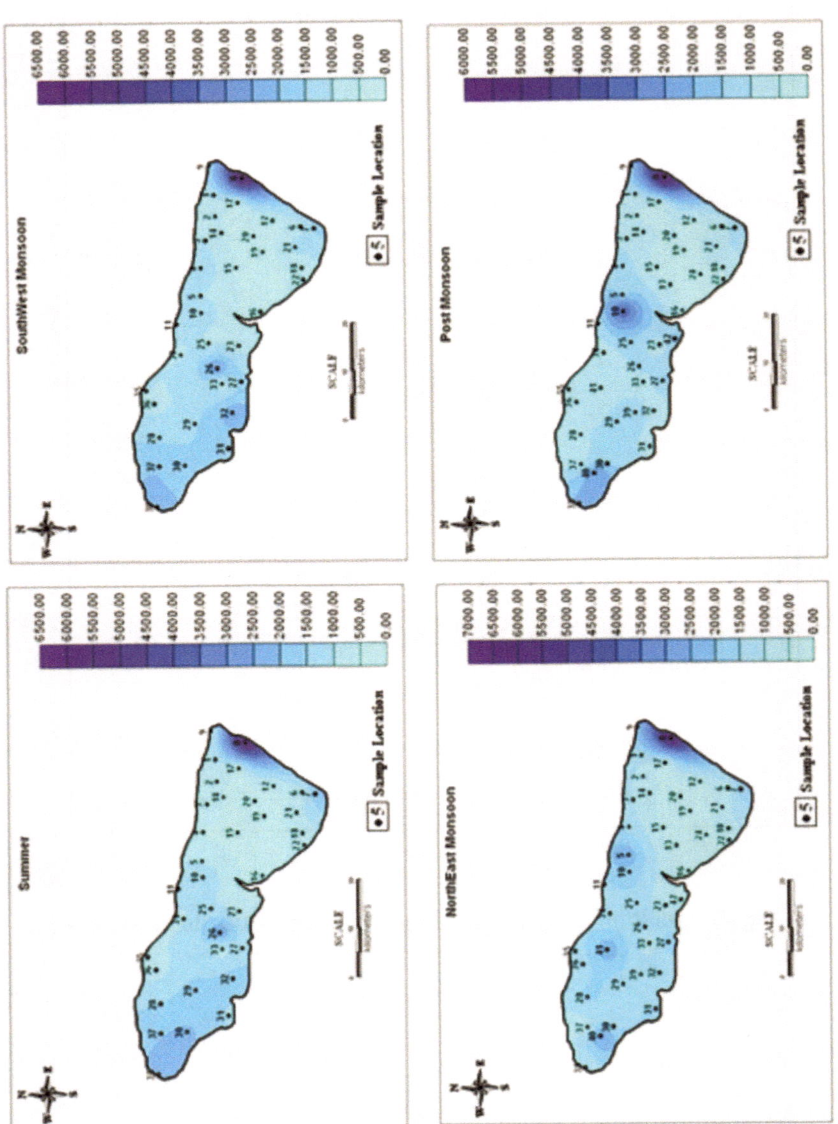

Fig. 3. Spatial distribution of EC in groundwater.

Chapter 11: Understanding the Seawater Intrusion Along the Coastal Aquifers Using Geochemical Techniques

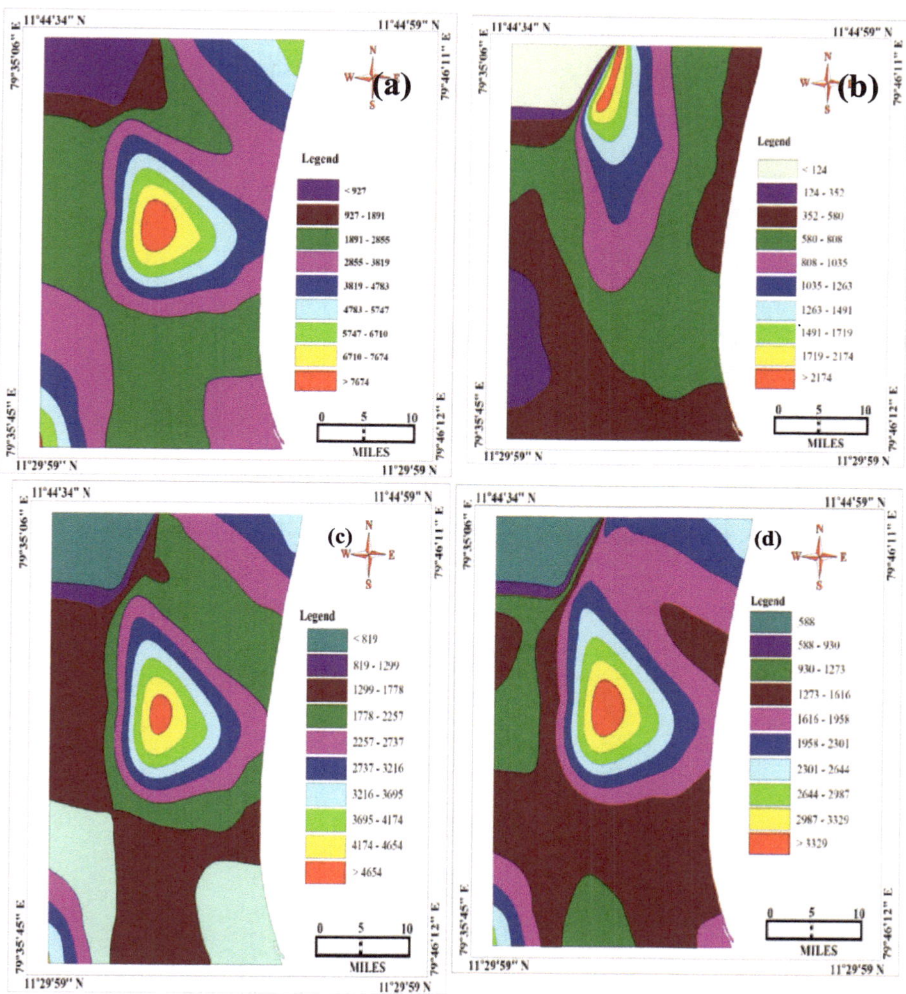

Fig. 2. Spatial distribution of EC (μs/cm) for a) PRM and b) SWM c) NEM and d) POM

Chapter 12: Use of Geochemical Methods to Evaluate the Hydraulic Connection of Aquifers: A Case Study from Southeast Coast of Tamil Nadu, India

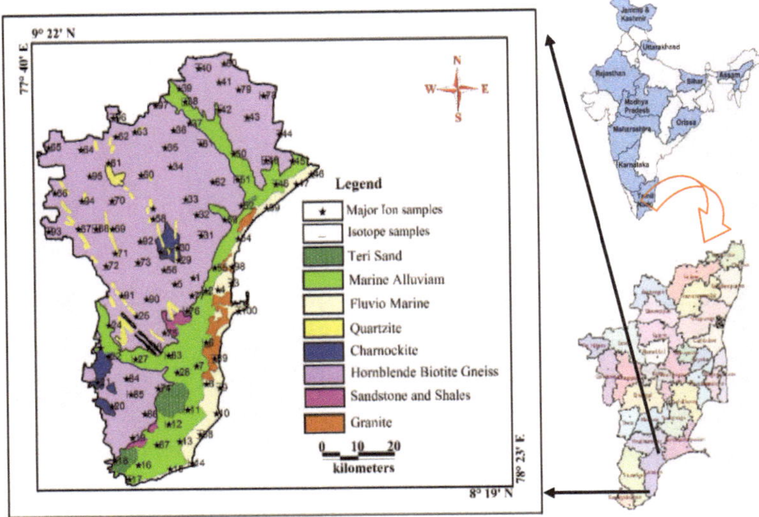

Fig. 1. Geology and Sample location map of the study area

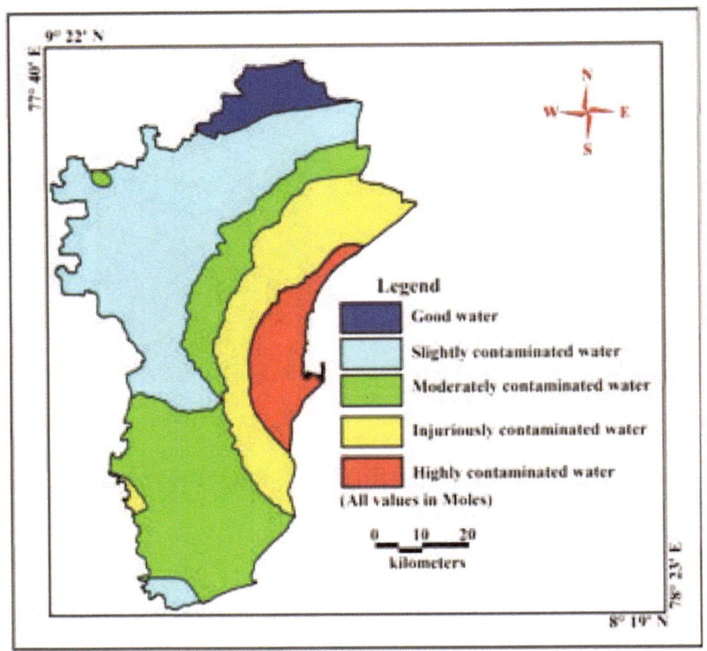

Fig. 4. Spatial distribution of Cl^-/HCO_3^- Moles ratio

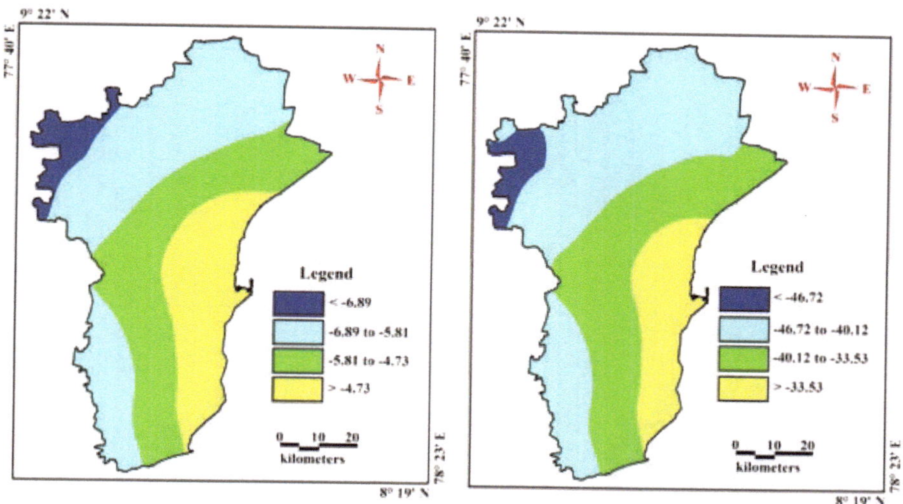

Fig. 6. Spatial variation of $\delta^{18}O$ and δD in the study area

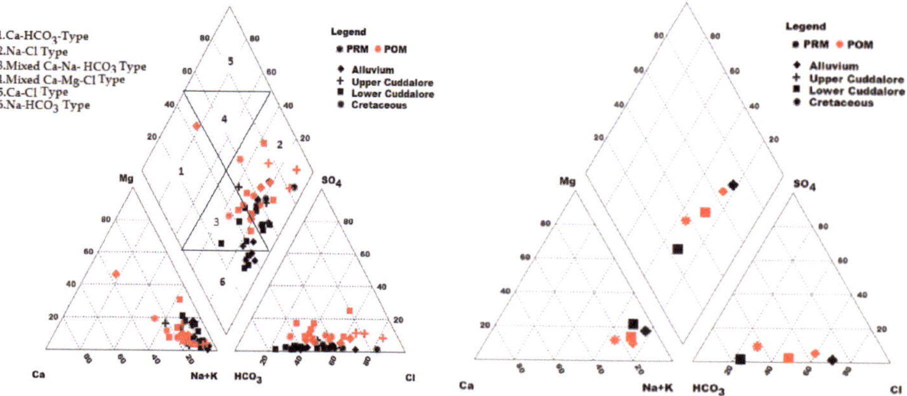

Fig. 3 (a) Piper facies diagram for groundwater samples for different season (b) Groundwater of higher DOC in different season and formation

Chapter 14: A Study on the Lithological Influence on Uranium Concentration in Groundwater of Hard Rock Aquifers of Madurai District, Tamil Nadu

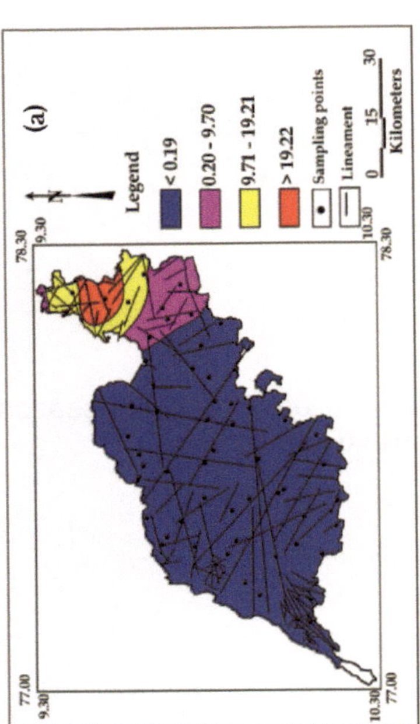

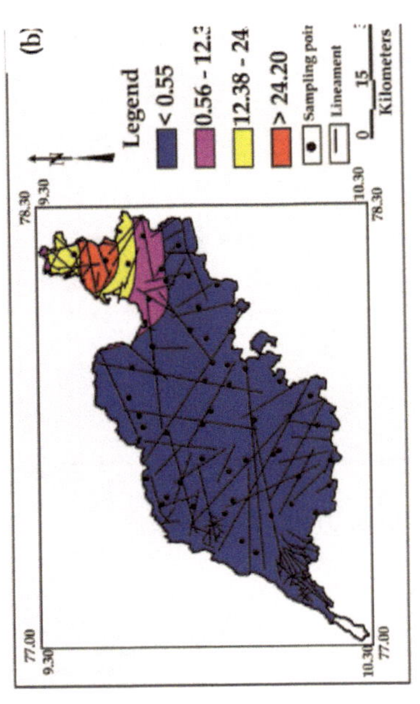

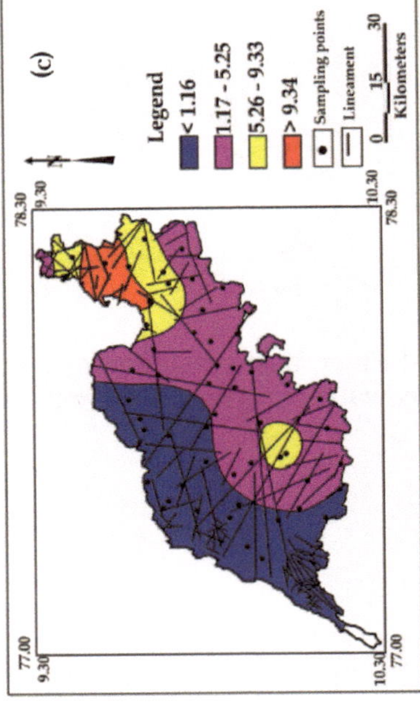

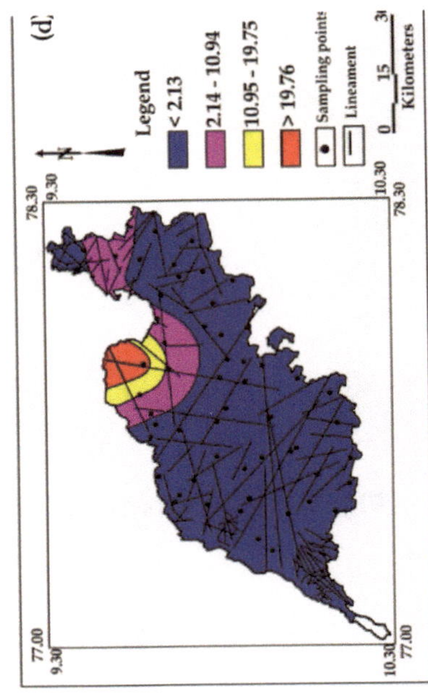

Fig. 6. Spatial distribution of U (ppb) and lineaments for all seasons a) PRM, b) SWM, c) NEM and d) POM Correlation matrix

Chapter 15: A Study on the Impact of Landuse Pattern on Groundwater in and Around Chidambaram Town

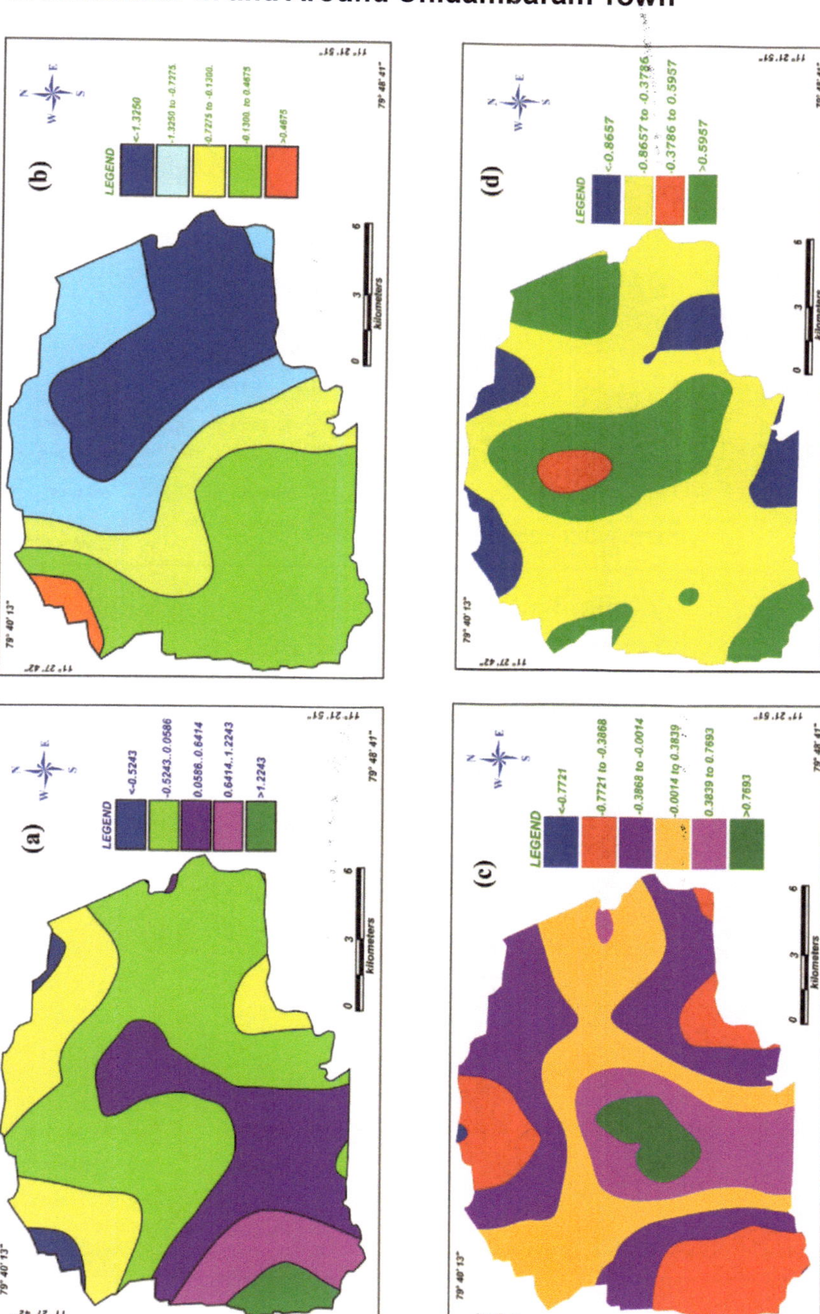

Fig. 4. Spatial distribution of factor scores (a) Factor-1; (b) Factor-2; (c) Factor-3 & (d) Factor-4

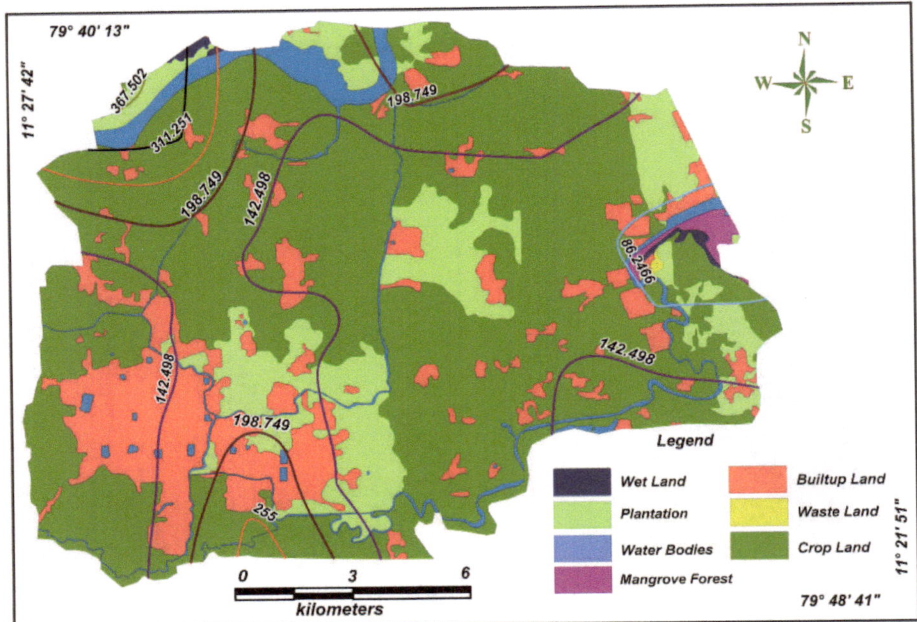

Fig. 5. FC overlay with LULC

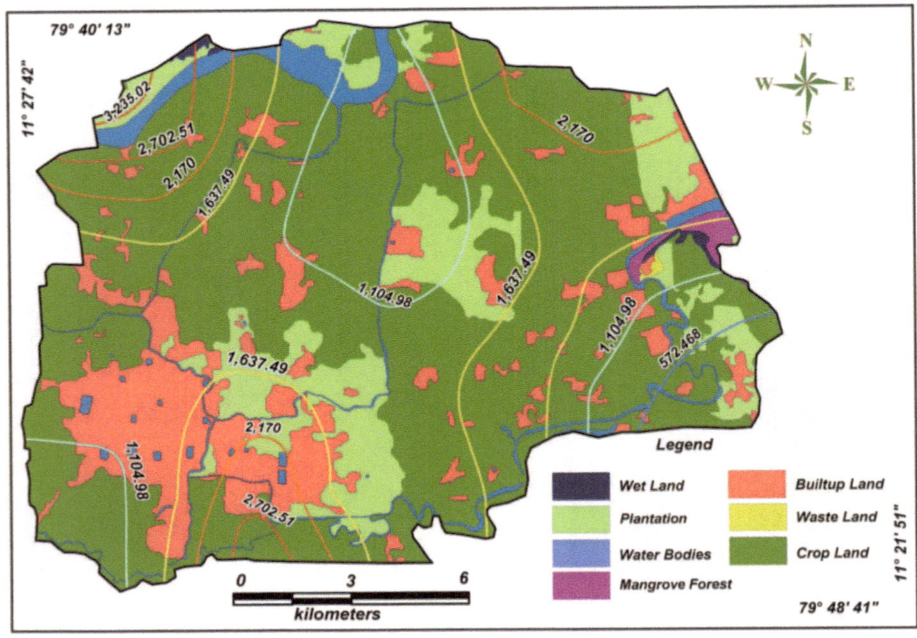

Fig. 6. TC overlay with LULC

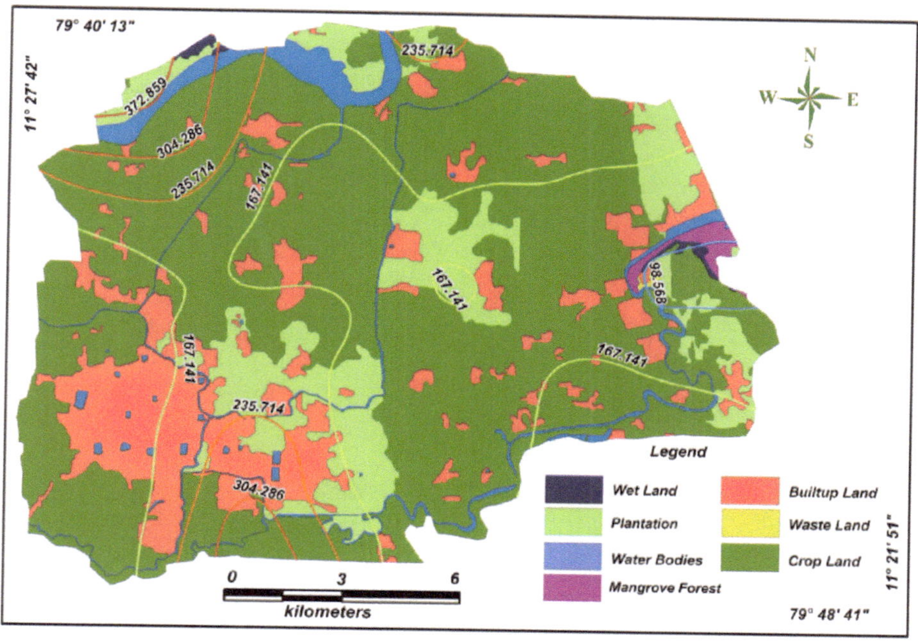

Fig. 7. TS overlay with LULC

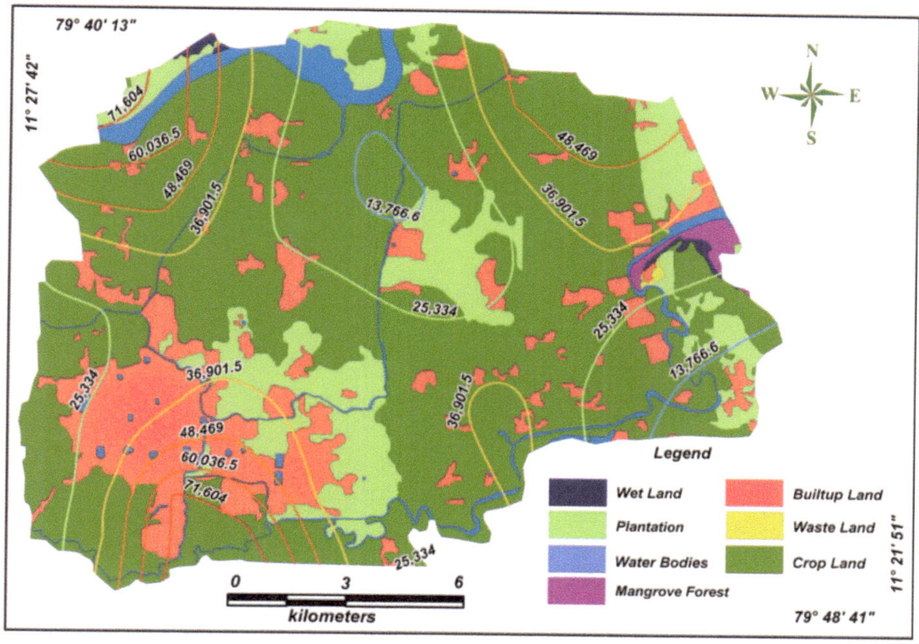

Fig. 8. TVS overlay with LULC

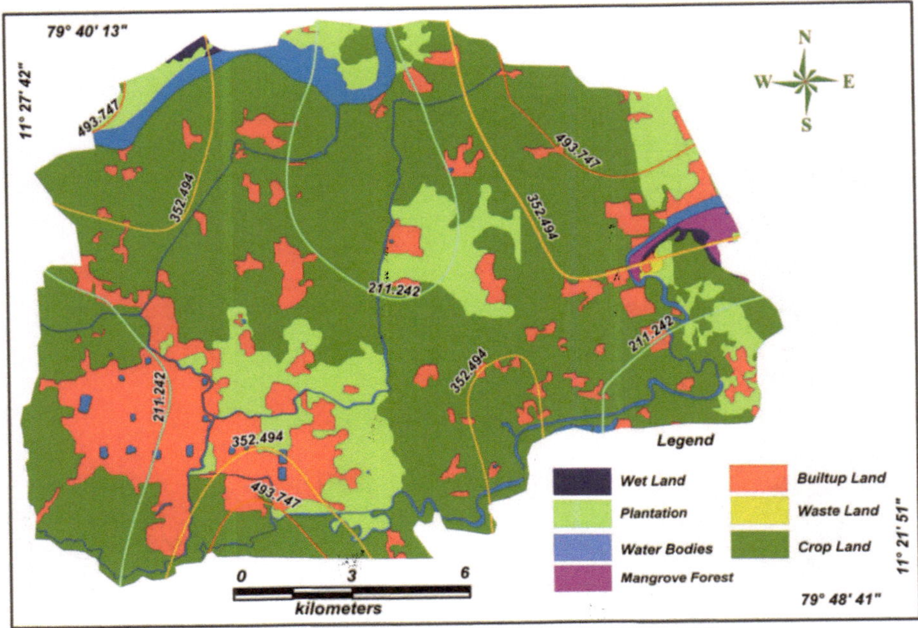

Fig. 9. VC overlay with LULC

Chapter 16: Identification of Spatial Extent of Fluoride Contamination in Groundwater of Dindigul District, Tamil Nadu - A Statistical Approach

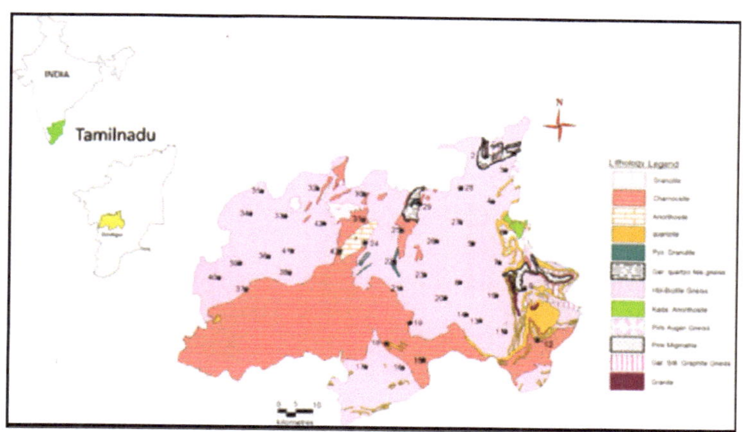

Fig. 1. Geological map of the Dindigul District with the sampling locations.

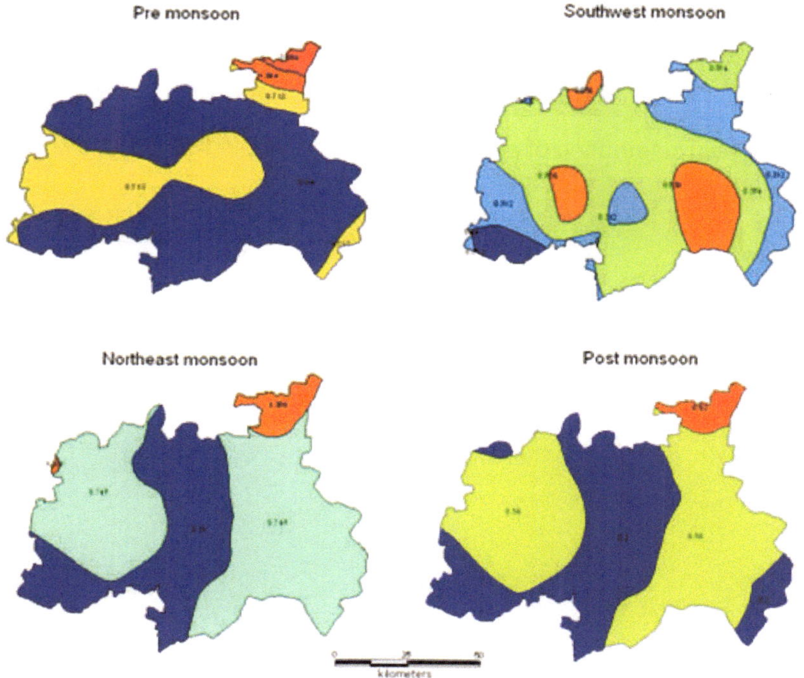

Fig. 2. Spatial and temporal distribution of F^- concentrations in the Dindigul district groundwater.

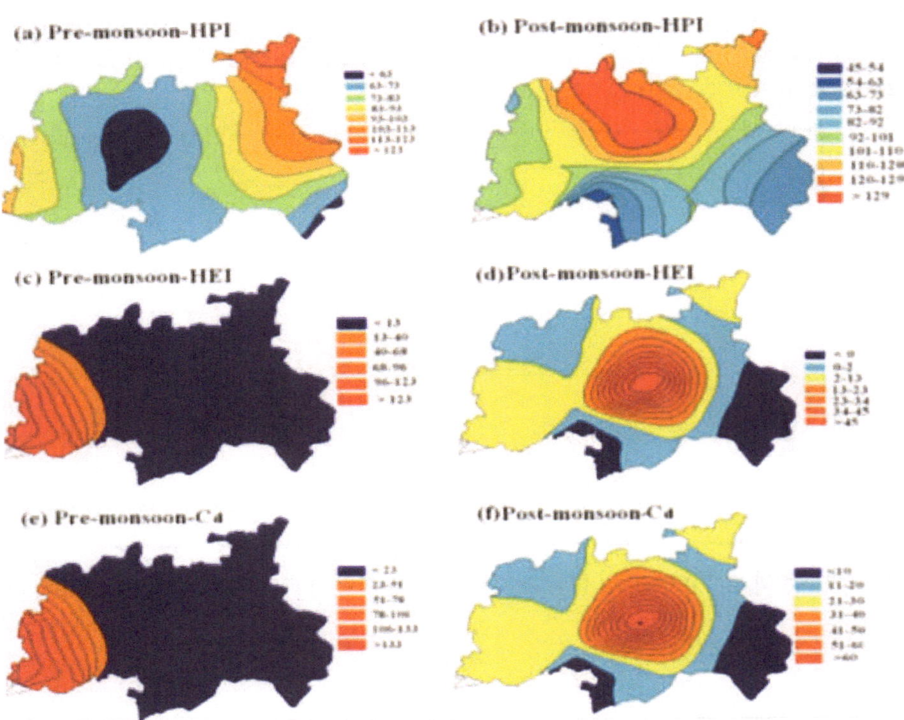

Fig. 4. Spatial distribution of pollution indices in the Dindigul groundwater.